# Springer Series in Operations Research and Financial Engineering

# Springer Series in Operations Research and Financial Engineering

*Altiok:* Performance Analysis of Manufacturing Systems
*Birge and Louveaux:* Introduction to Stochastic Programming
*Bonnans and Shapiro:* Perturbation Analysis of Optimization Problems
*Bramel, Chen, and Simchi-Levi:* The Logic of Logistics: Theory, Algorithms, and Applications for Logistics and Supply Chain Management (second edition)
*Dantzig and Thapa:* Linear Programming 1: Introduction
*Dantzig and Thapa:* Linear Programming 2: Theory and Extensions
*de Haan and Ferreira:* Extreme Value Theory: An Introduction
*Drezner (Editor):* Facility Location: A Survey of Applications and Methods
*Facchinei and Pang:* Finite-Dimensional Variational Inequalities and Complementarity Problems, Volume I
*Facchinei and Pang:* Finite-Dimensional Variational Inequalities and Complementarity Problems, Volume II
*Fishman:* Discrete-Event Simulation: Modeling, Programming, and Analysis
*Fishman:* Monte Carlo: Concepts, Algorithms, and Applications
*Haas:* Stochastic Petri Nets: Modeling, Stability, Simulation
*Klamroth:* Single-Facility Location Problems with Barriers
*Muckstadt:* Analysis and Algorithms for Service Parts Supply Chains
*Nocedal and Wright:* Numerical Optimization
*Olson:* Decision Aids for Selection Problems
*Pinedo:* Planning and Scheduling in Manufacturing and Services
*Pochet and Wolsey:* Production Planning by Mixed Integer Programming
*Resnick:* Heavy Tail Phenomena: Probabilistic and Statistical Modeling
*Whitt:* Stochastic-Process Limits: An Introduction to Stochastic-Process Limits and Their Application to Queues
*Yao (Editor):* Stochastic Modeling and Analysis of Manufacturing Systems
*Yao and Zheng:* Dynamic Control of Quality in Production-Inventory Systems: Coordination and Optimization
*Yeung and Petrosyan:* Cooperative Stochastic Differential Games

**Forthcoming**
*Muckstadt and Sapra:* Models and Solutions in Inventory Management

Sidney I. Resnick

# Heavy-Tail Phenomena

*Probabilistic and Statistical Modeling*

Springer

Sidney I. Resnick
Cornell University
School of Operations Research and
Industrial Engineering
Ithaca, NY 14853
U.S.A.
sirl@cornell.edu

*Series Editors:*

Thomas V. Mikosch
University of Copenhagen
Laboratory of Actuarial Mathematics
DK-1017 Copenhagen
Denmark
mikosh@act.ku.dk

Stephen M. Robinson
University of Wisconsin-Madison
Department of Industrial Engineering
Madison, WI 53706
U.S.A.
smrobins@facstaff.wise.edu

Sidney I. Resnick
Cornell University
School of Operations Research and
Industrial Engineering
Ithaca, NY 14853
U.S.A.
sirl@cornell.edu

Mathematics Subject Classification (2000): 60F17, 60F05, 60K25, 60K10, 60K30, 60G18, 60G51, 60G52, 60G55, 60G57, 60G70, 62G32, 62M10, 62E20, 62F10, 62F12, 62G30, 62G09, 62H20, 62P30, 62P05, 90B18, 90B15, 90B22, 68M20, 91B28, 91B30, 91B70

Library of Congress Control Number: 2006934621

ISBN-10: 0-387-24272-4 e-ISBN-10: 0-387-45024-6
ISBN-13: 978-0-387-24272-9 e-ISBN-13: 978-0-0387-45024-7

Printed on acid-free paper.



9 8 7 6 5 4 3 2 1

springer.com (JLS/Ham)

# Preface

"... was it heavy? Did it achieve total heaviosity?"
—Alvie (Woody Allen) to Annie (Diane Keaton) in *Annie Hall*, 1977.

Heavy-tail analysis is a branch of extreme-value theory devoted to studying phenomena governed by large movements rather than gradual ones. It encompasses both probability modeling as well as statistical inference. Its mathematical tools are based on regular variation, weak convergence of probability measures and random measures and point processes. Its applications are diverse, including the following:

- data networks, where the presence of heavy-tailed file sizes on network servers leads to long range dependence in the traffic rates;
- finance, where financial returns are heavy tailed and thus risk management calculations of *value-at-risk* require heavy-tailed methods;
- insurance, where the field of reinsurance is, by its nature, obsessed with very large values.

## The structure of the book

There is an introductory chapter to describe the flavor and applicability of the subject. Then there are two chapters termed *crash courses*: one on regular variation and the other on weak convergence. These chapters contain essential material that could have been relegated to appendices; however, you should go through them where they are placed in the book. If you know the material, move quickly. Otherwise, pay some attention to style and notation. In particular, note what goes on in Sections 3.4–3.6. Such chapters are, inevitably, a compromise between wanting the book to be self-contained and not wanting to duplicate at length what is standard in other excellent references.

Chapter 4 gets you into the heart of inference issues fairly quickly. The approach to inference is semiparametric and asymptotic in nature. This leads to a statistical theory that is different from classical contexts. We assume there is some structure out there at *asymptopia* and we are trying to infer what it is using a pitiful finite sample whose true model has not yet converged to the asymptotic model. Thus, maximum likelihood methods are not really available unless we simply assume from some threshold onwards that the asymptotic model holds. We give some diagnostics that help decide on values of parameters and when a heavy-tail model is appropriate.

Chapter 5 begins the probability treatment which is geared towards a dimensionless theory. It focuses on the Poisson process and stochastic processes derived from the Poisson process, including Lévy and extremal processes. We also give an introduction to data network modeling. Chapter 6 gives the dimensionless treatment of regular variation and its probabilistic equivalents. We survey weak convergence techniques and discuss why it is difficult to bootstrap heavy-tail phenomena. Chapter 7 exploits the weak convergence technology to discuss weak convergence of extremes to extremal processes and weak convergence of summation processes to Lévy limits. Special cases include sums of heavy-tailed iid random variables converging to $\alpha$-stable Lévy motion. We close the chapter with a unit on how weak convergence techniques can be used to study various transformations of regularly varying random vectors. We include Tauberian theory for Laplace transforms in this discussion.

Applied probability takes center stage in Chapter 8 which uses heavy-tail techniques to learn about the properties of three models. Two of the models are for data networks and the last one is a more traditional queueing model. We return to statistical issues in Chapter 9, discussing asymptotic normality for estimators and then moving to inference for multivariate heavy-tailed models. We include examples of analysis of exchange rate data, Internet data, telephone network data and insurance data. Finally, we close the chapter with a discussion of the much praised and vilified sample correlation function. There are some appendices devoted to notational conventions and a list of symbols and also a section which timidly discusses some useful software.

Each chapter contains exercises. Ignoring the exercises guarantees voyeur status.

# Acknowledgments

Several institutions provided support and allowed me to harrass their students with heavy lectures that eventually became the basis for this book. I gave a semester-long seminar at Cornell in the spring of 2002 which I thought would create the momentum for a quick completion of the writing task. However, having assumed the directorship of Cornell's superb School of Operations Research and Industrial Engineering, it was impossible to find the time to focus on the pleasures of writing. Upon finishing my term as Director in 2004, I spent a month at the University of Bern, giving a short course on heavy tails, and subsequently, during the summer of 2005, while serving as Eurandom Professor, I gave a modified set of lectures in Eindhoven, Netherlands.

Finally, in the fall of 2005, while visiting Columbia University I was able to obsessively focus on assembling all of my thoughts. I much appreciated Columbia's fine atmosphere and hospitality as well as the support and office space provided by Columbia's Departments of Statistics, Department of Industrial Engineering and Operations Research, and Columbia's Business School. While at Columbia I lectured from material written for the book and Soumik Pal and Frank Isaacs proved to be prize-winning fault finders.

The final polishing, indexing, and debugging was done while housed and supported by the University of North Carolina's Department of Statistics and Operations Research and also by SAMSI/NISS at Research Triangle Park. During this period, Bikramjit Das at Cornell supplied me with additional corrections and comments via Skype.

Some additional acknowledgments in random order:

For many years, the National Science Foundation and the National Security Agency have provided the research support necessary for ideas to flourish.

Springer/Birkhäuser continue to be a pleasure to work with, and Ann Kostant should be declared a national treasure. Finally, I want to express a big thank you to John Spiegelman, for working with me, tirelessly, often suggesting things I overlooked; he was altogether wonderful as well as amusing. And, as part of the Springer/Birkhäuser team, Elizabeth Loew continues to inspire total trust in handling my books.

Being on sabbatical is to be truly in a state of grace and I appreciate Cornell's support for the institution of sabbaticals. The opportunity to become a working member of Cornell in 1987 was a turning point in my adult professional life.

I have been blessed and supported by excellent colleagues, coworkers and students over the years who have helped me develop ideas about heavy tails. A partial list includes Laurens de Haan, Richard Davis, Holger Rootzen, Charles Goldie, Paul Feigin, Gennady Samorodnitsky, Catalin Stărică, Eric van den Berg, Krishanu Maulik, and Jan Heffernan. Two anonymous referees provided reviews chock full of useful, constructive criticism and suggestions.... And then there is Thomas Mikosch, who undertook to read the whole manuscript and made an uncountable number of a.e. useful and wise comments. Thomas displayed a genius for identifying inconsistencies and was only mildly sarcastic about my inept grammar, hyphenation, misplaced parentheses and creative spelling. ("Is that American spelling?") I, like, totally owe this guy a Coke! This is the second time [259] in my career I have been blessed by an effort whose helpfulness went way beyond what one has a right to expect from a colleague and friend.

Reminder to the blasé: TeX, LaTeX, and BibTeX are astonishingly useful and elegant, as is MikTeX.

Family history as described in literature:

(1987) I... thank Minna, Nathan, and Rachel Resnick for a cheery, happy family life. Minna and Rachel bought me the mechanical pencil that made this project possible, and Rachel generously shared her erasers with me as well as providing a backup mechanical pencil from her stockpile when the original died after 400 manuscript pages. I appreciate the fact that Nathan was only moderately aggressive about attacking my Springer-Verlag correspondence with a hole puncher [260].

(1992) Minna, Rachel, and Nathan Resnick provided a warm, loving family life and generously shared the home computer with me. They were also very consoling as I coped with two hard disk crashes and a monitor meltdown [262].

(1998) Rachel, who grew into a terrific adult, no longer needs to share her mechanical pencils with me. Nathan has stopped attacking my manuscripts with a hole puncher and gives ample evidence of the fine adult he will soon be. Minna is the ideal companion on the random path of life [264].

Time marches on and it is 2006. Nathan graduated Cornell, moved to New York City, and found a job, and Rachel (now a Director!) has married Randy. Nathan and Randy gang up on poor, defenseless me, which Minna claims I deserve. Minna and I, the artist and the ninja math geek, continue to explore a wonderful life together. As for mechanical pencils and home computers, I wrote this whole book on a laptop, the wonderous IBM (now Lenovo) ThinkPad. It is remarkable what one can accomplish while moving around with a good laptop and Internet connection.

Ithaca, NY

*Sidney Resnick*
June 2006

# Contents



## Part I Crash Courses

## Part II Statistics

## Part III Probability

# 1

# Introduction

## 1.1 Welcome

This is a survey of some of the mathematical, probabilistic and statistical tools used in heavy-tail analysis as well as some examples of their use. Heavy tails are characteristic of phenomena where the probability of a huge value is relatively big. Record-breaking insurance losses, financial log-returns, file sizes stored on a server, transmission rates of files are all examples of heavy-tailed phenomena. The modeling and statistics of such phenomena are tail dependent and much different than classical modeling and statistical analysis, which give primacy to central moments, averages, and the normal density, which has a wimpy, light tail.

An oversimplified view of heavy-tail analysis is that it rests on three subjects:

- *Mathematics*: The *theory of regularly varying functions* [26, 90, 102, 135, 144, 220, 260, 275] provides the right mathematical framework for heavy-tail analysis.
- *Probability theory and stochastic processes*: Heavy-tail analysis is a heavy consumer of weak convergence techniques [22, 23, 25, 301] since an organizing theme is that many limit relations giving approximations can be viewed as applications of almost surely continuous maps. It also requires knowledge of stochastic processes, such as point processes and random measures [65, 180, 230, 260], Brownian motion, Lévy processes, and stable processes [4, 19, 273, 274].
- *Statistics*: Are the data heavy tailed? Is a heavy-tailed model appropriate? How do you fit such a model to the data? Specialized techniques overlapping extreme-value theory [16, 90, 129, 260] are needed.

## 1.2 Survey

*Heavy-tail analysis* is an interesting and useful blend of mathematical analysis, probability, and stochastic processes and statistics. Heavy-tail analysis is the study of systems whose behavior is governed by large values which shock the system periodically. This is in contrast to many systems exhibiting stability whose behavior is determined largely by an averaging effect. In heavy-tailed analysis, typically the asymptotic behavior of descriptor variables is determined by the large values or merely a single large value.

Roughly speaking, a random variable $X$ has a heavy (right) tail if there exists a positive parameter $\alpha > 0$ such that

$$P[X > x] \sim x^{-\alpha}, \quad x \to \infty. \tag{1.1}$$

(Note that here and elsewhere that we use the notation

$$f(x) \sim g(x), \quad x \to \infty,$$

as shorthand for

$$\lim_{x \to \infty} \frac{f(x)}{g(x)} = 1,$$

for two real functions $f, g$. Similarly, $f(x) \sim g(x), x \to 0$, means the ratio approaches 1 as $x \to 0$.) Examples of such random variables are those with Cauchy, Pareto, $t$, $F$, or stable distributions. Stationary stochastic processes, such as the ARCH, GARCH, EGARCH, etc., which have been proposed as models for financial returns, typically have marginal distributions satisfying (1.1). It turns out that (1.1) is not quite the right mathematical setting for discussing heavy tails (that pride of place belongs to regular variation of real functions) but we will get to that in due course.

An elementary observation is that a heavy-tailed random variable has a relatively large probability of exhibiting a really large value, compared to random variables, which have exponentially bounded tails such as normal, Weibull, exponential, or gamma random variables. For a $N(0, 1)$ normal random variable $N$, with density $n(x)$, we have by Mill's ratio that

$$P[N > x] \sim \frac{n(x)}{x} \sim \frac{1}{x\sqrt{2\pi}} e^{-x^2/2}, \quad x \to \infty,$$

which has much weaker tail weight than suggested by (1.1).

There is a tendency to sometimes confuse the concept of a heavy-tail distribution with the concept of a distribution with infinite right support. (For a probability distribution $F$, the support is the smallest closed set $C$ such that $F(C) = 1$. For the exponential distribution with no translation, the support is $[0, \infty)$ and for the normal distribution, the support is $\mathbb{R}$.) The distinction is simple and exemplified by comparing a normally distributed random variable with one whose distribution is Pareto. Both have positive probability of achieving a value bigger than any preassigned threshold. However, the

Pareto random variable has, for large thresholds, a much bigger probability of exceeding the threshold. One cannot rule out heavy-tailed distributions by using the argument that everything in the world is bounded unless one agrees to rule out all distributions with unbounded support.

Much of classical statistics is often based on averages and moments. Try to imagine a statistical world in which you do not rely on moments since if (1.1) holds, moments above the $\alpha$th do not exist! This follows since

$$\int_0^\infty x^{\beta-1}P[X>x]dx \approx \int_1^\infty x^{\beta-1}x^{-\alpha}dx \begin{cases} <\infty & \text{if } \beta<\alpha, \\ =\infty & \text{if } \beta\geq\alpha, \end{cases}$$

where (in this case)

$$\int f \approx \int g$$

means both integrals either converge or diverge together. Much stability theory in stochastic modeling is expressed in terms of mean drifts, but what if the means do not exist. Descriptor variables in queueing theory are often in terms of means, such as mean waiting time, mean queue lengths, and so on. What if such expectations are infinite?

## 1.3 Context and examples

In this section, we outline scenarios where heavy-tailed analysis is used. The books [1, 16, 50, 90, 129, 209, 218, 238] contain other examples and application areas.

### 1.3.1 Data networks

Measurements on data networks often show empirical features that are surprising by the standards of classical queueing and telephone network models. Measurements often consist of data giving bitrate or packet rates. This means that a window resolution is selected (for example, 10 seconds, 1 second, 10 milliseconds, 1 millisecond, ...) and the number of bits or packets in adjacent time windows or slots is recorded. Significant examples include [118, 203, 305, 306].

Certain distinctive properties are common to many different data studies and such properties are termed *invariants* by network engineers. (In finance, the phrase *stylized fact* seems to be a synonym for *invariant*.) Here are some examples of *invariants* for network data:

- Heavy tails abound [204, 303, 304, 307] for such things as file sizes [6, 242], transmission rates, transmission durations [215, 267].

- The number of bits or packets per slot exhibits long-range dependence across time slots (e.g., [203, 305]). There is also a perception of self-similarity as the width of the time slot varies across a range of time scales exceeding a typical roundtrip time. See Section 5.2.3 (p. 126).
- Network traffic is bursty with rare but influential periods of very high transmission rates punctuating typical periods of modest activity.

Having observed empirical phenomena, there is an obligation to uncover relationships that explain the phenomena. An accepted network paradigm is that long-range dependence in traffic per time slot is caused by heavy tails of the file sizes of files stored on servers. This is discussed in Section 5.2 (p. 123), where a modeling explanation is provided for the relationship between long-range dependence and heavy tails.

An idealized data transmission model of a source destination pair is an alternating renewal *on/off* model, where constant-rate transmissions alternate with *off* periods. The *on* periods are random in length with a *heavy-tailed distribution*, and this leads to occasional large transmission lengths. Note that the constant transmission rate assumption means the transmission length is proportional to the size of the file being transmitted. This model provides one explanation of perceived *long-range dependence* in measured traffic rates. A competing model, which to some tastes is marginally more elegant, is the infinite-source Poisson model, to be discussed in Section 5.2.4 (p. 127).

*Example* 1.1. The Boston University study [52, 53, 63], now considered a classic, suggests self-similarity of web traffic stems from heavy-tailed file sizes. This means that we treat files as being randomly selected from a population and if $X$ represents a randomly selected file size, then the hypothesis of a heavy tail is

$$P[X > x] \sim x^{-\alpha}, \quad x \to \infty, \quad \alpha > 0, \tag{1.2}$$

where $\alpha$ is a shape parameter that must be statistically estimated. The BU study reports an overall estimate for a five-month measurement period (see [63]) of $\alpha = 1.05$. However, there is considerable month-to-month variation in these estimates and, for instance, the estimate for November 1994 in room 272 places $\alpha$ in the neighborhood of 0.66. Figure 1.1 gives the QQ and Hill plots [17, 165, 191, 252] of the file-size data for the month of November in the Boston University study. These are two graphical methods for estimating $\alpha$ and will be discussed in more detail in Section 4.6.1 (p. 97) and Section 4.4.2 (p. 85).

Extensive traffic measurements of *on* periods are reported in [305], where measured values of $\alpha$ were usually in the interval $(1, 2)$. Studies of sizes of files accessed on various servers by the Calgary study [6] report estimates of $\alpha$ from 0.4 to 0.6. So evidence exists which suggests values of $\alpha$ outside the range $(1, 2)$ should be considered. Also, as user

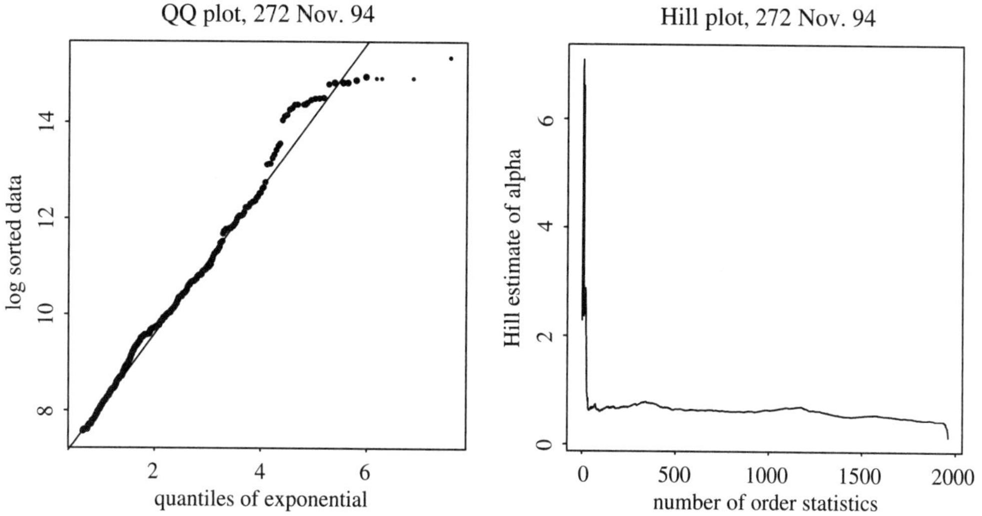


**Fig. 1.1.** QQ and Hill plots of November 1994 file lengths.

demands on the web grow and access speeds increase, there may be a drift toward heavier file-size distribution tails. However, this is a hypothesis that is currently untested.

### 1.3.2 Finance

In the study of financial returns of risky assets, it is empirically observed that "returns" possess notable features, which in the finance culture are termed *stylized facts*. This is similar to what we observed about network data sets, and *stylized facts* are to finance what *invariants* are to data networks.

What is a "return"? Suppose $\{S_i\}$ is the stochastic process representing the price of a speculative asset (stock, currency, derivative, commodity (corn, coffee, etc.)) at the $i$th measurement time. The return process is

$$\tilde{R}_i := (S_i - S_{i-1})/S_{i-1};$$

that is, the process giving the relative difference of prices. If the returns are small, then the differenced log-price process approximates the return process

$$R_i := \log S_i - \log S_{i-1} = \log \frac{S_i}{S_{i-1}} = \log\left(1 + \left(\frac{S_i}{S_{i-1}} - 1\right)\right)$$
$$\approx \frac{S_i}{S_{i-1}} - 1 = \tilde{R}_i$$

since for $|x|$ small,

$$\log(1+x) \sim x, \quad x \to 0,$$

by l'Hôpital's rule. So instead of studying the returns process $\{\tilde{R}_i\}$, the differenced log-price process $\{R_i\}$ is studied, and henceforth we refer to $\{R_i\}$ as the *returns process*. Recall that transforming a data set of positive observations by taking logarithms and then differencing is a common and comfortable procedure from time-series analysis [31], one which is often used to transform a nonstationary sequence to one that is plausibly modeled as stationary.

Empirically, the returns process often exhibits notable properties:

1. Heavy-tailed marginal distributions (but usually $\alpha > 2$, so the mean and variance exist).
2. Little or no correlation. However, by squaring or taking absolute values of the returns, one gets a highly correlated, even long-range-dependent process.
3. Dependence. (If the random variables were independent, so would the squares be independent, but squares are typically correlated.)

Hence one needs to model the data with a process that is stationary and has heavy-tailed marginal distributions and a dependence structure. This leads to the study of specialized models in economics with lots of acronyms like ARCH and GARCH. Estimation of the marginal distribution's shape parameter $\alpha$ is made more complex due to the fact that the observations are dependent.

Given $S_0$, there is a one-to-one correspondence between

$$\{S_0, S_1, \ldots, S_T\} \quad \text{and} \quad \{S_0, R_1, \ldots, R_T\}$$

since

$$\begin{aligned}\sum_{t=1}^{T} R_t &= (\log S_1 - \log S_0) + (\log S_2 - \log S_1) \\ &\quad + \cdots + (\log S_T - \log S_{T-1}) \\ &= \log S_T - \log S_0 = \log \frac{S_T}{S_0},\end{aligned}$$

so that

$$S_T = S_0 e^{\sum_{t=1}^{T} R_t}. \tag{1.3}$$

So why deal with returns rather than with the price process? Here are some reasons:

1. The returns are scale free and thus independent of the units as well as the size of the initial investment.

2. Returns have more attractive statistical properties than prices such as stationarity. Econometric models sometimes yield nonstationary price models but stationary returns.

Why deal with $\{R_t\}$ rather than $\{\tilde{R}_t\}$?

1. The process $\{R_t\}$ is nicely additive over time. It is easier to construct models for additive phenomena than for multiplicative ones (such as $1 + \tilde{R}_t = S_t/S_{t-1}$). One can recover $S_T$ from the returns by what is essentially an additive formula (1.3). (Additive is good!) Also, the $T$-day return process

$$R_T - R_1 = \log S_T - \log S_0$$

is additive. (Additive is good!)
2. The daily values of $\tilde{R}_t = S_t/S_{t-1} - 1$ satisfy

$$\frac{S_t}{S_{t-1}} - 1 \geq -1,$$

and for statistical modeling, it is a bit unnatural to have the variable bounded below by $-1$. For instance, one could not model such a process using a normal or two-sided stable density.
3. Certain economic facts are easily expressed by means of $\{R_t\}$. For example, if $S_t$ is the exchange rate of the US dollar against the British pound and $R_t = \log(S_t/S_{t-1})$, then $1/S_t$ is the exchange rate of pounds to dollars, and the return from the point of view of the British investor is

$$\log \frac{1/S_t}{1/S_{t-1}} = \log \frac{S_{t-1}}{S_t} = -\log \frac{S_t}{S_{t-1}},$$

which is minus the return for the American investor.
4. As mentioned, the operations of taking logarithms and differencing are standard time-series tools for coercing a data set into looking stationary. Both operations, as indicated, are easily undone. So there is a high degree of comfort with these operations.

Classical extreme-value theory, which subsumes heavy-tail analysis, uses techniques to estimate *value-at-risk* (or VaR), which is an extreme quantile of the profit-and-loss density, once the density is estimated. This is discussed further in Section 1.3.2 (p. 9) after Example 1.2 and also in Section 4.7 (p. 111).

*Example* 1.2 (*Standard & Poors* 500). We consider the data set *fm-poors.dat* in the package Xtremes [238], which gives the Standard & Poors 500 stock market index. The data, although somewhat old, are absolutely typical of many finance data sets; it is daily data from July 1962 to December 1987 but, of course, does not include days

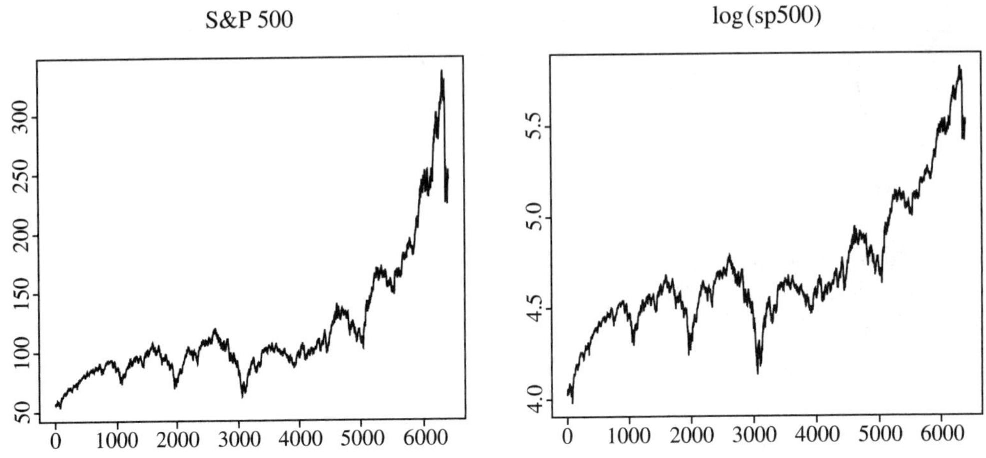


**Fig. 1.2.** Time-series plot of S&P 500 data (left) and log(S&P) (right).

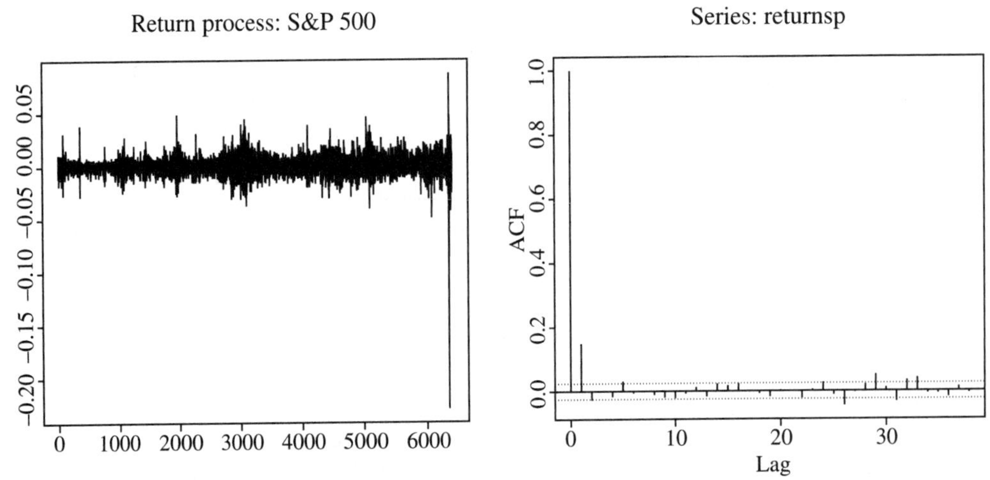


**Fig. 1.3.** Time-series plot of S&P 500 return data (left) and the sample autocorrelation function (right).

when the market is closed. In Figure 1.2, we display the time-series plots of the actual data for the index and the log of the data. Only someone delusional would conclude that these two series were stationary. On the left side of Figure 1.3, we exhibit the 6410 returns $\{R_t\}$ of the data by differencing at lag 1 the log(S&P) data. On the right side is the sample autocorrelation function. There is a large lag 1 correlation but otherwise few spikes are outside the 95% confidence window.

For a view of the *stylized facts* about these data, and to indicate the complexities of the dependence structure, we exhibit the autocorrelation function of the squared returns

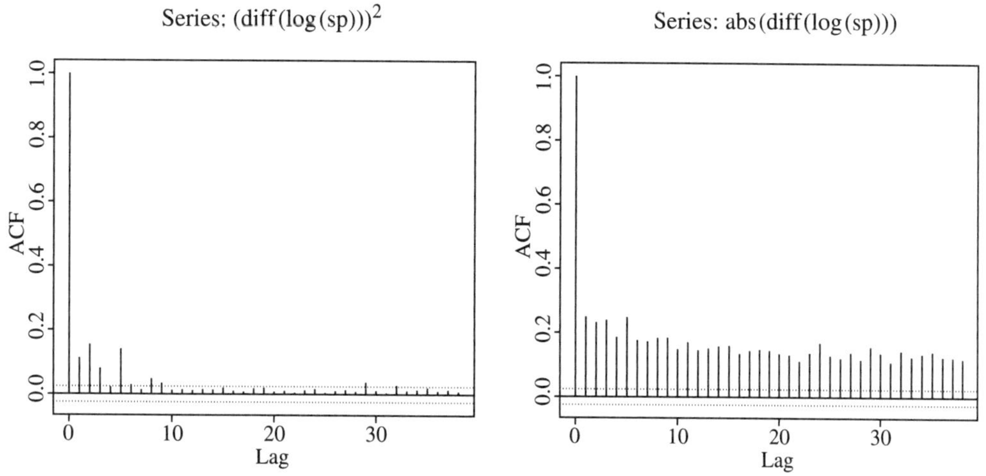


**Fig. 1.4.** (i) The sample autocorrelation function of the squared returns (left). (ii) The sample autocorrelation function of the absolute values of the returns (right).

in Figure 1.4 (left), and on the right, the autocorrelation function for the absolute value of the returns. Although there is little correlation in the original series, the iid hypothesis is obviously false.

One can compare the heaviness of the right and left tail of the marginal distribution of the process $\{R_t\}$ even if we do not believe that the process is iid. A reasonable assumption seems to be that the data can be modeled by a stationary, uncorrelated process, and we hope the standard exploratory extreme-value and heavy-tailed methods developed for iid processes still apply. We apply the QQ plotting technique to the data. (See Sections 4.6 (p. 97) and 11.1.2 (p. 366).) After playing a bit with the number of upper-order statistics used, we settled on $k = 200$ order statistics for the positive values (upper tail) which gives the slope estimate of $\hat{\alpha} = 3.61$. This is shown in the left side of Figure 1.5. On the right side of Figure 1.5 is the comparable plot for the left tail; here we applied the routine to abs(returns[returns $< 0$]), that is, to the absolute value of the negative data points in the log-return sample. After some experimentation, we obtained an estimate $\hat{\alpha} = 3.138$ using $k = 150$. Are the two tails symmetric, which is a common theoretical assumption? Unlikely!

### Value-at-risk

Extreme-value theory and heavy-tail analysis use techniques to estimate *value-at-risk* (VaR), which is an extreme quantile of the profit-and-loss density, once the density is estimated. Here is a rapid overview.

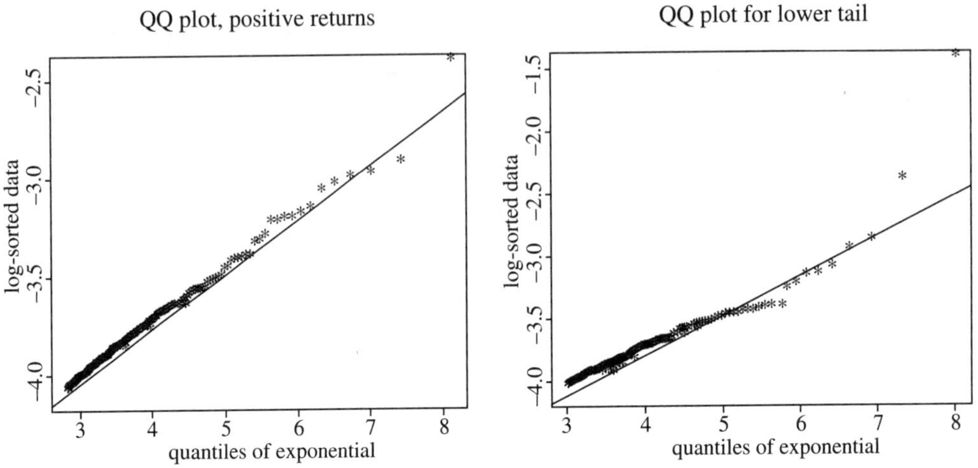


**Fig. 1.5.** Left: QQ plot and parfit estimate of $\alpha$ for the right tail using $k = 200$ upper-order statistics. Right: QQ plot and parfit estimate of $\alpha$ for the left tail using the absolute value of the negative values in the log-returns.

Financial institutions have to meet standards set by regulatory bodies designed to prevent overexposure to risks. Sufficient capital is required to withstand sudden dramatic unfavorable shifts in the market. A commonly used risk metric is *value-at-risk* (VaR), which is the point that is exceeded by a loss for the portfolio only with a specified low probability. This will just be a quantile of the loss distribution, which can be estimated from observable data.

The risk analysis is done in two stages:

- Express profit-and-loss in terms of returns.
- Statistically model the returns and compute the appropriate quantile.

*Representing market value in terms of returns.* Let $\{S_t\}$ be the price process of an asset such as a stock. Suppose at time 0 a decision is made to hold $h$ shares for the time horizon $t = 0, \ldots, T$. Then the market value of the asset at time $t$ is

$$V_t = hS_t, \quad t = 0, \ldots, T.$$

So $V_0$ is the initial value and $V_T$ is the final value at the end of the time horizon. The loss variable is the "loss" expressed in positive units:

$$L_t = -(V_t - V_0) = \begin{cases} |V_t - V_0| & \text{if } V_t - V_0 < 0, \\ -|V_t - V_0| & \text{if } V_t - V_0 > 0. \end{cases}$$

So if $L_t$ is negative, there is a profit.

We express this in terms of the return process $\{R_t\}$ defined in (1.3). Multiplying (1.3) by $h$, we have the $T$-period loss is

$$
\begin{aligned}
L_T = -h(S_T - S_0) &= -h\left(S_0 e^{\sum_{i=1}^T R_i} - S_0\right) \\
&= -hS_0\left(e^{\sum_{i=1}^T R_i} - 1\right) \\
&= V_0\left(1 - e^{\sum_{i=1}^T R_i}\right) \qquad (1.4) \\
&\approx V_0\left(-\sum_{i=1}^T R_i\right), \qquad (1.5)
\end{aligned}
$$

where the last approximation is tolerable provided $|\sum_{i=1}^T R_i|$ is small.

*Multivariate version.* Portfolios rarely contain a single asset, and typically diversification leads to portfolios being dependent on a large-dimensional vector of diverse asset returns. This is one reason for the increasing interest in multivariate heavy tails, which is the focus of this book.

Suppose a portfolio consists of $d$ assets with prices at time $t$ equal to $S_{t,1}, \ldots, S_{t,d}$, $t = 1, \ldots, T$. Let $h_j$ be the number of shares owned in asset $j$ during the period of observation so that the value of the $j$th asset at time $t$ is

$$V_{t,j} = h_j S_{t,j}, \quad j = 1, \ldots, d; \quad t = 0, \ldots, T.$$

The value of the total portfolio at $t$ is

$$V_t = \sum_{j=1}^d V_{t,j}, \quad t = 0, \ldots, T.$$

Let $\{R_{t,j}, t \geq 0\}$ be the return process for the $j$th asset. Also, we write

$$W_j = \frac{V_{0,j}}{V_0} = \frac{h_j S_{0,j}}{\sum_l h_l S_{0,l}}$$

to indicate how the portfolio is balanced at time 0. Define

$$
\begin{aligned}
L_T &= -(V_T - V_0) = -\sum_{j=1}^d (V_{T,j} - V_{0,j}) \\
&= -\sum_{j=1}^d \left(V_{0,j} e^{\sum_{t=1}^T R_{t,j}} - V_{0,j}\right) = -V_0 \sum_{j=1}^d \frac{V_{0,j}}{V_0}\left(e^{\sum_{t=1}^T R_{t,j}} - 1\right)
\end{aligned}
$$

$$= V_0 \sum_{j=1}^{d} W_j \left(1 - e^{\sum_{t=1}^{T} R_{t,j}}\right) \tag{1.6}$$

$$\approx V_0 \sum_{j=1}^{d} W_j \sum_{t=1}^{T} (-R_{t,j})$$

$$= V_0(W_1, \ldots, W_d) \begin{pmatrix} \sum_{t=1}^{T}(-R_{T,1}) \\ \vdots \\ \sum_{t=1}^{T}(-R_{T,d}) \end{pmatrix}. \tag{1.7}$$

*Definition and computation of VaR.* The value-at-risk, $\mathrm{VaR}(T, q)$, for the period $T$ is the $q$th quantile of the loss distribution defined by

$$P[L_T \leq \mathrm{VaR}(T, q)] = q. \tag{1.8}$$

For a single asset this is computed as follows. Define

$$F_T(x) = P\left[-\sum_{t=1}^{T} R_i \leq x\right],$$

which is basically the left tail of the $T$-period return variable. We claim

$$\mathrm{VaR}(T, q) = V_0(1 - e^{-F_T^{\leftarrow}(q)}), \tag{1.9}$$

which assumes $V_0$ is nonrandom and known. The reason for (1.9) is that

$$P[L_T \leq V_0(1 - e^{-F_T^{\leftarrow}(q)})] = P\left[V_0\left(1 - e^{\sum_{t=1}^{T} R_t}\right) \leq V_0(1 - e^{-F_T^{\leftarrow}(q)})\right]$$

$$= P\left[e^{\sum_{t=1}^{T} R_t} \geq e^{-F_T^{\leftarrow}(q)}\right] = P\left[\sum_{t=1}^{T} R_t \geq -F_T^{\leftarrow}(q)\right]$$

$$= P\left[-\sum_{t=1}^{T} R_t \leq F_T^{\leftarrow}(q)\right] = q.$$

Note that if we use the approximation $1 - e^{-F_T^{\leftarrow}(q)} \approx F_T^{\leftarrow}(q)$, then

$$\mathrm{VaR}(T, q) \approx V_0 F_T^{\leftarrow}(q).$$

However, this overestimates VaR since for $x > 0$, $1 - e^{-x} \leq x$.

The statistical problem is to estimate $\mathrm{VaR}(T, q)$ based on a sample of $T$-period returns. The empirical distribution function of the sample of $T$-period returns is an

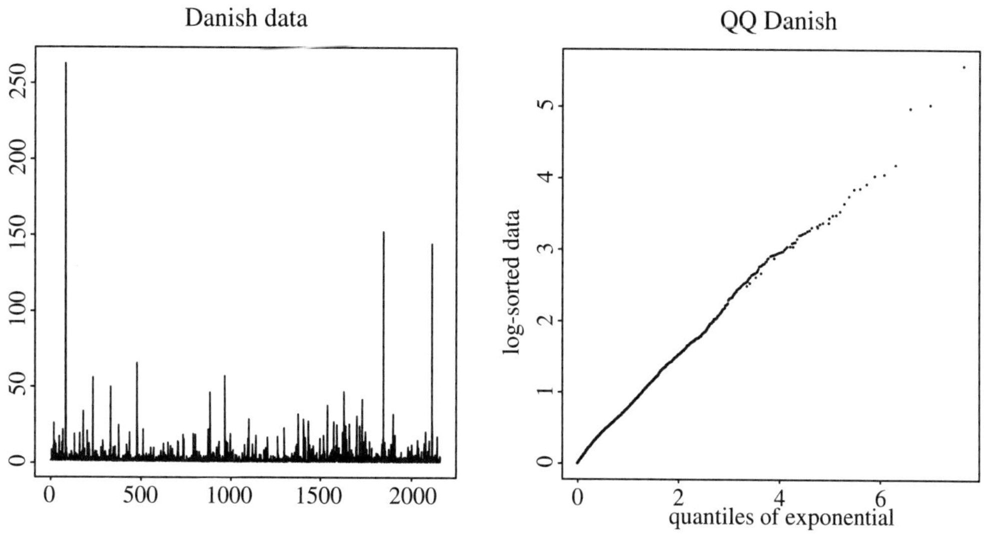


**Fig. 1.6.** Danish data (left) and QQ plot.

approximation of the true distribution of returns which is reasonably accurate in the center of the distribution. However, to estimate an extreme quantile such as VaR, we need a reasonable estimate not just in the center of the distribution but in the extreme tail. Thus extrapolation methods using *peaks-over-threshold* methods and asymptotic theory based on extreme-value and heavy-tail analysis must be used. See [50, 90, 129, 209, 218, 238]. We return to this point in Section 4.7 (p. 111).

### 1.3.3 Insurance and reinsurance

The general theme here is to model insurance claim sizes and frequencies so that premium rates may be set intelligently and risk to the insurance company quantified.

Smaller insurance companies sometimes pay for reinsurance or, more particularly, excess-of-loss (XL) insurance to a bigger company like Munich Re, Swiss Re, or Lloyd's of London. The excess claims over a certain contractually agreed threshhold are covered by the big insurance company. Such excess claims are by definition very large, so heavy-tail analysis is a natural tool to apply. What premium should the big insurance company charge to cover potential losses?

To convince you this might make a difference to somebody, note [129] that from 1970–1995, the two worst cumulative losses world wide were Hurrricane Andrew (my wife's cousin's yacht in Miami wound up on somebody's roof 30 miles to the north) and the Northridge earthquake in California. Losses in 1992 dollars were $16,000 and

$11,838 million dollars, respectively. (Note the unit is "millions of dollars.") The tally from Hurricane Katrina will undoubtedly exceed both these prior disasters.

As an example of data you might encounter, consider the Danish data on large fire insurance losses [219, 263]. Figure 1.6 gives a time-series plot of the 2156 Danish data consisting of losses of over one million Danish krone (DKK) and the right-hand plot is the QQ plot of these data, yielding a remarkably straight plot. The straight-line plot indicates the appropriateness of heavy-tail analysis. The data were collected from 1980–1990 inclusive and values adjusted for inflation to 1985 values.

# Part I

## Crash Courses

# 2

# Crash Course I: Regular Variation

The next two chapters are rapid overviews of two essential subjects: regular variation and weak convergence. This kind of material is sometimes relegated to appendices, which is an unloved practice requiring much paging forward and back. Readers who are familiar with these subjects will find these chapters reassuring collections of notation and basic results. Those readers with less familiarity should read through the chapters to gain some functionality with the topics without worrying about all details; they can return later to ponder details, get further references, and improve mastery as time and circumstances allow. Other treatments and more detail can be found in [26, 90, 102, 144, 260, 275].

The theory of regularly varying functions is the appropriate mathematical analysis tool for proper discussion of heavy-tail phenomena. We begin by reviewing some results from analysis starting with uniform convergence.

## 2.1 Preliminaries from analysis

### 2.1.1 Uniform convergence

If $\{f_n, n \geq 0\}$ are real-valued functions on $\mathbb{R}$ (or, in fact, any metric space), then $f_n$ converges uniformly on $A \subset \mathbb{R}$ to $f_0$ if

$$\sup_{x \in A} |f_0(x) - f_n(x)| \to 0 \tag{2.1}$$

as $n \to \infty$. The definition would still make sense if the range of $f_n$, $n \geq 0$, were a metric space, but then $|f_0(x) - f_n(x)|$ would have to be replaced by $d(f_0(x), f_n(x))$, where $d(\cdot, \cdot)$ is the metric. For functions on $\mathbb{R}$, the phrase *local uniform convergence* means that (2.1) holds for any compact interval $A$.

A very useful fact is that monotone functions converging pointwise to a continuous limit converge locally uniformly. (See [260, p. 1] for additional material.)

**Proposition 2.1.** *Suppose $U_n$, $n \geq 0$, are nondecreasing, real-valued functions on $\mathbb{R}$ and that $U_0$ is continuous. If for all $x$,*

$$U_n(x) \to U_0(x) \quad (n \to \infty),$$

*then $U_n \to U_0$ locally uniformly; i.e., for any $a < b$,*

$$\sup_{x \in [a,b]} |U_n(x) - U_0(x)| \to 0.$$

*Proof.* One proof of this fact is outlined as follows: If $U_0$ is continuous on $[a, b]$, then it is uniformly continuous. From the uniform continuity, for any $x$, there is an interval-neighborhood $O_x$ on which $U_0(\cdot)$ oscillates by less than a given $\epsilon$. This gives an open cover of $[a, b]$. Compactness of $[a, b]$ allows us to prune $\{O_x, x \in [a, b]\}$ to obtain a finite subcover $\{(a_i, b_i), i = 1, \ldots, K\}$. Using this finite collection and the monotonicity of the functions leads to the result: Given $\epsilon > 0$, there exists some large $N$ such that if $n \geq N$, then

$$\max_{1 \leq i \leq K} \left( |U_n(a_i) - U_0(a_i)| \bigvee |U_n(b_i) - U_0(b_i)| \right) < \epsilon \tag{2.2}$$

(by pointwise convergence). Observe that

$$\sup_{x \in [a,b]} |U_n(x) - U_0(x)| \leq \max_{1 \leq i \leq K} \sup_{x \in [a_i, b_i]} |U_n(x) - U_0(x)|. \tag{2.3}$$

For any $x \in [a_i, b_i]$, we have by monotonicity

$$\begin{aligned} U_n(x) - U_0(x) &\leq U_n(b_i) - U_0(a_i) \\ &\leq U_0(b_i) + \epsilon - U_0(a_i) \quad \text{(by (2.2))} \\ &\leq 2\epsilon, \end{aligned}$$

with a similar lower bound. This is true for all $i$, and hence we get uniform convergence on $[a, b]$. □

### 2.1.2 Inverses of monotone functions

Suppose $H : \mathbb{R} \mapsto (a, b)$ is a nondecreasing function on $\mathbb{R}$ with range $(a, b)$, where $-\infty \leq a < b \leq \infty$. With the convention that the infimum of an empty set is $+\infty$, we define the (left-continuous) inverse $H^{\leftarrow} : (a, b) \mapsto \mathbb{R}$ of $H$ as

$$H^{\leftarrow}(y) = \inf\{s : H(s) \geq y\}.$$

See Figure 2.1.

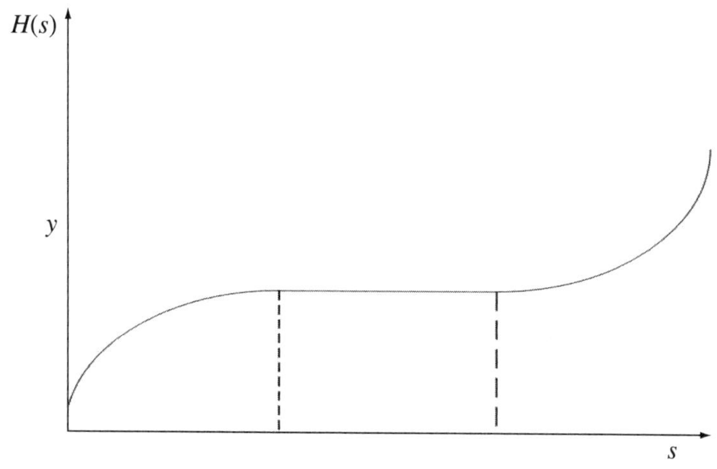


**Fig. 2.1.** The inverse at $y$ is the foot of the left dotted perpendicular.

In case the function $H$ is right continuous, we have the following desirable properties:

$$A(y) := \{s : H(s) \geq y\} \text{ is closed,} \tag{2.4}$$

$$H(H^{\leftarrow}(y)) \geq y, \tag{2.5}$$

$$H^{\leftarrow}(y) \leq t \quad \text{iff } y \leq H(t). \tag{2.6}$$

For (2.4), observe that if $s_n \in A(y)$ and $s_n \downarrow s$, then $y \leq H(s_n) \downarrow H(s)$, so $H(s) \geq y$ and $s \in A(y)$. If $s_n \uparrow s$ and $s_n \in A(y)$, then $y \leq H(s_n) \uparrow H(s-) \leq H(s)$ and $H(s) \geq y$, so $s \in A(y)$ again and $A(y)$ is closed. Since $A(y)$ is closed, $\inf A(y) \in A(y)$; that is, $H^{\leftarrow}(y) \in A(y)$ which means $H(H^{\leftarrow}(y)) \geq y$. This gives (2.5). Lastly, (2.6) follows from the definition of $H^{\leftarrow}$.

### 2.1.3 Convergence of monotone functions

For any function $H$ denote

$$\mathcal{C}(H) = \{x \in \mathbb{R} : H \text{ is finite and continuous at } x\}.$$

A sequence $\{H_n, n \geq 0\}$ of nondecreasing functions on $\mathbb{R}$ converges weakly to $H_0$ if as $n \to \infty$, we have

$$H_n(x) \to H_0(x)$$

for all $x \in \mathcal{C}(H_0)$. We will denote this by $H_n \to H_0$. No other form of convergence for monotone functions will be relevant. If $F_n$, $n \geq 0$, are probability distributions on $\mathbb{R}$, then a myriad of names give equivalent concepts: complete convergence, vague convergence, weak* convergence, narrow convergence. If $X_n$, $n \geq 0$, are random variables and $X_n$ has distribution function $F_n$, $n \geq 0$, then $X_n \Rightarrow X_0$ means $F_n \to F_0$. For the proof of the following, see [24], [260, p. 5], [264, p. 259].

**Proposition 2.2.** *If $H_n$, $n \geq 0$, are nondecreasing functions on $\mathbb{R}$ with range $(a, b)$ and $H_n \to H_0$, then $H_n^{\leftarrow} \to H_0^{\leftarrow}$ in the sense that for $t \in (a, b) \cap \mathcal{C}(H_0^{\leftarrow})$,*

$$H_n^{\leftarrow}(t) \to H_0^{\leftarrow}(t).$$

### 2.1.4 Cauchy's functional equation

Let $k(x)$, $x \in \mathbb{R}$, be a function that satisfies

$$k(x + y) = k(x) + k(y), \qquad x, y \in R.$$

If $k$ is measurable and bounded on a set of positive measure, then $k(x) = cx$ for some $c \in \mathbb{R}$. (See [275], [26, p. 4].)

## 2.2 Regular variation: Definition and first properties

The theory of regularly varying functions is an essential analytical tool for dealing with heavy tails, long-range dependence and domains of attraction. Roughly speaking, *regularly varying functions* are those functions which behave asymptotically like power functions. We will deal currently only with real functions of a real variable. Consideration of multivariate cases and probability concepts suggests recasting definitions in terms of vague convergence of measures, but we will consider this reformulation in Chapter 3.6 (p. 61) and Section 6.1.4 (p. 172).

**Definition 2.1.** A measurable function $U : \mathbb{R}_+ \mapsto \mathbb{R}_+$ is regularly varying at $\infty$ with index $\rho \in \mathbb{R}$ (written $U \in \mathrm{RV}_\rho$) if for $x > 0$,

$$\lim_{t\to\infty} \frac{U(tx)}{U(t)} = x^\rho.$$

We call $\rho$ the *exponent of variation.*

If $\rho = 0$, we call $U$ *slowly varying*. Slowly varying functions are generically denoted by $L(x)$. If $U \in \mathrm{RV}_\rho$, then $U(x)/x^\rho \in \mathrm{RV}_0$, and setting $L(x) = U(x)/x^\rho$, we see it is always possible to represent a $\rho$-varying function as $x^\rho L(x)$.

*Example* 2.1. The canonical $\rho$-varying function is $x^\rho$. The functions $\log(1+x)$, $\log\log(e+x)$ are slowly varying, as is $\exp\{(\log x)^\alpha\}$, $0<\alpha<1$. Any function $U$ such that $\lim_{x\to\infty} U(x) =: U(\infty)$ exists positive and finite is slowly varying. The following functions are not regularly varying: $e^x$, $\sin(x+2)$. Note that $[\log x]$ is slowly varying, but $\exp\{[\log x]\}$ is not regularly varying.

In probability applications, we are concerned with distributions whose tails are regularly varying. Examples are

$$1-F(x)=x^{-\alpha},\quad x\geq 1,\quad \alpha>0,$$

and the extreme-value distribution

$$\Phi_\alpha(x)=\exp\{-x^{-\alpha}\},\quad x\geq 0.$$

$\Phi_\alpha(x)$ has the property

$$1-\Phi_\alpha(x)\sim x^{-\alpha}\quad \text{as } x\to\infty.$$

A stable law (to be discussed later in Section 5.5.2 (p. 154)) with index $\alpha$, $0<\alpha<2$ has the property

$$1-G(x)\sim cx^{-\alpha},\quad x\to\infty,\quad c>0.$$

The Cauchy density $f(x)=(\pi(1+x^2))^{-1}$ has a distribution function $F$ with the property

$$1-F(x)\sim(\pi x)^{-1}.$$

If $N(x)$ is the standard normal distribution function, then $1-N(x)$ is not regularly varying nor is the tail of the Gumbel extreme-value distribution $1-\exp\{-e^{-x}\}$.

The definition of regular variation can be weakened slightly (cf. [102, 135, 260]).

**Proposition 2.3.**

(i) *A measurable function $U:\mathbb{R}_+\mapsto\mathbb{R}_+$ varies regularly if there exists a function $h$ such that for all $x>0$,*

$$\lim_{t\to\infty} U(tx)/U(t)=h(x).$$

*In this case $h(x)=x^\rho$ for some $\rho\in\mathbb{R}$ and $U\in RV_\rho$.*

(ii) *A monotone function $U:\mathbb{R}_+\mapsto\mathbb{R}_+$ varies regularly provided there are two sequences $\{\lambda_n\}$, $\{b_n\}$ of positive numbers satisfying*

$$b_n\to\infty,\qquad \lambda_n\sim\lambda_{n+1},\quad n\to\infty, \tag{2.7}$$

*and for all $x>0$,*

$$\lim_{n\to\infty}\lambda_n U(b_n x)=:\chi(x)\ \textit{exists positive and finite.} \tag{2.8}$$

*In this case $\chi(x)/\chi(1)=x^\rho$ and $U\in RV_\rho$ for some $\rho\in\mathbb{R}$.*

We frequently refer to (2.8) as the *sequential form of regular variation*. For probability purposes, it is most useful. Typically, $U$ is a distribution tail, $\lambda_n = n$, and $b_n$ is a distribution quantile.

*Proof.*

(i) The function $h$ is measurable since it is a limit of a family of measurable functions. Then for $x > 0$, $y > 0$,

$$\frac{U(txy)}{U(t)} = \frac{U(txy)}{U(tx)} \cdot \frac{U(tx)}{U(t)},$$

and letting $t \to \infty$ gives

$$h(xy) = h(y)h(x).$$

So $h$ satisfies the Hamel equation, which by change of variable can be converted to the Cauchy equation. Therefore, the form of $h$ is $h(x) = x^\rho$ for some $\rho \in \mathbb{R}$.

(ii) For concreteness assume $U$ is nondecreasing. Assume (2.7) and (2.8), and we show regular variation. Since $b_n \to \infty$, for each $t$ there is a finite $n(t)$ defined by

$$n(t) = \inf\{m : b_{m+1} > t\}$$

so that

$$b_{n(t)} \le t < b_{n(t)+1}.$$

Therefore, by monotonicity for $x > 0$,

$$\begin{aligned}&\left(\frac{\lambda_{n(t)+1}}{\lambda_{n(t)}}\right)\left(\frac{\lambda_{n(t)}U(b_{n(t)}x)}{\lambda_{n(t)+1}U(b_{n(t)+1})}\right)\\ &\quad\le \frac{U(tx)}{U(t)} \le \left(\frac{\lambda_{n(t)}}{\lambda_{n(t)+1}}\right)\left(\frac{\lambda_{n(t)+1}U(b_{n(t)+1}x)}{\lambda_{n(t)}U(b_{n(t)})}\right).\end{aligned}$$

Now let $t \to \infty$ and use (2.7) and (2.8) to get $\lim_{t\to\infty} \frac{U(tx)}{U(t)} = 1\frac{\chi(x)}{\chi(1)}$. Regular variation follows from part (i). □

*Remark* 2.1. Proposition 2.3(ii) remains true if we only assume (2.8) holds on a dense set. This is relevant to the case where $U$ is nondecreasing and $\lambda_n U(b_n x)$ converges weakly.

### 2.2.1 A maximal domain of attraction

Suppose $\{X_n, n \geq 1\}$ are iid with common distribution function $F(x)$. The extreme is

$$M_n = \bigvee_{i=1}^{n} X_i = \max\{X_1, \dots, X_n\}.$$

One of the extreme-value distributions is

$$\Phi_\alpha(x) := \exp\{-x^{-\alpha}\}, \quad x > 0, \quad \alpha > 0.$$

What are conditions on $F$, called *domain of attraction conditions*, so that there exists $b_n > 0$ such that

$$P[b_n^{-1} M_n \leq x] = F^n(b_n x) \to \Phi_\alpha(x) \tag{2.9}$$

weakly? How do you characterize the normalization sequence $\{b_n\}$?

Set $x_0 = \sup\{x : F(x) < 1\}$ which is called the right endpoint of $F$. We first check that (2.9) implies $x_0 = \infty$. Otherwise, if $x_0 < \infty$, we get from (2.9) that for $x > 0$, $b_n x \to x_0$; i.e., $b_n \to x_0 x^{-1}$. Since $x > 0$ is arbitrary, we get $b_n \to 0$, whence $x_0 = 0$. But then for $x > 0$, $F^n(b_n x) = 1$, which violates (2.9). Hence $x_0 = \infty$.

Furthermore, $b_n \to \infty$ since otherwise on a subsequence $n'$, $b_{n'} \leq K$ for some $K < \infty$. Then, since $F(K) < 1$,

$$0 < \Phi_\alpha(1) = \lim_{n' \to \infty} F^{n'}(b_{n'}) \leq \lim_{n' \to \infty} F^{n'}(K) = 0,$$

which is a contradiction.

In (2.9), take logarithms to get for $x > 0$, $\lim_{n\to\infty} n(-\log F(b_n x)) = x^{-\alpha}$. Now use the relation $-\log(1-z) \sim z$ as $z \to 0$ and (2.9) is equivalent to

$$\lim_{n \to \infty} n(1 - F(b_n x)) = x^{-\alpha}, \quad x > 0. \tag{2.10}$$

From (2.10) and Proposition 2.3, we get

$$1 - F(x) \sim x^{-\alpha} L(x), \quad x \to \infty \tag{2.11}$$

for some $\alpha > 0$. To characterize $\{b_n\}$, set $U(x) = 1/(1 - F(x))$, and (2.10) is the same as

$$U(b_n x)/n \to x^\alpha, \quad x > 0;$$

inverting, we find via Proposition 2.2 that

$$\frac{U^{\leftarrow}(ny)}{b_n} \to y^{1/\alpha}, \quad y > 0. \tag{2.12}$$

So $U^{\leftarrow}(n) = (1/(1-F))^{\leftarrow}(n) \sim b_n$, and this determines $b_n$ by the convergence-to-types theorem. See [135, 260, 264].

Conversely, if (2.11) holds, define $b_n = U^{\leftarrow}(n)$ as previously. Then

$$\lim_{n\to\infty} \frac{1-F(b_n x)}{1-F(b_n)} = x^{-\alpha},$$

and we recover (2.10) provided $1-F(b_n) \sim n^{-1}$ or what is the same provided $U(b_n) \sim n$, i.e., $U(U^{\leftarrow}(n)) \sim n$. Recall from (2.6) that $z < U^{\leftarrow}(n)$ iff $U(z) < n$, and setting $z = U^{\leftarrow}(n)(1-\varepsilon)$ and then $z = U^{\leftarrow}(n)(1+\varepsilon)$, we get

$$\frac{U(U^{\leftarrow}(n))}{U(U^{\leftarrow}(n)(1+\varepsilon))} \le \frac{U(U^{\leftarrow}(n))}{n} \le \frac{U(U^{\leftarrow}(n))}{U(U^{\leftarrow}(n)(1-\varepsilon))}.$$

Let $n \to \infty$, remembering $U = 1/(1-F) \in RV_\alpha$. Then

$$(1+\varepsilon)^{-\alpha} \le \liminf_{n\to\infty} n^{-1} U(U^{\leftarrow}(n)) \le \limsup_{n\to\infty} U(U^{\leftarrow}(n)) \le (1-\varepsilon)^{-\alpha},$$

and since $\varepsilon > 0$ is arbitrary, the desired result follows.

## 2.3 Regular variation: deeper results; Karamata's theorem

There are several deeper results that give the theory power and utility: uniform convergence; Karamata's theorem, which says that a regularly varying function integrates the way you expect a power function to integrate; and finally the Karamata representation theorem.

### 2.3.1 Uniform convergence

The first useful result is the uniform convergence theorem.

**Proposition 2.4.** *If $U \in RV_\rho$ for $\rho \in \mathbb{R}$, then*

$$\lim_{t\to\infty} U(tx)/U(t) = x^\rho$$

*locally uniformly in $x$ on $(0,\infty)$. If $\rho < 0$, then uniform convergence holds on intervals of the form $(b,\infty)$, $b > 0$. If $\rho > 0$, uniform convergence holds on intervals $(0,b]$ provided $U$ is bounded on $(0,b]$ for all $b > 0$.*

If $U$ is monotone the result already follows from the discussion in Section 2.1.1, since we have a family of monotone functions converging to a continuous limit. For detailed discussion, see [26, 102, 144, 275].

### 2.3.2 Integration and Karamata's theorem

The next set of results examines the integral properties of regularly varying functions [26, 102, 181, 183, 275]. For purposes of integration, a $\rho$-varying function behaves roughly like $x^\rho$. We assume all functions are locally integrable, and since we are interested in behavior at $\infty$, we assume integrability on intervals including 0 as well.

**Theorem 2.1 (Karamata's theorem).**

(a) *Suppose* $\rho \geq -1$ *and* $U \in \mathrm{RV}_\rho$. *Then* $\int_0^x U(t)dt \in \mathrm{RV}_{\rho+1}$ *and*

$$\lim_{x\to\infty} \frac{xU(x)}{\int_0^x U(t)dt} = \rho + 1. \tag{2.13}$$

*If* $\rho < -1$ *(or if* $\rho = -1$ *and* $\int_x^\infty U(s)ds < \infty$*), then* $U \in \mathrm{RV}_\rho$ *implies that* $\int_x^\infty U(t)dt$ *is finite,* $\int_x^\infty U(t)dt \in \mathrm{RV}_{\rho+1}$*, and*

$$\lim_{x\to\infty} \frac{xU(x)}{\int_x^\infty U(t)dt} = -\rho - 1. \tag{2.14}$$

(b) *If* $U$ *satisfies*

$$\lim_{x\to\infty} \frac{xU(x)}{\int_0^x U(t)dt} = \lambda \in (0, \infty), \tag{2.15}$$

*then* $U \in \mathrm{RV}_{\lambda-1}$. *If* $\int_x^\infty U(t)dt < \infty$ *and*

$$\lim_{x\to\infty} \frac{xU(x)}{\int_x^\infty U(t)dt} = \lambda \in (0, \infty), \tag{2.16}$$

*then* $U \in \mathrm{RV}_{-\lambda-1}$.

What Theorem 2.1 emphasizes is that for the purposes of integration, the slowly varying function can be passed from inside to outside the integral. For example, the way to remember and interpret (2.13) is to write $U(x) = x^\rho L(x)$ and then observe that

$$\int_0^x U(t)dt = \int_0^x t^\rho L(t)dt;$$

now pass the $L(t)$ in the integrand outside as a factor $L(x)$ to get

$$\begin{aligned} &\sim L(x)\int_0^x t^\rho dt = L(x)x^{\rho+1}/(\rho+1) \\ &= xx^\rho L(x)/(\rho+1) = xU(x)/(\rho+1), \end{aligned}$$

which is equivalent to the assertion (2.13).

*Proof.*

(a) For certain values of $\rho$, uniform convergence suffices after writing, for instance,

$$\frac{\int_0^x U(s)ds}{xU(x)} = \int_0^1 \frac{U(sx)}{U(x)}ds.$$

If we wish to proceed using elementary concepts, consider the following approach, which follows [102].

If $\rho > -1$, we show that $\int_0^\infty U(t)dt = \infty$. From $U \in \mathrm{RV}_\rho$, we have

$$\lim_{s\to\infty} U(2s)/U(s) = 2^\rho > 2^{-1}$$

since $\rho > -1$. Therefore, there exists $s_0$ such that $s > s_0$ necessitates $U(2s) > 2^{-1}U(s)$. For $n$ with $2^n > s_0$, we have

$$\int_{2^{n+1}}^{2^{n+2}} U(s)ds = 2\int_{2^n}^{2^{n+1}} U(2s)ds > \int_{2^n}^{2^{n+1}} U(s)ds,$$

and so setting $n_0 = \inf\{n : 2^n > s_0\}$ gives

$$\int_{s_0}^\infty U(s)ds \ge \sum_{n:2^n>s_0} \int_{2^{n+1}}^{2^{n+2}} U(s)ds > \sum_{n\ge n_0} \int_{2^{n_0+1}}^{2^{n_0+2}} U(s)ds = \infty.$$

Thus for $\rho > -1$, $x > 0$, and any $N < \infty$, we have

$$\int_0^t U(sx)ds \sim \int_N^t U(sx)ds, \quad t \to \infty,$$

since $U(sx)$ is a $\rho$-varying function of $s$. For fixed $x$ and given $\varepsilon$, there exists $N$ such that for $s > N$,

$$(1-\varepsilon)x^\rho U(s) \le U(sx) \le (1+\varepsilon)x^\rho U(s),$$

and thus

$$\begin{aligned}
\limsup_{t\to\infty} \frac{\int_0^{tx} U(s)ds}{\int_0^t U(s)ds} &= \limsup_{t\to\infty} \frac{x\int_0^t U(sx)ds}{\int_0^t U(s)ds} \\
&= \limsup_{t\to\infty} \frac{x\int_N^t U(sx)ds}{\int_N^t U(s)ds} \\
&\le \limsup_{t\to\infty} x^{\rho+1}(1+\varepsilon)\frac{\int_N^t U(s)ds}{\int_N^t U(s)ds} \\
&= (1+\varepsilon)x^{\rho+1}.
\end{aligned}$$

An analogous argument applies for lim inf, and thus we have proved

$$\int_0^x U(s)ds \in \mathrm{RV}_{\rho+1}$$

when $\rho > -1$.

In case $\rho = -1$, then either $\int_0^\infty U(s)ds < \infty$, in which case $\int_0^x U(s)ds \in \mathrm{RV}_{-1+1} = \mathrm{RV}_0$, or $\int_0^\infty U(s)ds = \infty$ and the previous argument is applicable. So we have checked that for $\rho \geq -1$, $\int_0^x U(s)ds \in \mathrm{RV}_{\rho+1}$.

We now focus on proving (2.13) when $U \in \mathrm{RV}_\rho$, $\rho \geq -1$. Define the function

$$b(x) := xU(x) / \int_0^x U(t)dt, \tag{2.17}$$

so that integrating $b(x)/x$ leads to the representations

$$\begin{aligned} \int_0^x U(s)ds &= c \exp\left\{\int_1^x t^{-1}b(t)dt\right\}, \\ U(x) &= cx^{-1}b(x) \exp\left\{\int_1^x t^{-1}b(t)dt\right\}. \end{aligned} \tag{2.18}$$

We must show $b(x) \to \rho + 1$. Observe first that

$$\begin{aligned} \liminf_{x\to\infty} 1/b(x) &= \liminf_{x\to\infty} \frac{\int_0^x U(t)dt}{xU(x)} \\ &= \liminf_{x\to\infty} \int_0^1 \frac{U(sx)}{U(x)} ds. \end{aligned}$$

Now make a change of variable $s = x^{-1}t$, and by Fatou's lemma this is

$$\begin{aligned} &\geq \int_0^1 \liminf_{x\to\infty} (U(sx)/U(x))ds \\ &= \int_0^1 s^\rho ds = \frac{1}{\rho+1}, \end{aligned}$$

and we conclude that

$$\limsup_{x\to\infty} b(x) \leq \rho + 1. \tag{2.19}$$

If $\rho = -1$, then $b(x) \to 0$ as desired, so now suppose $\rho > -1$.

We observe the following properties of $b(x)$:

(i) $b(x)$ is bounded on a semi-infinite neighborhood of $\infty$ (by (2.19)).

(ii) $b$ is slowly varying since $xU(x) \in RV_{\rho+1}$ and $\int_0^x U(s)ds \in RV_{\rho+1}$.

(iii) We have

$$b(xt) - b(x) \to 0$$

as $x \to \infty$, and the convergence is uniformly bounded for $t$ in finite intervals.

The last statement follows since by slow variation,

$$\lim_{x\to\infty} (b(xt) - b(x))/b(x) = 0$$

and the denominator is ultimately bounded.

From (iii) and dominated convergence

$$\lim_{x\to\infty} \int_1^s t^{-1}(b(xt) - b(x))dt = 0,$$

and the left side may be rewritten to obtain

$$\lim_{x\to\infty} \left\{ \int_1^s t^{-1} b(xt)dt - b(x)\log s \right\} = 0. \tag{2.20}$$

From (2.18)

$$c\exp\left\{ \int_1^x t^{-1} b(t)dt \right\} = \int_0^x U(s)ds \in RV_{\rho+1},$$

and from the regular variation property

$$\begin{aligned}(\rho+1)\log s &= \lim_{x\to\infty} \log\left( \frac{\int_0^{xs} U(t)dt}{\int_0^x U(t)dt} \right) \\ &= \lim_{x\to\infty} \int_x^{xs} t^{-1} b(t)dt = \lim_{x\to\infty} \int_1^s t^{-1} b(xt)dt;\end{aligned}$$

combining this with (2.20) leads to the desired conclusion that $b(x) \to \rho + 1$.

(b) We suppose (2.15) holds and check $U \in RV_{\lambda-1}$. Set

$$b(x) = xU(x) / \int_0^x U(t)dt,$$

so that $b(x) \to \lambda$. From (2.18)

$$\begin{aligned}U(x) &= cx^{-1} b(x) \exp\left\{ \int_1^x t^{-1} b(t)dt \right\} \\ &= cb(x)\exp\left\{ \int_1^x t^{-1}(b(t) - 1)dt \right\},\end{aligned}$$

and since $b(t) - 1 \to \lambda - 1$, $U$ satisfies the definition of being $(\lambda - 1)$-varying as can be checked from the definition. (See Corollary 2.1.) □

### 2.3.3 Karamata's representation

Theorem 2.1 leads in a straightforward way to what has been called the *Karamata representation* of a regularly varying function.

**Corollary 2.1 (the Karamata representation).**

(i) *The function $L$ is slowly varying iff $L$ can be represented as*

$$L(x) = c(x) \exp\left\{\int_1^x t^{-1}\varepsilon(t)dt\right\}, \quad x > 0, \tag{2.21}$$

*where $c : \mathbb{R}_+ \mapsto \mathbb{R}_+$, $\varepsilon : \mathbb{R}_+ \mapsto \mathbb{R}_+$, and*

$$\lim_{x\to\infty} c(x) = c \in (0, \infty), \tag{2.22}$$

$$\lim_{t\to\infty} \varepsilon(t) = 0. \tag{2.23}$$

(ii) *A function $U : \mathbb{R}_+ \mapsto \mathbb{R}_+$ is regularly varying with index $\rho$ iff $U$ has the representation*

$$U(x) = c(x) \exp\left\{\int_1^x t^{-1}\rho(t)dt\right\}, \tag{2.24}$$

*where $c(\cdot)$ satisfies* (2.22) *and $\lim_{t\to\infty} \rho(t) = \rho$. (This is obtained from* (i) *by writing $U(x) = x^\rho L(x)$ and using the representation for $L$.)*

*Proof.* If $L$ has a representation (2.21), then it must be slowly varying since for $x > 1$,

$$\lim_{t\to\infty} L(tx)/L(t) = \lim_{t\to\infty} (c(tx)/c(t)) \exp\left\{\int_t^{tx} s^{-1}\varepsilon(s)ds\right\}.$$

Given $\varepsilon$, there exists $t_0$ by (2.23) such that

$$-\varepsilon < \varepsilon(t) < \varepsilon, \quad t \geq t_0,$$

so that

$$-\varepsilon \log x = -\varepsilon \int_t^{tx} s^{-1}ds \leq \int_t^{tx} s^{-1}\varepsilon(s)ds \leq \varepsilon \int_t^{tx} s^{-1}ds = \varepsilon \log x.$$

Therefore, $\lim_{t\to\infty} \int_t^{tx} s^{-1}\varepsilon(s)ds = 0$ and $\lim_{t\to\infty} L(tx)/L(t) = 1$.

Conversely, suppose $L \in RV_0$. In a matter similar to (2.17), define

$$b(x) := xL(x)/\int_0^x L(s)ds,$$

and by Karamata's theorem, $b(x) \to 1$ as $x \to \infty$. Note that

$$L(x) = x^{-1} b(x) \int_0^x L(s) ds.$$

Set $\varepsilon(x) = b(x) - 1$, so $\varepsilon(x) \to 0$ and

$$\begin{aligned}
\int_1^x t^{-1} \varepsilon(t) dt &= \int_1^x \left( L(t) / \int_0^t L(s) ds \right) dt - \log x \\
&= \int_1^x d \left( \log \int_0^t L(s) ds \right) - \log x \\
&= \log \left( x^{-1} \int_0^x L(s) ds / \int_0^1 L(s) ds \right),
\end{aligned}$$

whence

$$\begin{aligned}
\exp \left\{ \int_1^x t^{-1} \varepsilon(t) dt \right\} &= x^{-1} \int_0^x L(s) ds / \int_0^1 L(s) ds \\
&= L(x) / \left( b(x) \int_0^1 L(s) ds \right), \qquad (2.25)
\end{aligned}$$

and the representation follows with

$$c(x) = b(x) \int_0^1 L(s) ds.$$

□

*Example* 2.2. The Cauchy density

$$F'(x) = \frac{1}{2\pi} \left( \frac{1}{1+x^2} \right), \quad x \in \mathbb{R},$$

satisfies

$$F'(x) \sim \frac{1}{2\pi} x^{-2}, \quad x \to \infty,$$

and hence

$$1 - F(x) \sim \frac{1}{2\pi} x^{-1}, \quad x \to \infty.$$

### 2.3.4 Differentiation

The previous results describe the asymptotic properties of the indefinite integral of a regularly varying function. We now describe what happens when a $\rho$-varying function is differentiated.

**Proposition 2.5.** *Suppose* $U : R_+ \mapsto R_+$ *is absolutely continuous with density* $u$ *so that*

$$U(x) = \int_0^x u(t)dt.$$

(a) (von Mises [293]) *If*

$$\lim_{x\to\infty} xu(x)/U(x) = \rho, \tag{2.26}$$

*then* $U \in \mathrm{RV}_\rho$.

(b) (Landau [196]) *If* $U \in \mathrm{RV}_\rho$, $\rho \in \mathbb{R}$, *and* $u$ *is monotone, then* (2.26) *holds, and if* $\rho \neq 0$, *then* $|u|(x) \in \mathrm{RV}_{\rho-1}$. (*See* [260, 275] *and* [102, *pp.* 23 *and* 109].)

*Proof.*

(a) Set

$$b(x) = xu(x)/U(x)$$

and as before we find that

$$U(x) = U(1)\exp\left\{\int_1^x t^{-1}b(t)dt\right\}$$

so that $U$ satisfies the representation theorem for a $\rho$-varying function.

(b) Suppose $u$ is nondecreasing. An analogous proof works in the case $u$ is nonincreasing. Let $0 < a < b$ and observe that

$$(U(xb) - U(xa))/U(x) = \int_{xa}^{xb} u(y)dy/U(x).$$

By monotonicity we get

$$u(xb)x(b-a)/U(x) \geq (U(xb)-U(xa))/U(x) \geq u(xa)x(b-a)/U(x). \tag{2.27}$$

From (2.27) and the fact that $U \in \mathrm{RV}_\rho$, we conclude that

$$\limsup_{x\to\infty} xu(xa)/U(x) \leq (b^\rho - a^\rho)/(b-a) \tag{2.28}$$

for any $b > a > 0$. So let $b \downarrow a$, which is tantamount to taking a derivative. Then (2.28) becomes

$$\limsup_{x\to\infty} xu(xa)/U(x) \leq \rho a^{\rho-1} \tag{2.29}$$

for any $a > 0$. Similarly, from the left-hand equality in (2.27) after letting $a \uparrow b$, we get

$$\liminf_{x\to\infty} xu(xb)/U(x) \geq \rho b^{\rho-1} \tag{2.30}$$

for any $b > 0$. Then (2.26) results by setting $a = 1$ in (2.29) and $b = 1$ in (2.30). □

## 2.4 Regular variation: Further properties

For the following list of properties, it is convenient to define *rapid variation* or regular variation with index $\infty$. We say $U : \mathbb{R}_+ \mapsto \mathbb{R}_+$ is regularly varying with index $\infty$ ($U \in \mathrm{RV}_\infty$) if for every $x > 0$,

$$\lim_{t\to\infty} \frac{U(tx)}{U(t)} = x^\infty := \begin{cases} 0 & \text{if } x < 1, \\ 1 & \text{if } x = 1, \\ \infty & \text{if } x > 1. \end{cases}$$

Similarly, $U \in \mathrm{RV}_{-\infty}$ if

$$\lim_{t\to\infty} \frac{U(tx)}{U(t)} = x^{-\infty} := \begin{cases} \infty & \text{if } x < 1, \\ 1 & \text{if } x = 1, \\ 0 & \text{if } x > 1. \end{cases}$$

The following proposition, modeled after [102] (see also [90]), collects useful properties of regularly varying functions.

**Proposition 2.6.**

(i) *If* $U \in \mathrm{RV}_\rho$, $-\infty \le \rho \le \infty$, *then*

$$\lim_{x\to\infty} \log U(x)/\log x = \rho$$

*so that*

$$\lim_{x\to\infty} U(x) = \begin{cases} 0 & \text{if } \rho < 0, \\ \infty & \text{if } \rho > 0. \end{cases}$$

(ii) (Potter bounds) *Suppose* $U \in \mathrm{RV}_\rho$, $\rho \in \mathbb{R}$. *Take* $\varepsilon > 0$. *Then there exists* $t_0$ *such that for* $x \ge 1$ *and* $t \ge t_0$,

$$(1-\varepsilon)x^{\rho-\varepsilon} < \frac{U(tx)}{U(t)} < (1+\varepsilon)x^{\rho+\varepsilon}. \tag{2.31}$$

(iii) *If* $U \in \mathrm{RV}_\rho$, $\rho \in \mathbb{R}$, *and* $\{a_n\}$, $\{b_n\}$ *satisfy* $0 < b_n \to \infty$, $0 < a_n \to \infty$, *and* $b_n \sim ca_n$ *as* $n \to \infty$ *for* $0 < c < \infty$, *then* $U(b_n) \sim c^\rho U(a_n)$. *If* $\rho \ne 0$, *the result also holds for* $c = 0$ *or* $\infty$. *Analogous results hold with sequences replaced by functions.*

(iv) *If* $U_1 \in \mathrm{RV}_{\rho_1}$ *and* $U_2 \in \mathrm{RV}_{\rho_2}$, $\rho_2 < \infty$, *and* $\lim_{x\to\infty} U_2(x) = \infty$, *then*

$$U_1 \circ U_2 \in \mathrm{RV}_{\rho_1\rho_2}.$$

(v) *Suppose $U$ is nondecreasing, $U(\infty) = \infty$, and $U \in \mathrm{RV}_\rho$, $0 \le \rho \le \infty$. Then*

$$U^{\leftarrow} \in \mathrm{RV}_{\rho^{-1}}.$$

(vi) *Suppose $U_1$, $U_2$ are nondecreasing and $\rho$-varying, $0 < \rho < \infty$. Then for $0 \le c \le \infty$,*

$$U_1(x) \sim cU_2(x), \quad x \to \infty,$$

*iff*

$$U_1^{\leftarrow}(x) \sim c^{-\rho^{-1}} U_2^{\leftarrow}(x), \quad x \to \infty.$$

(vii) *If $U \in \mathrm{RV}_\rho$, $\rho \neq 0$, then there exists a function $U^*$ that is absolutely continuous, strictly monotone, and*

$$U(x) \sim U(x)^*, \quad x \to \infty.$$

*Proof.*

(i) We give the proof for the case $0 < \rho < \infty$. Suppose $U$ has Karamata representation

$$U(x) = c(x) \exp\left\{\int_1^x t^{-1}\rho(t)dt\right\},$$

where $c(x) \to c > 0$ and $\rho(t) \to \rho$. Then

$$\log U(x)/\log x = o(1) + \int_1^x t^{-1}\rho(t)dt \Big/ \int_1^x t^{-1}dt \to \rho.$$

(ii) Using the Karamata representation,

$$U(tx)/U(t) = (c(tx)/c(t)) \exp\left\{\int_1^x s^{-1}\rho(ts)ds\right\},$$

and the result is apparent since we may pick $t_0$ so that $t > t_0$ implies that $\rho - \varepsilon < \rho(ts) < \rho + \varepsilon$ for $s > 1$.

(iii) If $c > 0$, then from the uniform convergence property in Proposition 2.4,

$$\lim_{n\to\infty} \frac{U(b_n)}{U(a_n)} = \lim_{n\to\infty} \frac{U(a_n(b_n/a_n))}{U(a_n)} = \lim_{t\to\infty} \frac{U(tc)}{U(t)} = c^\rho.$$

(iv) Again by uniform convergence, for $x > 0$,

$$\begin{aligned}\lim_{t\to\infty} \frac{U_1(U_2(tx))}{U_1(U_2(t))} &= \lim_{t\to\infty} \frac{U_1(U_2(t)(U_2(tx)/U_2(t)))}{U_1(U_2(t))} \\ &= \lim_{y\to\infty} \frac{U_1(yx^{\rho_2})}{U_1(y)} = x^{\rho_2\rho_1}.\end{aligned}$$

(v) Let $U_t(x) = U(tx)/U(t)$, so that if $U \in \mathrm{RV}_\rho$ and $U$ is nondecreasing, then $(0 < \rho < \infty)$

$$U_t(x) \to x^\rho, \quad t \to \infty,$$

which implies by Proposition 2.2 that

$$U_t^{\leftarrow}(x) \to x^{\rho^{-1}}, \quad t \to \infty;$$

that is,

$$\lim_{t\to\infty} U^{\leftarrow}(xU(t))/t = x^{\rho^{-1}}.$$

Therefore,

$$\lim_{t\to\infty} U^{\leftarrow}(xU(U^{\leftarrow}(t)))/U^{\leftarrow}(t) = x^{\rho^{-1}}.$$

This limit holds locally uniformly since monotone functions are converging to a continuous limit. Now $U \circ U^{\leftarrow}(t) \sim t$ as $t \to \infty$, and if we replace $x$ by $xt/U \circ U^{\leftarrow}(t)$ and use uniform convergence, we get

$$\lim_{t\to\infty} \frac{U^{\leftarrow}(tx)}{U^{\leftarrow}(t)} = \lim_{t\to\infty} \frac{U^{\leftarrow}((xt/U \circ U^{\leftarrow}(t))U \circ U^{\leftarrow}(t))}{U^{\leftarrow}(t)}$$
$$= \lim_{t\to\infty} \frac{U^{\leftarrow}(xU \circ U^{\leftarrow}(t))}{U^{\leftarrow}(t)} = x^{\rho^{-1}},$$

which makes $U^{\leftarrow} \in \mathrm{RV}_{\rho^{-1}}$.

(vi) If $c > 0$, $0 < \rho < \infty$, we have for $x > 0$,

$$\lim_{t\to\infty} \frac{U_1(tx)}{U_2(t)} = \lim_{t\to\infty} \frac{U_1(tx)U_2(tx)}{U_2(tx)U_2(t)} = cx^\rho.$$

Inverting, we find for $y > 0$,

$$\lim_{t\to\infty} U_1^{\leftarrow}(yU_2(t))/t = (c^{-1}y)^{\rho^{-1}},$$

and so

$$\lim_{t\to\infty} U_1^{\leftarrow}(yU_2 \circ U_2^{\leftarrow}(t))/U_2^{\leftarrow}(t) = (c^{-1}y)^{\rho^{-1}},$$

and since $U_2 \circ U_2^{\leftarrow}(t) \sim t$,

$$\lim_{t\to\infty} U_1^{\leftarrow}(yt)/U_2^{\leftarrow}(t) = (c^{-1}y)^{\rho^{-1}}.$$

Set $y = 1$ to obtain the result.

(vii) For instance, if $U \in RV_\rho$, $\rho > 0$, define

$$U^*(t) = \int_1^t s^{-1} U(s) ds.$$

Then $s^{-1}U(s) \in RV_{\rho-1}$, and by Karamata's theorem,

$$U(x)/U^*(x) \to \rho.$$

$U^*$ is absolutely continuous, and since $U(x) \to \infty$ when $\rho > 0$, then $U^*$ is ultimately strictly increasing. □

## 2.5 Problems

**2.1.** Suppose

$$U(x) = 2\log x + \sin(\log x), \quad x > e.$$

Is $U(\cdot)$ regularly varying? If so, what is the index? What is the Karamata representation?

**2.2.** Give an example of a slowly varying function $L(x)$ such that $\lim_{x\to\infty} L(x)$ does not exist. (Would the Karamata representation be helpful?)

**2.3.** Verify that the following functions are slowly varying and give the Karamata representation:

1. $(1 + x^{-1}) \log x$; $x > e$.
2. $\exp\{(\log x)^\alpha\}$, $x > e$, $0 < \alpha < 1$.
3. $2 + \sin(\log\log x)$; $x > e^e$.
4. $\sum_{k=1}^{[x]} 1/k$; $x \ge 1$.

**2.4.** Check that the following functions are not regularly varying:

1. $2 + \sin(\log x)$; $x > e$.
2. $\exp\{[\log x]\}$; $x > e$.
3. $2 + \sin x$, $x > 0$.

Regarding item 2, is

$$\int_{u=e}^x \exp\{[\log u]\} du$$

regularly varying?

**2.5 (Variant of Karamata's theorem).** Suppose $F$ is a distribution on $\mathbb{R}_+$ and

$$1 - F(x) \sim x^{-\alpha} L(x), \quad x \to \infty.$$

1. For $\eta \geq \alpha$, show by integrating by parts or using Fubini's theorem that

$$\lim_{x\to\infty} \frac{\int_0^x u^\eta F(du)}{x^\eta(1-F(x))} = \frac{\alpha}{\eta - \alpha}.$$

2. For $\eta > 0$, show that

$$\lim_{x\to\infty} \frac{\int_x^\infty u^{-\eta} F(du)}{x^{-\eta}(1-F(x)} = \frac{\alpha}{\alpha + \eta}.$$

**2.6 (Variant of Potter's inequality [255]).** Let $Z$ be a nonnegative random variable with distribution $F$ such that $1-F$ is regularly varying with index $-\alpha, \alpha > 0$. If $\epsilon > 0$ is given, there exist constants $x_0 = x_0(\epsilon)$, $K = K(\epsilon) > 0$ such that for any $c > 0$, we have the following:

(i) *Tail ratio result*:

$$\frac{1-F(x/c)}{1-F(x)} \leq \begin{cases} (1+\epsilon)c^{\alpha+\epsilon} & \text{if } c \geq 1, x/c \geq x_0, \\ (1+\epsilon)c^{\alpha-\epsilon} & \text{if } c < 1, x \geq x_0. \end{cases}$$

(ii) *Expectation result*:

$$E(cZ \wedge x)^{\alpha+\epsilon} \leq \begin{cases} Kc^{\alpha+\epsilon}x^{\alpha+\epsilon}(1-F(x)) & \text{if } c \geq 1, x/c \geq x_0, \\ Kc^{\alpha-\epsilon}x^{\alpha+\epsilon}(1-F(x)) & \text{if } c < 1, x \geq x_0. \end{cases}$$

**2.7.** Suppose $\{N_n, n \geq 1\}$ is a sequence of nonnegative random variables such that

$$\frac{N_n}{n} \xrightarrow{P} N.$$

Assume $a(\cdot) \in \mathrm{RV}_\rho$ and $\mathbb{P}[N > 0] = 1$. Prove that

$$\frac{a(N_n)}{a(n)} \xrightarrow{P} N^\rho.$$

**2.8.** Prove that $L$ is slowly varying iff for all $x > 1$,

$$\lim_{t\to\infty} \frac{L(tx)}{L(t)} = 1.$$

If $L$ is monotone, it is enough to check the limit for one positive $x \neq 1$.

**2.9 (Relative stability of sums).** Prove the following are equivalent for iid nonnegative random variables $\{X_n, n \geq 1\}$.

1. There exist constants $b_n \to \infty$ such that as $n \to \infty$,

$$b_n^{-1} \sum_{i=1}^{n} X_i \xrightarrow{P} 1.$$

2. The Laplace transform

$$\phi(\lambda) := \mathbf{E}(e^{-\lambda X_1}), \quad \lambda > 0,$$

satisfies

$$\frac{1 - \phi(\tau^{-1})}{\tau^{-1}} \in \mathrm{RV}_0;$$

that is,

$$\frac{1 - \phi(\tau)}{\tau} = \int_0^\infty e^{-\tau x} \mathbb{P}[X_1 > x] dx$$

is slowly varying at 0.

3. The function

$$U(x) = \int_0^x P[X_1 > s] ds$$

is slowly varying at $\infty$. (This requires the use of a Tauberian theorem. See Section 7.3.3.)

Characterize the constant $b_n$ in terms of $\phi$. Verify that $b_n$ can also be characterized as follows: Set $H(x) = x/U(x)$ and then set

$$b_n = H^{\leftarrow}(n),$$

where $H^{\leftarrow}$ is the inverse function of $H$ satisfying $H(H^{\leftarrow}(x)) \sim x$.

**2.10 (Π-variation).** A measurable function $U : (0, \infty) \mapsto (0, \infty)$ is called $\Pi$-varying (written $U \in \Pi$) [26, 102] if there exists $g \in \mathrm{RV}_0$ such that for all $x > 0$,

$$\lim_{t \to \infty} \frac{U(tx) - U(t)}{g(t)} = \log x. \tag{2.32}$$

Call $g$ the auxiliary function. Sometimes we then write $U \in \Pi(g)$.

(a) Suppose

$$U(x) = \int_0^x u(s)ds, \quad x > 0, \quad u(\cdot) \in \mathrm{RV}_{-1}.$$

Show that $U \in \Pi$ [104].

(b) Suppose $U$ is nondecreasing. Show that $U \in \Pi$ iff there exists $a(n) \to \infty$ and

$$\frac{n}{a(n)} U(a(n)\cdot) \xrightarrow{v} L(\cdot),$$

where $L$ is the measure satisfying $L(a, b] = \log b/a, 0 < a < b < \infty$ [259].

**2.11 (More on $\Pi$-variation [96]).**

(a) Show that $U \in \Pi(g)$ iff $U \circ r \in \Pi$ for every $r \in \mathrm{RV}_1$. The auxiliary function of $U \circ r$ is $g \circ r$.

(b) If (2.32) holds except that the limit is $-\log x$, say, $U \in \Pi_-$, and if (2.32) holds, say, $U \in \Pi_+$, show that $U \in \Pi_+(g)$ iff $1/U \in \Pi_-$. The auxiliary function of $1/U$ is $g/U^2$.

(c) If $U \in \Pi(g)$ and $L_0 \in \mathrm{RV}_0$, then the product $U \cdot L_0 \in \Pi$ iff

$$\left(\frac{L_0(tx)}{L_0(t)} - 1\right)\frac{U(t)}{g(t)} \to 0 \quad (t \to \infty)$$

for all $x > 0$.

# 3

# Crash Course II: Weak Convergence; Implications for Heavy-Tail Analysis

Asymptotic properties of statistics in heavy-tailed analysis are clearly understood with an interpretation which comes from the modern theory of weak convergence of probability measures on metric spaces, as originally promoted in [22] and updated in [25]. Additionally, utilizing the power of weak convergence allows for a rather unified treatment of the one-dimensional and higher-dimensional cases of heavy-tailed phenomena.

## 3.1 Definitions

Let $\mathbb{S}$ be a complete, separable metric space with metric $d$ and let $\mathcal{S}$ be the Borel $\sigma$-algebra of subsets of $\mathbb{S}$ generated by open sets. Suppose $(\Omega, \mathcal{A}, \mathbb{P})$ is a probability space. A random element $X$ in $\mathbb{S}$ is a measurable map from such a space $(\Omega, \mathcal{A})$ into $(\mathbb{S}, \mathcal{S})$.

With a random variable, a point $\omega \in \Omega$ is mapped into a real-valued member of $\mathbb{R}$. With a random element, a point $\omega \in \Omega$ is mapped into an element of the metric space $\mathbb{S}$. Some common examples of this paradigm are given in Table 3.1.

Given a sequence $\{X_n, n \geq 0\}$ of random elements of $\mathbb{S}$, there is a corresponding sequence of distributions on $\mathcal{S}$,

$$P_n = \mathbb{P} \circ X_n^{-1} = \mathbb{P}[X_n \in \cdot], \quad n \geq 0.$$

$P_n$ is called the *distribution* of $X_n$. Then $X_n$ converges weakly to $X_0$ (written $X_n \Rightarrow X_0$ or $P_n \Rightarrow P_0$) if whenever $f \in C(\mathbb{S})$, the class of bounded, continuous, real-valued functions on $\mathbb{S}$, we have

$$\mathbf{E}f(X_n) = \int_{\mathbb{S}} f(x) P_n(dx) \to \mathbf{E}f(X_0) = \int_{\mathbb{S}} f(x) P_0(dx).$$

Recall that the definition of weak convergence of random variables in $\mathbb{R}$ is given in terms of one-dimensional distribution functions, which does not generalize nicely to

| **Metric space $\mathbb{S}$** | **Random element $X$ is a...** |
|---|---|
| $\mathbb{R}$ | random variable |
| $\mathbb{R}^d$ | random vector |
| $\mathbb{R}^\infty$ | random sequence |
| $C[0,\infty)$, the space of real-valued continuous functions on $[0,\infty)$ | random process with continuous paths |
| $D[0,\infty)$, the space of real-valued, right-continuous functions on $[0,\infty)$ with finite left limits existing on $(0,\infty)$ | right-continuous random process with jump discontinuities |
| $M_p(\mathbb{E})$, the space of point measures on a nice space $\mathbb{E}$ | stochastic point process on $\mathbb{E}$ |
| $M_+(\mathbb{E})$, the space of Radon measures on a nice space $\mathbb{E}$ | random measure on $\mathbb{E}$ |

**Table 3.1.** Various metric spaces and random elements.

higher dimensions. The definition in terms of integrals of test functions $f \in C(\mathbb{S})$ is very flexible and well defined for any metric space $\mathbb{S}$.

## 3.2 Basic properties of weak convergence

### 3.2.1 Portmanteau theorem

The basic *Portmanteau theorem* [22, p. 11], [25] says the following are equivalent:

$$X_n \Rightarrow X_0. \tag{3.1}$$

$$\lim_{n\to\infty} \mathbb{P}[X_n \in A] = \mathbb{P}[X_0 \in A] \quad \forall A \in \mathcal{S} \text{ such that } \mathbb{P}[X_0 \in \partial A] = 0. \tag{3.2}$$

Here $\partial A$ denotes the boundary of the set $A$.

$$\limsup_{n\to\infty} \mathbb{P}[X_n \in F] \le \mathbb{P}[X_0 \in F] \quad \forall \text{ closed } F \in \mathcal{S}. \tag{3.3}$$

$$\liminf_{n\to\infty} \mathbb{P}[X_n \in G] \ge \mathbb{P}[X_0 \in G] \quad \forall \text{ open } G \in \mathcal{S}. \tag{3.4}$$

$$\mathbf{E}f(X_n) \to \mathbf{E}f(X_0) \quad \forall f \text{ that are bounded and uniformly continuous.} \tag{3.5}$$

Although it may seem comfortable to express weak convergence of probability measures in terms of sets, it is mathematically simplest to rely on integrals with respect to test functions as given, for instance, in (3.5).

### 3.2.2 Skorohod's theorem

*Skorohod's theorem* [23, Proposition 0.2] is a nice way to think about weak convergence since, for certain purposes, it allows one to replace convergence in distribution with almost sure convergence. In a theory which relies heavily on continuity, this is a big advantage. Almost sure convergence, being pointwise, is very well suited to continuity arguments.

Let $\{X_n, n \geq 0\}$ be random elements of the metric space $(\mathbb{S}, \mathcal{S})$, and suppose the domain of each $X_n$ is $(\Omega, \mathcal{A}, \mathbb{P})$. Let

$$([0, 1], \mathcal{B}[0, 1], \mathbb{LEB}(\cdot))$$

be the usual probability space on $[0, 1]$, where $\mathbb{LEB}(\cdot)$ is Lebesgue measure or length and $\mathcal{B}[0, 1]$ is the Borel subsets of $[0, 1]$. We call this space the uniform probability space. Skorohod's theorem expresses that $X_n \Rightarrow X_0$ iff there exist random elements $\{X_n^*, n \geq 0\}$ in $\mathbb{S}$ defined on the uniform probability space, such that

$$X_n \stackrel{d}{=} X_n^* \quad \text{for each } n \geq 0$$

and

$$X_n^* \to X_0^* \quad \text{a.s.}$$

The second statement means

$$\mathbb{LEB}\left\{t \in [0, 1] : \lim_{n\to\infty} d(X_n^*(t), X_0^*(t)) = 0\right\} = 1.$$

Almost sure convergence always implies convergence in distribution, so Skorohod's theorem provides a partial converse. To see why almost sure convergence implies weak convergence is easy. With $d(\cdot, \cdot)$ as the metric on $\mathbb{S}$, we have $d(X_n, X_0) \to 0$ almost surely, and for any $f \in C(\mathbb{S})$, we get by continuity that $f(X_n) \to f(X_0)$ almost surely. Since $f$ is bounded, by dominated convergence we get $\mathbf{E}f(X_n) \to \mathbf{E}f(X_0)$.

Recall that in one dimension, Skorohod's theorem has an easy proof. If $X_n \Rightarrow X_0$ and $X_n$ has distribution function $F_n$, then

$$F_n \to F_0, \quad n \to \infty.$$

Thus, by Proposition 2.2, $F_n^{\leftarrow} \to F_0^{\leftarrow}$. Then with $U$, the identity function on $[0, 1]$ (so that $U$ is uniformly distributed),

$$X_n \stackrel{d}{=} F_n^{\leftarrow}(U) =: X_n^*, \quad n \geq 0,$$

and

$$\begin{aligned}\mathbb{LEB}[X_n^* \to X_0^*] &= \mathbb{LEB}\{t \in [0,1] : F_n^{\leftarrow}(t) \to F_0^{\leftarrow}(t)]\\ &\geq \mathbb{LEB}(\mathcal{C}(F_0^{\leftarrow})) = 1,\end{aligned}$$

since the set of discontinuities of the monotone function $F_0^{\leftarrow}(\cdot)$ is countable and hence has Lebesgue measure 0.

The power of weak convergence theory comes from the fact that once a basic convergence result has been proved, many corollaries emerge with little effort, often using only continuity. Suppose $(\mathbb{S}_i, d_i), i = 1, 2$, are two metric spaces and $h : \mathbb{S}_1 \mapsto \mathbb{S}_2$ is continuous. If $\{X_n, n \geq 0\}$ are random elements in $(\mathbb{S}_1, \mathcal{S}_1)$ and $X_n \Rightarrow X_0$, then $h(X_n) \Rightarrow h(X_0)$ as random elements in $(\mathbb{S}_2, \mathcal{S}_2)$. Justification is straightforward: Let $f_2 \in C(\mathbb{S}_2)$, and we must show that $\mathbf{E} f_2(h(X_n)) \to \mathbf{E} f_2(h(X_0))$. But $f_2(h(X_n)) = f_2 \circ h(X_n)$, and since $f_2 \circ h \in C(\mathbb{S}_1)$, the result follows from the definition of $X_n \Rightarrow X_0$ in $\mathbb{S}_1$.

If $\{X_n\}$ are random variables that converge, then letting $h(x) = x^2$ or $\arctan x$ or ... yields additional convergences for free.

### 3.2.3 Continuous mapping theorem

The function $h$ used in the previous paragraphs need not be continuous everywhere, and, in fact, many of the maps $h$ that we will wish to use are definitely not continuous everywhere. For a function $h : \mathbb{S}_1 \mapsto \mathbb{S}_2$, define the discontinuity set of $h$ as

$$D(h) := \{s_1 \in \mathbb{S}_1 : h \text{ is discontinuous at } s_1\}.$$

Similarly, define

$$\mathcal{C}(h) =: \{s_1 \in \mathbb{S}_1 : h \text{ is continuous at } s_1\}.$$

**Theorem 3.1 (continuous mapping theorem).** *Let $(\mathbb{S}_i, d_i)$, $i = 1, 2$, be two metric spaces, and suppose $\{X_n, n \geq 0\}$ are random elements of $(\mathbb{S}_1, \mathcal{S}_1)$ and $X_n \Rightarrow X_0$. If $h : \mathbb{S}_1 \mapsto \mathbb{S}_2$ satisfies*

$$\mathbb{P}[X_0 \in D(h)] = \mathbb{P}[X_0 \in \{s_1 \in \mathbb{S}_1 : h \text{ is discontinuous at } s_1\}] = 0,$$

*then*

$$h(X_n) \Rightarrow h(X_0)$$

*in $\mathbb{S}_2$.*

*Proof.* For a traditional proof, see [22, p. 30]. This result is an immediate consequence of Skorohod's theorem. If $X_n \Rightarrow X_0$, then there exist almost surely convergent random elements of $\mathbb{S}_1$ defined on the unit interval, denoted $X_n^*$, such that

$$X_n^* \stackrel{d}{=} X_n, \quad n \geq 0.$$

Then it follows that

$$\mathbb{LEB}[h(X_n^*) \to h(X_0^*)] \geq \mathbb{LEB}[X_0^* \notin D(h)].$$

Since $X_0 \stackrel{d}{=} X_0^*$, we get

$$\mathbb{LEB}[X_0^* \notin D(h)] = \mathbb{P}[X_0 \notin D(h)] = 1,$$

and therefore $h(X_n^*) \to h(X_0^*)$ almost surely. Since almost sure convergence implies convergence in distribution, $h(X_n^*) \Rightarrow h(X_0^*)$. Since for every $n \geq 0$, we have $h(X_n) \stackrel{d}{=} h(X_n^*)$, the result follows. □

### 3.2.4 Subsequences and Prohorov's theorem

Often to prove weak convergence, subsequence arguments are used and the following is necessary. A family $\Pi$ of probability measures on a complete, separable metric space is *relatively compact* or *sequentially compact* if every sequence $\{P_n\} \subset \Pi$ contains a weakly convergent subsequence. Note that the family of all measures can be metrized so that this notion of relative compactness coincides with the metric definition and expresses the Bolzano–Weierstrass equivalence of compactness and sequential compactness. See [25].

Relative compactness is theoretically useful but hard to check in practice, so we need a workable criterion. Call the family $\Pi$ tight (and by abuse of language we will refer to the corresponding random elements also as a tight family) if for any $\varepsilon$, there exists a compact $K_\varepsilon \in \mathcal{S}$ such that

$$P(K_\varepsilon) > 1 - \varepsilon \quad \text{for all } P \in \Pi.$$

This is the kind of condition that precludes probability mass from escaping from the state space. Prohorov's theorem [25] guarantees that when $\mathbb{S}$ is separable and complete, tightness of $\Pi$ is the same as relative compactness. Tightness can be checked, although it is seldom easy.

## 3.3 Some useful metric spaces

It pays to spend a bit of time remembering details of examples of metric spaces that will be useful. To standardize notation, we set

$$\begin{aligned}\mathcal{F}(\mathbb{S}) &= \text{closed subsets of } \mathbb{S},\\ \mathcal{G}(\mathbb{S}) &= \text{open subsets of } \mathbb{S},\\ \mathcal{K}(\mathbb{S}) &= \text{compact subsets of } \mathbb{S}.\end{aligned}$$

### 3.3.1 $\mathbb{R}^d$, finite-dimensional Euclidean space

We set

$$\mathbb{R}^d := \{(x_1, \dots, x_d) : x_i \in \mathbb{R}, i = 1, \dots, d\} = \mathbb{R} \times \mathbb{R} \times \cdots \times \mathbb{R}.$$

The metric is defined by

$$d(\boldsymbol{x}, \boldsymbol{y}) = \sqrt{\sum_{i=1}^{d} (x_i - y_i)^2}$$

for $\boldsymbol{x}, \boldsymbol{y} \in \mathbb{R}^d$. Convergence of a sequence in this space is equivalent to componentwise convergence.

Define an interval

$$(\boldsymbol{a}, \boldsymbol{b}] = \{\boldsymbol{x} \in \mathbb{R}^d : a_i < x_i \le b_i, i = 1, \dots, d\}.$$

A probability measure $P$ on $\mathbb{R}^d$ is determined by its distribution function

$$F(\boldsymbol{x}) := P(-\infty, \boldsymbol{x}],$$

and a sequence of probability measures $\{P_n, n \ge 0\}$ on $\mathbb{R}^d$ converges to $P_0$ iff

$$F_n(\boldsymbol{x}) \to F_0(\boldsymbol{x}) \quad \forall \boldsymbol{x} \in \mathcal{C}(F_0).$$

Note that this statement equates convergence in distribution of a sequence of random vectors with weak convergence of their distribution functions. While this is concrete, it is seldom useful since multivariate distribution functions are usually awkward to deal with in practice.

Also, recall $K \in \mathcal{K}(\mathbb{R}^d)$ iff $K$ is closed and bounded.

### 3.3.2 $\mathbb{R}^\infty$, sequence space

Define

$$\mathbb{R}^\infty := \{(x_1, x_2, \dots) : x_i \in \mathbb{R}, i \geq 1\} = \mathbb{R} \times \mathbb{R} \times \cdots .$$

The metric can be defined by

$$d(\boldsymbol{x}, \boldsymbol{y}) = \sum_{i=1}^{\infty} (|x_i - y_i| \wedge 1) 2^{-i},$$

for $\boldsymbol{x}, \boldsymbol{y} \in \mathbb{R}^\infty$. This gives a complete, separable metric space where convergence of a family of sequences means coordinatewise convergence; that is,

$$\boldsymbol{x}(n) \to \boldsymbol{x}(0) \text{ iff } x_i(n) \to x_i(0) \forall i \geq 1.$$

The topology $\mathcal{G}(\mathbb{R}^\infty)$ can be generated by basic neighborhoods of the form

$$N_d(\boldsymbol{x}) = \left\{ \boldsymbol{y} : \bigvee_{i=1}^{d} |x_i - y_i| < \epsilon \right\}$$

as we vary $d$, the center $\boldsymbol{x}$, and $\epsilon$.

A set $A \subset \mathbb{R}^\infty$ is relatively compact iff every one-dimensional section is bounded, that is, iff for any $i \geq 1$,

$$\{x_i : \boldsymbol{x} \in A\} \text{ is bounded.}$$

For more details, see [25, 106, 116].

### 3.3.3 $C[0, 1]$ and $C[0, \infty)$, continuous functions

The metric on $C[0, M]$, the space of real-valued continuous functions with domain $[0, M]$ is the uniform metric

$$d_M(x(\cdot), y(\cdot)) = \sup_{0 \leq t \leq M} |x(t) - y(t)| =: \|x(\cdot) - y(\cdot)\|_M,$$

and the metric on $C[0, \infty)$ is

$$d(x(\cdot), y(\cdot)) = \sum_{n=1}^{\infty} \frac{d_n(x, y) \wedge 1}{2^n},$$

where we interpret $d_n(x, y)$ as the $C[0, n]$ distance of $x$ and $y$ restricted to $[0, n]$. The metric on $C[0, \infty)$ induces the topology of local uniform convergence.

For $C[0, 1]$ (or $C[0, M]$), we have that every function is uniformly continuous since a continuous function on a compact set is always uniformly continuous. Uniform continuity can be expressed by the modulus of continuity, which for $x \in C[0, 1]$ is defined by

$$\omega_x(\delta) = \sup_{|t-s|<\delta} |x(t) - x(s)|, \quad 0 < \delta < 1.$$

Then uniform continuity means

$$\lim_{\delta \to 0} \omega_x(\delta) = 0.$$

The Arzelà–Ascoli theorem [106, 280] expresses the fact that a uniformly bounded equicontinuous family of functions in $C[0, 1]$ has a uniformly convergent subsequence; that is, this family is relatively compact or has compact closure. Thus a set $A \subset C[0, 1]$ is relatively compact iff

(i) $A$ is uniformly bounded; that is,

$$\sup_{0 \le t \le 1} \sup_{x \in A} |x(t)| < \infty, \tag{3.6}$$

and

(ii) $A$ is equicontinuous; that is,

$$\lim_{\delta \downarrow 0} \sup_{x \in A} \omega_x(\delta) = 0.$$

Since the functions in a compact family vary in a controlled way, (3.6) can be replaced by

$$\sup_{x \in A} |x(0)| < \infty. \tag{3.7}$$

Compare this result with the compactness characterization in $\mathbb{R}^\infty$, where relative compactness meant that each one-dimensional section was bounded. Here, a family $A$ of continuous functions is relatively compact if each one-dimensional section is bounded in a uniform way *and* equicontinuity is present.

### 3.3.4 $D[0, 1]$ and $D[0, \infty)$

Start by considering $D[0, 1]$, the space of right-continuous functions on $[0, 1)$ that have finite left limits on $(0, 1]$. Minor changes allow us to consider $D[0, M]$ for any $M > 0$.

In the uniform topology, two functions $x(\cdot)$ and $y(\cdot)$ are close if their graphs are uniformly close. In the Skorohod topology on $D[0, 1]$, we consider $x$ and $y$ close if after deforming the time scale of one of them, for example, $y$, the resulting graphs are close. Consider the following simple example:

$$x_n(t) = 1_{[0,\frac{1}{2}+\frac{1}{n})}(t), \quad x(t) = 1_{[0,\frac{1}{2})}(t). \tag{3.8}$$

The uniform distance is always 1, but a time deformation allows us to consider the functions to be close. (Various metrics and their applications to functions with jumps are considered in detail in [301].)

Define time deformations

$$\Lambda = \{\lambda : [0,1] \mapsto [0,1] : \lambda(0) = 0, \lambda(1) = 1,$$
$$\lambda(\cdot) \text{ is continuous, strictly increasing}\}. \tag{3.9}$$

Let $e(t) \in \Lambda$ be the identity transformation and denote the uniform distance between $x$ and $y$ as

$$\|x - y\| := \sup_{0 \le t \le 1} |x(t) - y(t)|.$$

The Skorohod metric $d(x, y)$ between two functions $x, y \in D[0, 1]$ is

$$d(x, y) = \inf\{\epsilon > 0 : \exists \lambda \in \Lambda \text{ such that } \|\lambda - e\| \vee \|x - y \circ \lambda\| \le \epsilon\},$$
$$= \inf_{\lambda \in \Lambda} \|\lambda - e\| \vee \|x - y \circ \lambda\|.$$

Simple consequences of the definitions:

1. Given a sequence $\{x_n\}$ of functions in $D[0, 1]$, we have $d(x_n, x_0) \to 0$ iff there exist $\lambda_n \in \Lambda$ and

$$\|\lambda_n - e\| \to 0, \quad \|x_n \circ \lambda_n - x_0\| \to 0. \tag{3.10}$$

2. From the definition, we always have

$$d(x, y) \le \|x - y\|, \quad x, y \in D[0, 1]$$

since one choice of $\lambda$ is the identity, but this may not give the infimum. Therefore, uniform convergence always implies Skorohod convergence. The converse is false; see (3.8).

3. If $d(x_n, x_0) \to 0$ for $x_n \in D[0, 1]$, $n \ge 0$, then for all $t \in \mathcal{C}(x_0)$, we have pointwise convergence,

$$x_n(t) \to x_0(t).$$

To see this, suppose (3.10) holds. Then

$$\|\lambda_n - e\| = \|\lambda_n^{\leftarrow} - e\| \to 0.$$

Thus

$$|x_n(t) - x_0(t)| \leq |x_n(t) - x_0 \circ \lambda_n^{\leftarrow}(t)| + |x_0 \circ \lambda_n^{\leftarrow}(t) - x_0(t)|$$
$$\leq \|x_n \circ \lambda_n - x_0\| + o(1)$$

since $x$ is continuous at $t$ and $\lambda_n^{\leftarrow} \to e$.

4. If $d(x_n, x_0) \to 0$ and $x_0 \in C[0, 1]$, then uniform convergence holds.

   If (3.10) holds, then as in item 3 we have for each $t \in [0, 1]$,

$$|x_n(t) - x_0(t)| \leq \|x_n \circ \lambda_n - x_0\| + \|x_0 - x_0 \circ \lambda_n\| \to 0,$$

   and hence

$$\|x_n(t) - x_0(t)\| \to 0.$$

**The space $D[0, \infty)$.** Denote the restriction of $x \in D[0, \infty)$ to the interval $[0, s]$ by $r_s x(\cdot)$, where

$$r_s x(t) = x(t), \quad 0 \leq t \leq s.$$

Let $d_s$ be the Skorohod metric on $D[0, s]$ and define $d_\infty$, the Skorohod metric on $D[0, \infty)$, by

$$d_\infty(x, y) = \int_0^\infty e^{-s}(d_s(r_s x, r_s y) \wedge 1) ds.$$

The impact of this is that Skorohod convergence on $D[0, \infty)$ reduces to convergence on finite intervals since $d_\infty(x_n, x_0) \to 0$ iff for any $s \in \mathcal{C}(x_0)$, we have $d_s(r_s x_n, r_s x_0) \to 0$. For more detail, see [25, 208, 260, 300, 301].

### 3.3.5 Radon measures and point measures; vague convergence

#### Spaces of measures

Suppose $\mathbb{E}$ is a nice space. The technical meaning of *nice* is that $\mathbb{E}$ should be a locally compact topological space with countable base; often it is safe to think of $\mathbb{E}$ as a finite-dimensional Euclidean space or $\mathbb{R}^d$. The case $d = 1$ is important but $d > 1$ is also very useful. When it comes time to construct point processes, $\mathbb{E}$ will be the space in which our points live. We assume $\mathbb{E}$ comes with a $\sigma$-field $\mathcal{E}$, which can be the $\sigma$-field generated by the open sets or, equivalently, the rectangles of $\mathbb{E}$.

How can we model a random distribution of points in $\mathbb{E}$? One way is to specify random elements $\{X_n\}$ in $\mathbb{E}$ and then to define the corresponding stochastic point process to be the counting function whose value at the region $A \in \mathcal{E}$ is the number of random elements $\{X_n\}$ that fall in $A$. This is intuitively appealing but has some technical drawbacks, and it is mathematically preferable to focus on counting functions rather than on points.

A measure $\mu : \mathcal{E} \mapsto [0, \infty]$ is an assignment of positive numbers to sets in $\mathcal{E}$ such that

1. $\mu(\emptyset) = 0$ and $\mu(A) \geq 0$ for all $A \in \mathcal{E}$;
2. if $\{A_n, n \geq 1\}$ are mutually disjoint sets in $\mathcal{E}$, then the $\sigma$-additivity property holds:

$$\mu\left(\bigcup_{i=1}^{\infty} A_i\right) = \sum_{i=1}^{\infty} \mu(A_i).$$

The measure $\mu$ is called *Radon* if

$$\mu(K) < \infty \quad \forall K \in \mathcal{K}(\mathbb{E}).$$

Thus compact sets are known to have finite $\mu$-mass. Knowing where the measure is required to be finite helps us to keep track of infinities in a useful way and prevents illegal operations like $\infty - \infty$.

Define

$$M_+(\mathbb{E}) = \{\mu : \mu \text{ is a nonnegative measure on } \mathcal{E} \text{ and } \mu \text{ is Radon}\}. \tag{3.11}$$

The space $M_+(\mathbb{E})$ can be made into a complete separable metric space under what is called the *vague* metric. For now, instead of describing the metric, we will describe the notion of convergence consistent with the metric.

## Convergence concept

The way we defined convergence of *probability* measures was by means of test functions. We integrate a test function that is bounded and continuous on the metric space, and if the resulting sequence of numbers converges, then we have weak convergence. However, with infinite measures in $M_+(\mathbb{E})$, we cannot just integrate a bounded function to get something finite. However, we know our measures are also Radon, and this suggests using functions that vanish on complements of compact sets. So define

$$C_K^+(\mathbb{E}) = \{f : \mathbb{E} \mapsto \mathbb{R}_+ : f \text{ is continuous with compact support}\}.$$

For a function to have compact support means that it vanishes off a compact set.

The notion of convergence in $M_+(\mathbb{E})$: If $\mu_n \in M_+(\mathbb{E})$ for $n \geq 0$, then $\mu_n$ converges vaguely to $\mu_0$, written $\mu_n \stackrel{v}{\rightarrow} \mu_0$, if for all $f \in C_K^+(\mathbb{E})$, we have

$$\mu_n(f) := \int_{\mathbb{E}} f(x)\mu_n(dx) \rightarrow \mu_0(f) := \int_{\mathbb{E}} f(x)\mu_0(dx)$$

as $n \rightarrow \infty$.

*Example* 3.1 (*trivial but mildly illuminating example*). Suppose $\mathbb{E}$ is some finite-dimensional Euclidean space with metric $d(\cdot,\cdot)$, and define for $\boldsymbol{x} \in \mathbb{E}$ and $A \in \mathcal{E}$,

$$\epsilon_{\boldsymbol{x}}(A) = \begin{cases} 1 & \text{if } \boldsymbol{x} \in A, \\ 0 & \text{if } \boldsymbol{x} \in A^c. \end{cases}$$

Then

$$\mu_n := \epsilon_{\boldsymbol{x}_n} \stackrel{v}{\to} \mu_0 := \epsilon_{\boldsymbol{x}_0}$$

in $M_+(\mathbb{E})$ iff

$$\boldsymbol{x}_n \to \boldsymbol{x}_0$$

in the metric on $\mathbb{E}$.

To see this, suppose that $\boldsymbol{x}_n \to \boldsymbol{x}_0$ and $f \in C_K^+(\mathbb{E})$. Then

$$\mu_n(f) = f(\boldsymbol{x}_n) \to f(\boldsymbol{x}_0) = \mu_0(f),$$

since $f$ is continuous and the points are converging. Conversely, suppose that $\boldsymbol{x}_n \not\to \boldsymbol{x}_0$. Define $\phi : \mathbb{R} \mapsto [0, 1]$ by

$$\phi(t) = \begin{cases} 1 & \text{if } t < 0, \\ 1 - t & \text{if } 0 \le t \le 1, \\ 0 & \text{if } t > 1. \end{cases}$$

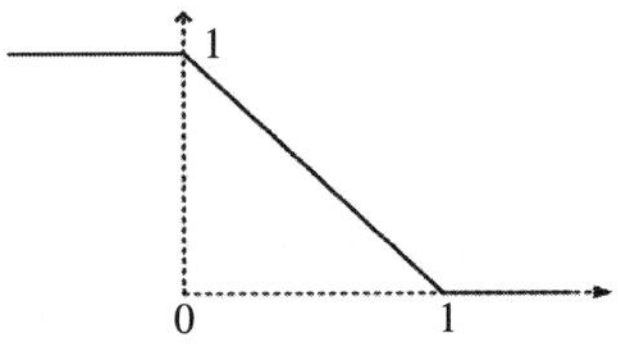


There exists a subsequence $\{n'\}$ such that $d(\boldsymbol{x}_{n'}, \boldsymbol{x}_0) > \epsilon$. Define

$$f(\boldsymbol{y}) = \phi(d(\boldsymbol{x}_0, \boldsymbol{y})/\epsilon),$$

so that $f \in C_K^+(\mathbb{E})$. Then

$$|f(\boldsymbol{x}_{n'}) - f(\boldsymbol{x}_0)| = |0 - 1| \not\to 0,$$

and then we have $\mu_n(f) \not\to \mu_0(f)$.

**Point measures.** A point measure $m$ is an element of $M_+(\mathbb{E})$ of the form

$$m = \sum_i \epsilon_{x_i}. \tag{3.12}$$

Built into this definition is the understanding that $m(\cdot)$ is Radon: $m(K) < \infty$ for $K \in \mathcal{K}(\mathbb{E})$. Think of $\{x_i\}$ as the atoms and $m$ as the function that counts how many atoms fall in a set. The set $M_p(\mathbb{E})$ is the set of all Radon point measures of the form (3.12). This turns out to be a closed subset of $M_+(\mathbb{E})$.

**The vague topology; more on $M_+(\mathbb{E})$ (and hence, more on $M_p(\mathbb{E})$)**

We can specify open sets, a topology (a system of open sets satisfying closure properties), and then a notion of "distance" in $M_+(\mathbb{E})$. Define a *basis* set to be a subset of $M_+(\mathbb{E})$ of the form

$$\{\mu \in M_+(\mathbb{E}) : \mu(f_i) \in (a_i, b_i), i = 1, \dots, d\}, \tag{3.13}$$

where $f_i \in C_K^+(\mathbb{E})$ and $0 \le a_i \le b_i$. Now imagine varying the choices of the integer $d$, functions $f_1, \dots, f_d$, and endpoints $a_1, \dots, a_d; b_1, \dots, b_d$. Unions of basis sets form the class of open sets constituting the vague topology.

The topology is metrizable as a complete, separable metric space, and we can put a metric $d(\cdot,\cdot)$ on the space, which yields the same open sets. The metric $d(\cdot,\cdot)$ can be specified as follows: There exists *some* sequence of functions $f_i \in C_K^+(\mathbb{E})$ and for $\mu_1, \mu_2 \in M_+(\mathbb{E})$,

$$d(\mu_1, \mu_2) = \sum_{i=1}^{\infty} \frac{|\mu_1(f_i) - \mu_2(f_i)| \wedge 1}{2^i}. \tag{3.14}$$

*An interpretation*: If $\mu \in M_+(\mathbb{E})$, then $\mu$ is determined by our knowledge of $\{\mu(f), f \in C_K^+(\mathbb{E})\}$. This may seem reasonable, and we will see why this is true shortly in Lemma 3.1 (p. 52). Think of $\mu$ as an object with components $\{\mu(f), f \in C_K^+(\mathbb{E})\}$, where we imagine $\mu(f)$ as the $f$th component of $\mu$. Then (3.14) indicates, in fact, that it is enough to have a countable set of components to determine $\mu$, and we can think about $\mu$ being represented as

$$\mu = \{\mu(f_i), i \ge 1\}. \tag{3.15}$$

So we measure distance in $M_+(\mathbb{E})$ as if the objects were in $\mathbb{R}^\infty$.

This analogy makes plausible the following characterization of compactness: A subset $M \subset M_+(\mathbb{E})$ is vaguely relatively compact iff

$$\sup_{\mu \in M} \mu(f) < \infty \quad \forall f \in C_K^+(\mathbb{E}). \tag{3.16}$$

To show compactness implies that (3.16) is easy and helps us digest the concepts. Suppose $M$ is relatively compact. For $f \in C_K^+(\mathbb{E})$, define the projection onto the $f$th component $T_f : M_+(\mathbb{E}) \mapsto [0, \infty)$ by

$$T_f(\mu) = \mu(f).$$

Then $T_f$ is continuous since $\mu_n \xrightarrow{v} \mu$ implies that

$$T_f(\mu_n) = \mu_n(f) \to \mu(f) = T_f(\mu).$$

For fixed $f \in C_K^+(\mathbb{E})$, we note that

$$\sup_{\mu \in M} \mu(f) = \sup_{\mu \in M} T_f(\mu) = \sup_{\mu \in M^-} T_f(\mu)$$

since the supremum of a continuous function on $M$ must be the same as the supremum on the closure $M^-$.

If $M$ is relatively compact, then the closure $M^-$ is compact. Since $T_f$ is continuous on $M_+(\mathbb{E})$, $T_f(M^-)$ is a compact subset of $[0, \infty)$. (Continuous images of compact sets are compact.) Compact sets in $[0, \infty)$ are bounded, so

$$\infty > \sup T_f(M^-) = \sup\{T_f(\mu), \mu \in M^-\} = \sup_{\mu \in M^-} \{\mu(f)\}.$$

Why emphasize integrals of test functions rather than measures of sets? Proofs are a bit simpler with this formulation and it is easier to capitalize on continuity arguments. One can always formulate parallel definitions and concepts with sets using a variant of Urysohn's lemma. See [116, p. 47], [280, p. 135], [180], [260, p. 141].

**Lemma 3.1.**

(a) *Suppose $K \in \mathcal{K}(\mathbb{E})$. There exists $K_n \in \mathcal{K}(\mathbb{E})$, $K_n \downarrow K$, and there exist $f_n \in C_K^+(\mathbb{E})$ with $\{f_n\}$ nonincreasing such that*

$$1_K \le f_n \le 1_{K_n} \downarrow 1_K. \tag{3.17}$$

(b) *Suppose $G \in \mathcal{G}(\mathbb{E})$, and $G$ is relatively compact. There exist open, relatively compact $G_n \uparrow G$ and $f_n \in C_K^+(\mathbb{E})$ with $\{f_n\}$ nondecreasing such that*

$$1_G \ge f_n \ge 1_{G_n} \uparrow 1_G. \tag{3.18}$$

From Lemma 3.1, comes a Portmanteau theorem.

**Theorem 3.2.** *Let $\mu, \mu_n \in M_+(\mathbb{E})$. The following are equivalent:*

(a) $\mu_n \stackrel{v}{\to} \mu$.
(b) *$\mu_n(B) \to \mu(B)$ for all relatively compact $B$ satisfying $\mu(\partial B) = 0$.*
(c) *For all $K \in \mathcal{K}(\mathbb{E})$, we have*

$$\limsup_{n\to\infty} \mu_n(K) \le \mu(K),$$

*and for all $G \in \mathcal{G}(\mathbb{E})$ that are relatively compact, we have*

$$\liminf_{n\to\infty} \mu_n(G) \ge \mu(G).$$

## 3.4 How to prove weak convergence

We outline some tools useful for proving weak convergence.

### 3.4.1 Methods in spaces useful for heavy-tail analysis

Here is an outline of what it takes to prove weak convergence in some spaces of immediate interest:

1. In $\mathbb{R}^d$, we can show that random vectors $\{\boldsymbol{X}_n, n \geq 0\}$ converge weakly,

$$\boldsymbol{X}_n \Rightarrow \boldsymbol{X}_0,$$

   by any of the following methods:

   (a) Show convergence of the finite-dimensional distributions

$$\mathbb{P}[\boldsymbol{X}_n \leq \boldsymbol{x}] \to \mathbb{P}[\boldsymbol{X}_0 \leq \boldsymbol{x}]$$

   at continuity points of the limit. Sometimes this can even be done by showing convergence of the joint densities when they exist.

   (b) Show convergence of the characteristic functions

$$\mathbf{E}e^{i\boldsymbol{t}\cdot\boldsymbol{X}_n} \to \mathbf{E}e^{i\boldsymbol{t}\cdot\boldsymbol{X}_0}$$

   for $\boldsymbol{t} \in \mathbb{R}^d$.

   (c) Reduce the problem to one dimension and prove that

$$\boldsymbol{t} \cdot \boldsymbol{X}_n \Rightarrow \boldsymbol{t} \cdot \boldsymbol{X}_0,$$

   which works because of item 1(b). This is called the Cramér–Wold device [24, 25].

   (d) If $\boldsymbol{X}_n \geq \boldsymbol{0}$, show that Laplace transforms converge,

$$\mathbf{E}e^{-\boldsymbol{\lambda}\cdot\boldsymbol{X}_n} \to \mathbf{E}e^{-\boldsymbol{\lambda}\cdot\boldsymbol{X}_0}$$

   for $\boldsymbol{\lambda} > \boldsymbol{0}$. See [135, 302].

2. In $\mathbb{R}^\infty$, random sequences $\{\boldsymbol{X}_n, n \geq 0\}$ of the form

$$\boldsymbol{X}_n = (X_n^{(1)}, X_n^{(2)}, \ldots)$$

   satisfy

$$X_n \Rightarrow X_0$$

if we show for any $d > 0$ that

$$(X_n^{(1)}, X_n^{(2)}, \dots, X_n^{(d)}) \Rightarrow (X_0^{(1)}, X_0^{(2)}, \dots, X_0^{(d)})$$

in $\mathbb{R}^d$.

3. In $M_+(\mathbb{E})$, random measures $\{\xi_n(\cdot), n \geq 0\}$ converge weakly,

$$\xi_n \Rightarrow \xi_0,$$

iff for any family $\{h_j\}$ with $h_j \in C_K^+(\mathbb{E})$, we have

$$(\xi_n(h_j), j \geq 1) \Rightarrow (\xi_0(h_j), j \geq 1)$$

in $\mathbb{R}^\infty$. It would suffice to prove this for the family of functions alluded to in (3.14) (p. 51). In practice, one assumes a sequence $\{h_j\}$ and proves $\mathbb{R}^\infty$ convergence; this reduces to proving $\mathbb{R}^d$-convergence by item 2, and often this can be reduced to one-dimensional convergence.

### 3.4.2 Donsker's theorem

The most famous result in the basic theory of weak convergence is Donsker's theorem, which informs us that a random walk with suitable time and space scaling looks roughly like a Brownian motion. There are many results that can be based on Donsker's theorem using methods outlined in the next section. For a classical proof of Donsker's theorem using convergence of the finite-dimensional distributions plus tightness, see [23, 25].

**Theorem 3.3.** *Suppose $\{\xi_j, j \geq 1\}$ are iid random variables satisfying*

$$\mathbf{E}(\xi_j) = 0 \quad \textit{and} \quad \mathrm{Var}(\xi_j) = 1.$$

*Define*

$$S_0 = 0, \qquad S_n = \sum_{i=1}^{n} \xi_i, \quad n \geq 1.$$

*Then in $D[0, \infty)$,*

$$\frac{S_{[n\cdot]}}{\sqrt{n}} \Rightarrow W(\cdot),$$

*where $W(\cdot)$ is a standard Brownian motion, that is, a continuous path process with stationary independent increments, $W(0) = 0$, and $W(1)$ has a standard normal distribution.*

## 3.5 New convergences from old

Since proving tightness is no picnic, when one has a basic weak convergence result, it is desirable to milk it for all its worth. The continuous mapping theorem is one way to accomplish this, but there are other ways to do this as well. The Slutsky or *converging together* lemmas are a simple approximation method. The idea is that we want to prove that $\{Y_n\}$ converges. If we already know that some approximation $\{X_n\}$ converges and $\{Y_n\}$ is close to $\{X_n\}$, then it should be the case that $\{Y_n\}$ also converges. A second theme is to build convergence in a product space from convergences in factor spaces.

### 3.5.1 Slutsky approximations

There are two approximation results.

**Theorem 3.4 (Slutsky's theorem).** *Suppose* $\{X, X_n, Y_n, n \geq 1\}$ *are random elements of a metric space* $(\mathbb{S}, \mathcal{S})$ *with metric* $d(\cdot,\cdot)$. *If* $X_n \Rightarrow X$ *and* $d(X_n, Y_n) \xrightarrow{P} 0$, *then* $Y_n \Rightarrow X$.

*Proof.* Let $f : \mathbb{S} \mapsto \mathbb{R}$ be real-valued, bounded, and uniformly continuous; this will be sufficient by (3.5) (p. 40). Define the modulus of continuity

$$\omega_\delta(f) = \sup_{d(x,y)\leq\delta} |f(x) - f(y)|.$$

Because $f$ is uniformly continuous,

$$\omega_\delta(f) \to 0, \quad \delta \to 0. \tag{3.19}$$

From the Portmanteau theorem (Section 3.2 (p. 40)), it suffices to show that $\mathbf{E}f(Y_n) \to \mathbf{E}f(X)$. To do this, observe that

$$\begin{aligned}
&|\mathbf{E}f(Y_n) - \mathbf{E}f(X)| \\
&\quad\leq |\mathbf{E}f(Y_n) - \mathbf{E}f(X_n)| + |\mathbf{E}f(X_n) - \mathbf{E}f(X)| \\
&\quad= \mathbf{E}|f(Y_n) - f(X_n)|1_{[d(Y_n,X_n)\leq\delta]} \\
&\qquad + 2\sup_x |f(x)|\mathbb{P}[d(Y_n, X_n) > \delta] + o(1),
\end{aligned}$$

where the $o(1)$ term results from $X_n \Rightarrow X$. The above is bounded by

$$\leq o(1) + \omega_\delta(f) + (\text{const})\mathbb{P}[d(Y_n, X_n) > \delta].$$

The last probability goes to 0 by assumption. Let $\delta \to 0$ and use (3.19). □

Slutsky's theorem is sometimes called the first converging together result. Here is the generalization that is especially useful for truncation arguments.

**Theorem 3.5 (second converging together theorem).** *Suppose that* $\{X_{Mn}, X_M, Y_n, X; n \geq 1, M \geq 1\}$ *are random elements of the metric space* $(\mathbb{S}, \mathcal{S})$ *and are defined on a common domain. Assume for each* $M$*, as* $n \to \infty$*,*

$$X_{Mn} \Rightarrow X_M,$$

*and as* $M \to \infty$*,*

$$X_M \Rightarrow X.$$

*Suppose further that for all* $\epsilon > 0$*,*

$$\lim_{M\to\infty} \limsup_{n\to\infty} \mathbb{P}[d(X_{Mn}, Y_n) > \epsilon] = 0. \tag{3.20}$$

*Then as* $n \to \infty$*, we have*

$$Y_n \Rightarrow X.$$

*Proof.* For any bounded, uniformly continuous function $f : \mathbb{S} \mapsto \mathbb{R}$, we must show that

$$\lim_{n\to\infty} \mathbf{E} f(Y_n) = \mathbf{E} f(X).$$

Without loss of generality, we may, for the sake of neatness, suppose that

$$\sup_{x\in\mathbb{S}} |f(x)| \leq 1.$$

Now write

$$|\mathbf{E}f(Y_n) - \mathbf{E}f(X)| \leq \mathbf{E}|f(Y_n) - f(X_{Mn})| + |\mathbf{E}f(X_{Mn}) - f(X_M)| \\ + |\mathbf{E}f(X_M) - f(X)|,$$

so that

$$\begin{aligned}
&\limsup_{n\to\infty} |\mathbf{E}f(Y_n) - \mathbf{E}f(X)| \\
&\quad\leq \lim_{M\to\infty} \limsup_{n\to\infty} \mathbf{E}|f(Y_n) - f(X_{Mn})| + 0 + 0 \\
&\quad\leq \lim_{M\to\infty} \limsup_{n\to\infty} \mathbf{E}|f(Y_n) - f(X_{Mn})| 1_{[d(Y_n, X_{Mn}) \leq \epsilon]} \\
&\qquad + \lim_{M\to\infty} \limsup_{n\to\infty} \mathbf{E}|f(Y_n) - f(X_{Mn})| 1_{[d(Y_n, X_{Mn}) > \epsilon]} \\
&\quad\leq \sup\{|f(x) - f(y)| : d(x, y) \leq \epsilon\} \\
&\qquad + \lim_{M\to\infty} \limsup_{n\to\infty} 2\mathbb{P}[d(Y_n, X_{Mn}) > \epsilon] \\
&\quad\leq \omega_\epsilon(f) + 0 \to 0
\end{aligned}$$

as $\epsilon \to 0$. □

### 3.5.2 Combining convergences

For scaling arguments involving random change of time, we need the following simple result. It is indicative of a body of results that allow conclusion of joint convergence from existence of marginal convergences.

**Proposition 3.1.** *Let $\mathbb{E}$ and $\mathbb{E}'$ be two complete separable metric spaces, and suppose $\{\xi_n, n \geq 0\}$ and $\{\eta_n, n \geq 0\}$ are random elements of $\mathbb{E}$ and $\mathbb{E}'$, respectively, defined on the same probability space. Suppose*

$$\xi_n \Rightarrow \xi_0$$

*in $\mathbb{E}$ and*

$$\eta_n \xrightarrow{P} e_0',$$

*where $e_0'$ is a fixed point of $\mathbb{E}'$; that is, $e_0'$ is nonrandom. Then we have jointly in $\mathbb{E} \times \mathbb{E}'$,*

$$(\xi_n, \eta_n) \Rightarrow (\xi_0, e_0')$$

*as $n \to \infty$.*

*Remark* 3.1. Weak convergence on product spaces $\mathbb{E} \times \mathbb{E}'$ deserves some comments. (Full treatment is found, for example, in [25, 301].) If $\mathbb{E}$ and $\mathbb{E}'$ are complete, separable metric spaces with metrics $d$ and $d'$, then $\mathbb{E} \times \mathbb{E}'$ is a complete, separable metric space with metric (for example)

$$d_{\text{prod}}((e_1, e_1'), (e_2, e_2')) = d(e_1, e_2) + d'(e_1', e_2').$$

*Proof.* Referring to Slutsky's theorem, Theorem 3.4 (p. 55), set

$$X_n := (\xi_n, e_0') \in \mathbb{E} \times \mathbb{E}',$$

and

$$Y_n := (\xi_n, \eta_n) \in \mathbb{E} \times \mathbb{E}'.$$

Then

$$d_{\text{prod}}(X_n, Y_n) = d(\xi_n, \xi_n) + d'(\eta_n, e_0') \xrightarrow{P} 0.$$

Furthermore, suppose $f \in C(\mathbb{E} \times \mathbb{E}')$ is bounded and continuous on $\mathbb{E} \times \mathbb{E}'$. We have as $n \to \infty$,

$$|\mathbf{E}(f(X_n)) - \mathbf{E}(f(\xi_0, e_0'))| = |\mathbf{E}(f(\xi_n, e_0')) - \mathbf{E}(f(\xi_0, e_0'))| \to 0$$

since $f(\cdot, e_0') \in C(\mathbb{E})$. Thus $X_n \Rightarrow (\xi_0, e_0')$ and the desired conclusion follows from Slutsky's theorem, Theorem 3.4. □

### 3.5.3 Inversion techniques

There are two convenient results for getting new convergences from old when the converging processes are nondecreasing. We outline these in a form that will be needed. More detailed results are contained in, for example, [300, 301].

*Remark* 3.2. A small technical point that we intend to overlook: In the next two sections, we will consider the map

$$x \mapsto x^{\leftarrow},$$

where $x(\cdot)$ is a nondecreasing function in $D[0, \infty)$. Since inverses were defined to be left-continuous in Section 2.1.2 (p. 18), we have $x^{\leftarrow} \in D_{\text{left}}[0, \infty)$, the space of left-continuous functions on $[0, \infty)$ with finite right limits on $(0, \infty)$. The space $D_{\text{left}}[0, \infty)$ can be metrized by the Skorohod metric, just as we did with $D[0, \infty)$. We will allow ourselves the luxury of ignoring the difference between $D[0, \infty)$ and $D_{\text{left}}[0, \infty)$.

#### Inverses

**Proposition 3.2.**

(a) *If $x_n \in D[0, \infty)$ is nondecreasing, $x_n(0) = 0$ and $x_n \to x_0$ in $D[0, \infty)$, where $x_0$ is continuous, strictly increasing, then*

$$x_n^{\leftarrow} \to x_0^{\leftarrow}$$

*locally uniformly and in $D[0, \infty)$.*

(b) *Suppose $\xi_n$ is a stochastic process with nondecreasing paths in $D[0, \infty)$ such that $\xi_n(0) = 0$, and*

$$\xi_n \xrightarrow{P} \xi_0, \tag{3.21}$$

*in $D[0, \infty)$. If almost all paths of $\xi_0$ are continuous and strictly increasing, then*

$$\xi_n^{\leftarrow} \xrightarrow{P} \xi_0^{\leftarrow}. \tag{3.22}$$

*The result holds true if $\xrightarrow{P}$ is replaced by $\Rightarrow$, and then, in fact, we have*

$$(\xi_n, \xi_n^{\leftarrow}) \Rightarrow (\xi_0, \xi_0^{\leftarrow}) \tag{3.23}$$

*in $D[0, \infty) \times D[0, \infty)$.*

*Proof.*

(a) We have

$$x_n^{\leftarrow}(t) \to x_0^{\leftarrow}(t)$$

pointwise by inversion. This gives monotone functions converging to a continuous limit and hence convergence is locally uniform. Local uniform convergence implies convergence in the Skorohod metric.

(b) Let $d(\cdot, \cdot)$ be the Skorohod metric on $D[0, \infty)$ and (3.21) expresses the fact that

$$d(\xi_n, \xi_0) \xrightarrow{P} 0; \tag{3.24}$$

we need to show that

$$d(\xi_n^{\leftarrow}, \xi_0^{\leftarrow}) \xrightarrow{P} 0. \tag{3.25}$$

We use the subsequence characterization of convergence in probability (see [264, Section 6.3] or [24]). Given a subsequence $\{n''\}$, it suffices to find a further subsequence $\{n'\} \subset \{n''\}$ such that

$$d(\xi_{n'}^{\leftarrow}, \xi_0^{\leftarrow}) \xrightarrow{\text{a.s.}} 0.$$

From (3.24), pick $\{n'\}$ such that

$$d(\xi_{n'}, \xi_0) \xrightarrow{\text{a.s.}} 0.$$

Then for almost all $\omega$,

$$\xi_{n'}(t, \omega) \to \xi_0(t, \omega) \quad \forall t \geq 0,$$

and so by inverting the monotone functions

$$\xi_{n'}^{\leftarrow}(t, \omega) \to \xi_0^{\leftarrow}(t, \omega) \quad \forall t \geq 0.$$

Since $\xi_{n'}^{\leftarrow}(t, \omega)$ is monotone in $t$ and $\xi_0^{\leftarrow}(t, \omega)$ is continuous in $t$, the convergence is locally uniform in $t$, as required. □

### Vervaat's lemma

The next little gem is useful when considering asymptotic normality of estimators. See [289, 290].

**Proposition 3.3.**

(a) *Suppose for each $n$ that $x_n \in D[0, \infty)$ is a nondecreasing function and, furthermore, that $x_0 \in C[0, \infty)$. If $c_n \to \infty$ and*

$$c_n(x_n(t) - t) \to x_0(t) \quad (n \to \infty) \tag{3.26}$$

*locally uniformly, then also*

$$c_n(x_n^{\leftarrow}(t) - t) \to -x_0(t) \quad (n \to \infty) \tag{3.27}$$

*locally uniformly.*

(b) *Suppose $X_n$ is a sequence of $D[0,\infty)$ valued random elements and $X_0$ has continuous paths. Denote the identity by $e(t) = t$. If $X_n$ has nondecreasing paths and if $c_n \to \infty$, then*

$$c_n(X_n - e) \Rightarrow X_0 \quad (n \to \infty)$$

*in $D[0,\infty)$ implies that*

$$c_n(X_n^{\leftarrow} - e) \Rightarrow -X_0 \quad (n \to \infty)$$

*in $D[0,\infty)$. In fact, we also have*

$$c_n(X_n - e, X_n^{\leftarrow} - e) \Rightarrow (X_0, -X_0) \tag{3.28}$$

*in $D[0,\infty) \times D[0,\infty)$.*

*Proof.*

(a) Suppose (3.26) holds. Since $c_n \to \infty$, we have pointwise convergence $x_n(t) \to t$. Due to Proposition 2.2 (p. 20), $x_n^{\leftarrow}(t) \to t$, and applying Proposition 2.1 (p. 18), we conclude that convergence is locally uniform.
For the purpose of getting a contradiction, suppose (3.27) fails. Then there exist $T > 0$, $\epsilon > 0$, and $n' \to \infty$ such that

$$\sup_{0 \le t \le T} |c_{n'}(x_{n'}^{\leftarrow}(t) - t) + x_0(t)| > 2\epsilon,$$

so that there exist $\{t_{n'}\} \subset [0, T]$ satisfying

$$|c_{n'}(x_{n'}^{\leftarrow}(t_{n'}) - t_{n'}) + x_0(t_{n'})| > \epsilon.$$

Either
(a) $c_{n'}(x_{n'}^{\leftarrow}(t_{n'}) - t_{n'}) + x_0(t_{n'}) > \epsilon$
or
(b) $c_{n'}(x_{n'}^{\leftarrow}(t_{n'}) - t_{n'}) + x_0(t_{n'}) < -\epsilon$, that is, $c_{n'}(t_{n'} - x_{n'}^{\leftarrow}(t_{n'})) - x_0(t_{n'}) > \epsilon$.

If (b) is true (otherwise, a similar argument applies), examine (3.26) on a subsequence $\{x_{n'}^{\leftarrow}(t_{n'})\}$ and write the inequality using (2.5) (p. 19) as

$$\begin{aligned} c_{n'}(x_{n'}(x_{n'}^{\leftarrow}(t_{n'})) - x_{n'}^{\leftarrow}(t_{n'})) - x_0(x_{n'}^{\leftarrow}(t_{n'})) \\ \geq c_{n'}(t_{n'} - x_{n'}^{\leftarrow}(t_{n'})) - x_0(t_{n'}) + x_0(t_{n'}) - x_0(x_{n'}^{\leftarrow}(t_{n'})) \\ \geq \epsilon + o(1). \end{aligned}$$

This is a contradiction to the local uniform convergence in (3.26).

(b) Since

$$\xi_n := c_n(X_n - e) \Rightarrow X_0$$

in $D[0, \infty)$, we have from Skorohod's theorem (see Section 3.2.2 (p. 41)) that there exist $\tilde{\xi}$, $\tilde{X}_0$ defined on $[0, 1]$ and

$$\tilde{\xi}_n \overset{\text{a.s.}}{\to} \tilde{X}_0$$

in $D[0, \infty)$. Define

$$\tilde{X}_n := \frac{\tilde{\xi}_n}{c_n} + e.$$

Then $\tilde{X}_n$ is almost surely nondecreasing since $X_n \overset{d}{=} \tilde{X}_n$. Since

$$c_n(\tilde{X}_n - e) \overset{\text{a.s.}}{\to} \tilde{X}_0,$$

we get from part (a) that

$$c_n(\tilde{X}_n^{\leftarrow} - e) \overset{\text{a.s.}}{\to} -\tilde{X}_0,$$

and, in fact, in $D[0, \infty) \times D[0, \infty)$

$$c_n(\tilde{X}_n - e, \tilde{X}_n^{\leftarrow} - e) \overset{\text{a.s.}}{\to} (\tilde{X}_0, -\tilde{X}_0).$$

The rest follows since for each $n$

$$c_n(X_n - e, X_n^{\leftarrow} - e) \overset{d}{=} c_n(\tilde{X}_n - e, \tilde{X}_n^{\leftarrow} - e)$$

and

$$(X_0, -X_0) \overset{d}{=} (\tilde{X}_0, -\tilde{X}_0).$$

□

## 3.6 Vague convergence and regular variation

Regular variation of distribution tails can be reformulated in terms of vague convergence and with this reformulation, the generalization to higher dimensions is effortless. Here we discuss the reformulation in one dimension. We will see implications of the reformulation in Chapter 4.

**Vague convergence on $(0, \infty]$**

**Theorem 3.6.** *Suppose $X_1$ is a nonnegative random variable with distribution function $F(x)$. Set $\bar{F} = 1 - F$. The following are equivalent:*

(i) $\bar{F} \in \mathrm{RV}_{-\alpha}$, $\alpha > 0$.
(ii) *There exists a sequence $\{b_n\}$ with $b_n \to \infty$ such that*

$$\lim_{n\to\infty} n\bar{F}(b_n x) = x^{-\alpha}, \quad x > 0.$$

(iii) *There exists a sequence $\{b_n\}$ with $b_n \to \infty$ such that*

$$\mu_n(\cdot) := n\mathbb{P}\left[\frac{X_1}{b_n} \in \cdot\right] \xrightarrow{v} \nu_\alpha(\cdot) \tag{3.29}$$

*in $M_+(0, \infty]$, where $\nu_\alpha(x, \infty] = x^{-\alpha}$.*

*Remark* 3.3. Here are three remarks on Theorem 3.6.

(a) If any of (i), (ii), or (iii) is true, we may always define

$$b(t) = \left(\frac{1}{1-F}\right)^{\leftarrow}(t) = F^{\leftarrow}\left(1 - \frac{1}{t}\right) \tag{3.30}$$

and set $b_n = b(n)$. The quantity $b(t)$ is just a large quantile; it is the high level such that there is only probability $1/t$ that $X_1$ exceeds the level. Observe that if (i) holds, then

$$\bar{F} \in \mathrm{RV}_{-\alpha} \text{ implies } \frac{1}{1-F} \in \mathrm{RV}_\alpha \text{ implies } b(\cdot) = \left(\frac{1}{1-F}\right)^{\leftarrow}(\cdot) \in \mathrm{RV}_{1/\alpha}.$$

(b) Note in (iii) that the space $\mathbb{E} = (0, \infty]$ has 0 excluded and $\infty$ included. This is required since we need neighborhoods of $\infty$ to be relatively compact. Vague convergence only controls setwise convergence on relatively compact sets (with no mass on the boundary). With the usual topology on $[0, \infty)$, sets of the form $(x, \infty)$ are not bounded; yet consideration of $n\bar{F}(b_n x) = nP[X_1/b_n > x]$ requires considering exactly such sets. We need some topology which makes semi-infinite intervals compact. More on this later when we discuss the *one-point uncompactification* in Section 6.1.3 (p. 170). If it helps, think of $(0, \infty]$ as the homeomorphic stretching of $(0, 1]$ or as the homeomorphic image of $[0, \infty)$ under the map $x \mapsto 1/x$, which takes $[0, \infty) \mapsto (0, \infty]$.

(c) *Preview of things to come*: Note that if $\{X_j, j \geq 1\}$ is an iid sequence of nonnegative random variables with common distribution $F$, then the measure $\mu_n$ defined in (3.29) is also the mean measure of the empirical measure

$$\mu_n(\cdot) = \mathbf{E}\left(\sum_{i=1}^{n} \epsilon_{X_i/b(n)}(\cdot)\right).$$

of the scaled sample. The convergence of $\mu_n$ is equivalent to convergence of the sequence of empirical measures to a limiting Poisson process.

*Proof.* The equivalence of (i) and (ii) is part (ii) of Proposition 2.3 (p. 21).

(ii) $\to$ (iii). Let $f \in C_K^+((0,\infty])$ and we must show that

$$\mu_n(f) := n\mathbf{E}f\left(\frac{X_1}{b_n}\right) = \int f(x)n\mathbb{P}\left[\frac{X_1}{b_n} \in dx\right] \to \nu_\alpha(f).$$

Since $f$ has compact support, the support of $f$ is contained in $(\delta, \infty]$ for some $\delta > 0$. We know that

$$\mu_n(x,\infty] \to x^{-\alpha} = \nu_\alpha(x,\infty] \quad \forall x > 0. \tag{3.31}$$

On $(\delta, \infty]$, define

$$P_n(\cdot) = \frac{\mu_n}{\mu_n(\delta,\infty]} \tag{3.32}$$

so that $P_n$ is a probability measure on $(\delta, \infty]$. Then for $y \in (\delta, \infty]$,

$$P_n(y,\infty] \to P(y,\infty] = \frac{y^{-\alpha}}{\delta^{-\alpha}}.$$

In $\mathbb{R}$, convergence of distribution functions (or tails) is equivalent to weak convergence, so $\{P_n\}$ converges weakly to $P$. Since $f$ is bounded and continuous on $(\delta, \infty]$, we get from weak convergence that

$$P_n(f) \to P(f);$$

that is,

$$\frac{\mu_n(f)}{\mu_n(\delta,\infty]} \to \frac{\nu_\alpha(f)}{\delta^{-\alpha}}.$$

In light of (3.31), this implies

$$\mu_n(f) \to \nu_\alpha(f),$$

as required.

(iii) $\to$ (ii). Since

$$\mu_n \xrightarrow{v} \nu_\alpha,$$

we have

$$\mu_n(x,\infty] \to \nu_\alpha(x,\infty] \quad \forall x > 0$$

since $(x,\infty]$ is relatively compact and

$$\nu_\alpha(\partial(x,\infty]) = \nu_\alpha(\{x\}) = 0. \qquad \square$$

## 3.7 Problems

**3.1.** Suppose for $n \geq 0$ that $\mu_n \in M_+(\mathbb{E})$. Show that $\mu_n \xrightarrow{v} \mu_0$ in $M_+(\mathbb{E})$ iff for all $f \in C_K^+(\mathbb{E})$,

$$\mu_n(1 - e^{-f}) \to \mu_0(1 - e^{-f}).$$

**3.2.** Suppose $\{\xi_n, n \geq 0\}$ are random elements of $M_+([0,\infty))$ and that

$$\xi_n \Rightarrow \xi_0.$$

If $t$ satisfies

$$P[\xi_0(\{t\}) = 0] = 1,$$

does

$$\xi_n[0,t] \Rightarrow \xi_0[0,t]$$

in $[0,\infty)$?

*Hint*: Is the map $T_t : M_+[0,\infty) \mapsto \mathbb{R}_+$ defined by

$$T_t(\mu) = \mu[0,t]$$

continuous? Almost surely continuous?

**3.3.** Show that the transformations in (a) and (b) are vaguely continuous:

(a) $T_1 : M_p(\mathbb{E}) \times M_p(\mathbb{E}) \mapsto M_p(\mathbb{E})$ defined by

$$T_1(m_1, m_2) = m_1 + m_2.$$

(b) $T_2 : M_+(\mathbb{E}) \times (0,\infty) \mapsto M_+(\mathbb{E})$ defined by

$$T_2(\mu, \lambda) = \lambda\mu.$$

Define the scaling function $T_3 : M_+((0,\infty]) \times (0,\infty) \mapsto M_+(0,\infty]$ by

$$T_3(\mu, \lambda) = \mu(\lambda(\cdot)).$$

Is $T_3$ continuous?

**3.4.** Suppose $x_n$, $n \geq 0$, are points of $\mathbb{E}$ and $c_n$, $n \geq 0$, are positive constants. Then

$$c_n \epsilon_{x_n} \xrightarrow{v} c_0 \epsilon_{x_0}$$

as $n \to \infty$ iff

$$c_n \to c_0 \quad \text{and} \quad x_n \to x_0.$$

**3.5.** In $M_+[0,\infty)$, prove that

$$\frac{1}{n} \sum_{i=1}^{\infty} \epsilon_{i/n} \xrightarrow{v} \mathbb{LEB}(\cdot).$$

**3.6.** If $K \in \mathcal{K}(\mathbb{E})$ is compact, prove that

$$\{\mu \in M_+(\mathbb{E}) : \mu(K) < t\}$$

is open in $M_+(\mathbb{E})$.

**3.7.** Assume that $\mathbb{E}_i$, $i = 1, 2$, are two nice spaces and that $\mathbb{E}_2$ is compact. Suppose for $n \geq 0$ that $m_n \in M_p(\mathbb{E}_1 \times \mathbb{E}_2)$ and $m_n \xrightarrow{v} m_0$ in $M_p(\mathbb{E}_1 \times \mathbb{E}_2)$. Conclude that

$$m_n(\cdot \times E_2) \xrightarrow{v} m_0(\cdot \times \mathbb{E}_2)$$

in $M_p(\mathbb{E}_1)$.

**3.8.** Suppose $\mathbb{E}$ and $\mathbb{E}'$ are two nice spaces with $\mathbb{E}$ compact and suppose $T : \mathbb{E} \mapsto \mathbb{E}'$ is continuous on an open subset $G$ of $\mathbb{E}$. Then if $m \in M_p(\mathbb{E})$ is a point measure with support contained in $G$, the mapping $\hat{T} : M_p(E) \mapsto M_p(\mathbb{E}')$ defined by

$$\hat{T}\left(\sum_i \epsilon_{x_i}\right) = \sum_i \epsilon_{Tx_i}$$

is continuous at $m$ [76].

**3.9.** Suppose $\mathbb{E}_1$, $\mathbb{E}_2$, $\mathbb{E}_2'$ are nice spaces with $\mathbb{E}_2$ compact. Assume that $T : \mathbb{E}_2 \mapsto \mathbb{E}_2'$ is continuous on an open subset $G_2$ of $\mathbb{E}_2$. If $m \in M_p(\mathbb{E}_1 \times \mathbb{E}_2)$ has the property $m(\mathbb{E}_1 \times G_2^c) = 0$, then

$$\hat{T} : M_p(\mathbb{E}_1 \times \mathbb{E}_2) \mapsto M_p(\mathbb{E}_1 \times \mathbb{E}_2'),$$

defined by

$$\hat{T}\left(\sum_i \epsilon_{(t_i, x_i)}\right) = \sum_i \epsilon_{(t_i, Tx_i)},$$

is continuous at $m$ [76].

**3.10.** (a) Suppose the random vectors $\boldsymbol{X}_n$ and $\boldsymbol{Y}_n$ in $\mathbb{R}^d$ are independent for each $n$ and that $\boldsymbol{X}_n \Rightarrow \boldsymbol{X}$ and $\boldsymbol{Y}_n \Rightarrow \boldsymbol{Y}$. Show that in $\mathbb{R}^{2d}$, we have

$$(\boldsymbol{X}_n, \boldsymbol{Y}_n) \Rightarrow (\boldsymbol{X}, \boldsymbol{Y}),$$

where $\boldsymbol{X}, \boldsymbol{Y}$ are independent.
(b) Show also that

$$\boldsymbol{X}_n + \boldsymbol{Y}_n \Rightarrow \boldsymbol{X} + \boldsymbol{Y}$$

in $\mathbb{R}^d$.

**3.11.** Let $\{X_n\}$ be a sequence of random variables such that $\mathbf{E}X_n = m$ and $\mathrm{Var}(X_n) = \sigma_n^2 > 0$ for all $n$, where $\sigma_n^2 \to 0$ as $n \to \infty$. Define

$$Z_n = \sigma_n^{-1}(X_n - m),$$

and let $f$ be a function with nonzero derivative $f'(m)$ at $m$.

1. Show that $X_n - m \Rightarrow 0$.
2. If
$$Y_n = \frac{f(X_n) - f(m)}{\sigma_n f'(m)},$$
show that $Y_n - Z_n \Rightarrow 0$.
3. Show that if $Z_n$ converges in probability or in distribution, then so does $Y_n$.
4. If $S_n$ is binomially distributed with parameters $n$ and $p$ and $f'(p) \neq 0$, use the preceding results to determine the asymptotic distribution of $f(S_n/n)$.

**3.12.** If $f$ is bounded and upper semicontinuous, show that $P_n \Rightarrow P$ implies that

$$\limsup_{n\to\infty} P_n(f) \le P(f).$$

**3.13.** Suppose the family of measures $\Pi$ is defined by

$$\Pi = \{\epsilon_x(\cdot), x \in A\},$$

where $A \subset S$ and

$$\epsilon_x(B) = \begin{cases} 1 & \text{if } x \in B, \\ 0 & \text{if } x \in B^c. \end{cases}$$

Show $\Pi$ is relatively compact iff $A^-$ is compact in $S$.

**3.14.** If the sequence of random variables $\{|X_n|^\delta\}$ is uniformly integrable for some $\delta > 0$, then $\{|X_n|\}$ is tight. In particular, the condition

$$\sup_n E(|X_n|^{\delta+\eta}) < \infty$$

for some $\eta > 0$ is sufficient for tightness.

**3.15 (Second-order regular variation [90, 101, 235]).** A function $U : (0,\infty) \mapsto (0,\infty)$ is second-order regularly varying with first-order parameter $\gamma > 0$ and second-order parameter $\rho \leq 0$ (written $U \in 2\,\mathrm{RV}(\gamma,\rho)$) if there exists a function $A(t) \to 0$ which is ultimately of constant sign, and such that

$$\lim_{t\to\infty} \frac{\frac{U(tx)}{U(t)} - x^\gamma}{A(t)} = cx^\gamma \left(\frac{x^\rho - 1}{\rho}\right), \quad x > 0, \rho \leq 0, c \neq 0. \tag{3.33}$$

Now suppose $F$ is a distribution on $[0,\infty)$, and define

$$U = \left(\frac{1}{1-F}\right)^{\leftarrow} = F^{\leftarrow}\left(1 - \frac{1}{t}\right).$$

Show using Vervaat's lemma (Proposition 3.3 (p. 59)) that $U \in 2\,\mathrm{RV}(\gamma,\rho)$ is equivalent to

$$\lim_{t\to\infty} \frac{\frac{\bar F(tx)}{\bar F(t)} - x^{-1/\gamma}}{A\left(\frac{1}{1-F(t)}\right)} = c' x^{-1/\gamma}\left(\frac{x^{\rho/\gamma} - 1}{\rho\gamma}\right), \quad c' \neq 0, \quad x > 0. \tag{3.34}$$

**3.16 (More on second-order regular variation).** Verify the second-order regular variation for the following examples:

1. Suppose
$$1 - F(x) = x^{-1/\gamma} + cx^{-1/\delta},$$
where $c > 0$, $1/\delta > 1/\gamma$; that is, $\gamma > \delta$ [155, 297].

2. *Cauchy*:
$$F'(x) = \frac{1}{\pi(1+x^2)}$$
or
$$F(x) = \frac{1}{2} + \frac{1}{\pi}\arctan x, \quad x \in \mathbb{R}.$$
(Consider working with $U = (1/(1-F))^{\leftarrow}$ rather than $1 - F$, but feel free to experiment.)

3. Stable, for which you will need a series expansion. (See [135], for example.)

4. *Log-gamma*: An example of a log-gamma distribution is constructed by taking $X_1, X_2$ iid with standard exponential density and computing the distribution of $\exp\{X_1 + X_2\}$. For $x > 1$,

$$\begin{aligned} P[\exp\{X_1 + X_2\} > x] &= P[X_1 + X_2 > \log x] \\ &= \exp\{-\log x\} + \exp\{-\log x\} \log x \\ &= x^{-1}(1 + \log x) := 1 - F(x). \end{aligned}$$

   For this example, we have $\alpha = 1$, $\rho = 0$.

Finally, verify (quickly) that the Pareto is *not* second-order regularly varying. (This is more of an observation than anything else.)

**3.17 (Even more 2 RV).** Let $Z_1, Z_2$ be nonnegative iid random variables with common distribution $F$ satisfying $1 - F \in 2\,\mathrm{RV}(-\alpha, \rho)$. Then for $x > 0$,

$$\lim_{t\to\infty} \frac{\frac{P[Z_1 \vee Z_2 > tx]}{1-F(t)} - 2x^{-\alpha}}{A(t)} = 2H(x) - l x^{-2\alpha},$$

where $H(x) = cx^{-\alpha} \int_1^x u^{\rho-1} du$, $x > 0$, $c > 0$, if

$$\lim_{t\to\infty} \frac{1 - F(t)}{A(t)} = l, \quad |l| < \infty,$$

and if $|l| = \infty$,

$$\lim_{t\to\infty} \frac{\frac{P[Z_1 \vee Z_2 > tx]}{1-F(t)} - 2x^{-\alpha}}{1 - F(t)} = -x^{-2\alpha}.$$

(See [146] for this and harder results.)

**3.18.** Suppose $\{X_k, k \geq 0\}$ is a Markov chain with state space $\{0, 1, 2, \dots\}$ and transition matix $P = (p_{ij})$. Assume that

$$\pi' P = \pi' \quad \text{and} \quad p_{ij}^{(n)} \to \pi_j.$$

Show in $\mathbb{R}^\infty$ that as $n \to \infty$,

$$\{X_k, k \geq n\} \Rightarrow \{X_k^\#, k \geq 0\},$$

where $\{X_k^\#, k \geq 0\}$ is a stationary Markov chain with transition matrix $P$ and initial distribution $\pi$.

**3.19 (Second continuous mapping theorem).** Suppose $\mathbb{S}$ and $\mathbb{S}'$ are two complete, separable metric spaces and that we are given measurable maps $h_n : \mathbb{S} \mapsto \mathbb{S}'$. Let $\mathbb{D}$ be the set of $x \in \mathbb{S}$ such that $h_n(x_n) \not\to h(x)$ for some sequence $\{x_n\}$ converging to $x$. If $P_n, n \geq 0$, are probability measures on $(\mathbb{S}, \mathcal{S})$, and $P_n \Rightarrow P_0$, then

$$P_n \circ h_n^{-1} \Rightarrow P_0 \circ h_0^{-1},$$

provided $P_0(\mathbb{D}) = 0$ [25, p. 79]. (*Hint*: Modify the proof of Theorem 3.1 (p. 42).)

**3.20 (Combining independent convergences).** Let $\mathbb{E}$ and $\mathbb{E}'$, be two complete separable metric spaces and suppose $\{\xi_n, n \geq 0\}$ and $\{\eta_n, n \geq 0\}$ are random elements of $\mathbb{E}$ and $\mathbb{E}'$, respectively, defined on the same probability space. Suppose further, for each $n \geq 1$, that $\xi_n$ and $\eta_n$ are independent. Assume

$$\xi_n \Rightarrow \xi_0$$

in $\mathbb{E}$ and

$$\eta_n \Rightarrow \eta_0$$

in $\mathbb{E}'$. Then jointly in $\mathbb{E} \times \mathbb{E}'$ we have as $n \to \infty$ that the distribution of $(\xi_n, \eta_n)$ converges to a product measure whose factor distributions are the distributions of $\xi_0$ and $\eta_0$.

**3.21 (The supremum map).** Prove that the map

$$x \mapsto \sup_{0 \leq s \leq 1} x(s)$$

is continuous from $D[0, 1] \mapsto \mathbb{R}$.

**3.22 (Impossibility of Skorohod convergence).** Suppose that $X_n \in C[0, 1]$ and that $X_\infty \in D[0, 1] \setminus C[0, 1]$ and that the finite-dimensional distributions of $X_n$ converge to those of $X_\infty$. Argue that it is impossible for $X_n \Rightarrow X_\infty$ in the Skorohod topology.

# Part II

# Statistics

# 4

# Dipping a Toe in the Statistical Water

This material is designed to give immediate payoff for the previous two chapters. We give some estimators of the tail index, prove consistency, and evaluate the effectiveness of the estimation. We will return to statistical inference problems on several occasions, and the present chapter is a first experience with the statistical side of the subject. In particular, we will return to issues of asymptotic normality of the estimators in Chapter 9.1.

## 4.1 Statistical inference for heavy tails: This is a song about $\alpha$

How does one go about devising and using statistical methods for heavy tails? For the simplest formulation, suppose that one-dimensional data have been collected, and that fortune has smiled on us in that the data look stationary and even independent and identically distributed (iid).

The following are the initial steps in any heavy-tailed statistical analysis of one-dimensional data that are at least stationary:

- decide that a heavy-tailed model is appropriate, and then
- estimate the tail index $\alpha$ of the marginal distribution.

Various graphical and estimation techniques exist to help accomplish these steps: QQ estimation and plotting, Hill estimation and plotting, and Pickands estimation, to name just a few. There are also many techniques applicable from extreme-value methods [16, 50, 90, 129, 197, 238].

Suppose $X, X_1, \ldots, X_n$ have the same distribution $F(x)$ and that inference is to be based on $X_1, \ldots, X_n$. There are at least two competing heavy-tailed models and philosophies—although similar, they differ in important ways:

- Assume that $F$ has a Pareto right tail from some point on. This means that there there exist some $x_l > 0$, $c > 0$, and $\alpha > 0$ such that

$$\mathbb{P}[X > x] = cx^{-\alpha}, \quad x > x_l. \tag{4.1}$$

So we assume an exact Pareto tail from $x_l$ onwards. The form of the tail for $x < x_l$ may or may not be specified in this approach, depending on the purpose of the analysis.

- Assume that $F$ has a regularly varying right tail with index $-\alpha$,

$$\mathbb{P}[X > x] = 1 - F(x) = \bar{F}(x) = x^{-\alpha}L(x), \quad x > 0. \tag{4.2}$$

For the most part, we will assume the semiparametric assumption (4.2) of regular variation and focus on the problem of estimating the index of regular variation $\alpha$. The *Hill estimator* is a popular, though troubled, estimator of $1/\alpha$ and has a voluminous literature. A partial list of references is [57, 72, 100, 112, 155, 165, 212, 236, 252]. The Hill estimator is defined as follows: Assume for simplicity that observations $X_1, \dots, X_n$ are nonnegative. For $1 \le i \le n$, write $X_{(i)}$ for the $i$th largest value of $X_1, X_2, \dots, X_n$, so that

$$X_{(1)} \ge X_{(2)} \ge \dots \ge X_{(n)}.$$

Then Hill's estimator of $1/\alpha$ based on $k$ upper-order statistics is defined as

$$H_{k,n} := \frac{1}{k} \sum_{i=1}^{k} \log \frac{X_{(i)}}{X_{(k+1)}}. \tag{4.3}$$

The theory is most easily developed for the case in which $\{X_j, j \ge 1\}$ is iid, although applications often do not provide us with independent observations but rather with dependent, stationary data. So attention needs to be paid to applying the Hill estimator in non-iid cases.

## 4.2 Exceedances, thresholds, and the POT method

Why does the Hill estimator make intuitive sense? Suppose, temporarily, that instead of the semiparametric assumption (4.2), we assume that we have data from the more precisely specified iid Pareto parametric family

$$\bar{F}(x) := \mathbb{P}[X_i > x] = x^{-\alpha}, \quad x > 1, \alpha > 0. \tag{4.4}$$

Thus $F$ is a Pareto distribution with support $[1, \infty)$. Then the maximum-likelihood estimator of $1/\alpha$ is

$$\widehat{\alpha^{-1}} = \frac{1}{n}\sum_{i=1}^{n} \log X_i.$$

This follows readily since $\{\log X_i, 1 \le i \le n\}$ is a random sample from the distribution with tail

$$\mathbb{P}[\log X_1 > x] = \mathbb{P}[X_1 > e^x] = e^{-\alpha x}, \quad x > 0,$$

which is the exponential distribution tail. The mean of this exponential distribution is $\alpha^{-1}$ and the MLE is $\bar{X}$, which in this case is the given estimator.

But what if (4.4), a rather strong assumption, is implausible? A somewhat weaker assumption is to assume a Pareto tail from some point onwards, as in (4.1), rather than the exact model. This leads to the peaks-over-threshold (POT) method discussed in Section 4.2.3 (p. 77). First, some background.

### 4.2.1 Exceedances

Consider a precise definition of an exceedance. Given observations $x_1, \dots, x_n$ and a threshold $u$, we call an observation $x_j$ an *exceedance over* $u$ if $x_j > u$. In this case, $x_j - u$ is the *excess*.

Let $X_1, \dots, X_n$ be iid random variables and set

$$K_n = \sum_{j=1}^{n} 1_{(u,\infty)}(X_j) = \#\text{ of exceedances of } u \text{ in the first } n \text{ variables.}$$

This is a binomial random variable with success probability $p = \mathbb{P}[X_1 > u]$.

### 4.2.2 Exceedance times

Suppose $\{X_n, n \ge 1\}$ are iid and $u$ is a threshold. Define the *exceedance times* $\{\tau_j, j \ge 1\}$ by

$$\begin{aligned}
\tau_1 &= \inf\{j \ge 1 : X_j > u\}, \\
\tau_2 &= \inf\{j > \tau_1 : X_j > u\}, \\
&\vdots \\
\tau_r &= \inf\{j > \tau_{r-1} : X_j > u\}.
\end{aligned}$$

The sequence $\{X_{\tau_r}, r \ge 1\}$ are the *exceedances*.

**Subsequence principle**

If $\{X_n, n \geq 1\}$ is iid with common distribution $F$, then $\{X_{\tau_j}, j \geq 1\}$ is also iid, and

$$\mathbb{P}[X_{\tau_j} > x] = \bar{F}^{[u]}(x) := \mathbb{P}[X > x | X > u]$$
$$= \begin{cases} \frac{\bar{F}(x)}{\bar{F}(u)} & \text{for } x > u, \\ 1 & \text{for } x \leq u. \end{cases} \tag{4.5}$$

Note that $F^{[u]}(\cdot)$ is the conditional distribution of $X$ given that $X > u$. We sometimes write informally

$$X_{\tau_j} \stackrel{d}{=} X_1 | X_1 > u$$

to mean just this. Furthermore,

$$\tau_1, \tau_2 - \tau_1, \tau_3 - \tau_2, \ldots$$

are iid with

$$\mathbb{P}[\tau_1 > k] = \mathbb{P}[X_1 \leq u, \ldots, X_k \leq u] = (F(u))^k$$

and

$$\mathbb{P}[\tau_1 = k] = \mathbb{P}[X_1 \leq u, \ldots, X_{k-1} \leq u, X_k > u] = F^{k-1}(u)\bar{F}(u).$$

So $\tau_1$ has a geometric distribution with

$$\mathbb{P}[\tau_1 = k] = q^{k-1} p, \quad k = 1, 2, \ldots, \tag{4.6}$$

where

$$p = \bar{F}(u), \quad q = 1 - p = F(u).$$

Where does the distribution of $\{X_{\tau_j}, j \geq 1\}$ come from? This is a special case of an old result dating to P. Lévy and is sometimes called the *Découpage de Lévy* (see [260] and Problem 4.4). To quickly obtain the flavor of a partial proof, consider $X_{\tau_1}$. For $x > u$,

$$\mathbb{P}[X_{\tau_1} > x] = \sum_{k=1}^{\infty} \mathbb{P}[\tau_1 = k, X_{\tau_1} > x] = \sum_{k=1}^{\infty} \mathbb{P}[\tau_1 = k, X_k > x]$$
$$= \sum_{k=1}^{\infty} \mathbb{P}[X_1 \leq u, \ldots, X_{k-1} \leq u, X_k > u, X_k > x],$$

and for $x > u$ this is (with $q = F(u)$)

$$= \sum_{k=1}^{\infty} q^{k-1} \bar{F}(x) = \frac{\bar{F}(x)}{1-q} = \frac{\bar{F}(x)}{\bar{F}(u)}.$$

The moments of $\tau_1$ are easy to compute since $\tau_1$ is a geometrically distributed random variable. For threshold $u$,

$$E(\tau_1) = \frac{1}{p} = \frac{1}{1 - F(u)}, \tag{4.7}$$

and likewise,

$$\mathrm{Var}(\tau_1) = \frac{q}{p^2} = \frac{F(u)}{(\bar{F}(u))^2}.$$

### 4.2.3 Peaks over threshold

Suppose the model (4.1) is assumed in which the distribution tail is Pareto beyond $x_l$. Consider exceedances over level $x_l$ or the *peaks over the threshold* $x_l$. Then from (4.1) and (4.5), we have for $x > x_l$,

$$\mathbb{P}[X_{\tau_1} > x] = \frac{\bar{F}(x)}{\bar{F}(x_l)} = \frac{cx^{-\alpha}}{cx_l^{-\alpha}} = \left(\frac{x}{x_l}\right)^{-\alpha},$$

so that for $y > 1$

$$\mathbb{P}\left[\frac{X_{\tau_1}}{x_l} > y\right] = y^{-\alpha}, \quad y > 1. \tag{4.8}$$

*Conclusion*: Assuming that the distribution of the iid sample satisfies (4.1), that is, the distribution has a Pareto tail from $x_l$ onwards, means that the relative exceedances

$$\left\{\frac{X_{\tau_j}}{x_l}, j \geq 1\right\}$$

are an iid sample from a Pareto distribution with parameter $\alpha$ and support $[1, \infty)$. Assuming the $\{X_n\}$ are iid, applying the argument of Section 4.2 (p. 74) makes the Hill estimator the MLE estimator applied to the relative exceedances of level $x_l = X_{(k+1)}$, where we assume $k$-exceedances of level $x_l = X_{(k+1)}$. Relying on exceedances is the peaks-over-threshold (POT) method.

What is possible if we assume only the regular variation assumption (4.2)? Relative exceedances of $x_l$ now have the distribution tail (cf. (4.8))

$$\mathbb{P}\left[\frac{X_{\tau_1}}{x_l} > y\right] = \frac{\bar{F}(x_l y)}{\bar{F}(x_l)} \approx y^{-\alpha}, \quad y > 0, \quad x_l \text{ large}, \tag{4.9}$$

which is only approximately a Pareto tail. One way to proceed is to pretend the approximate equality given by (4.9) is an actual equality. This has the advantage that the method of maximum-likelihood estimation is available, and this method is a powerful, off-the-shelf technology. However, there is no obvious way to quantify the errors introduced by a misspecified model if one assumes (4.1) when it is false. The other way to proceed is to prove asymptotic properties of estimators based on regular variation assumptions and refinements.

Exceedances and the POT method will be revisited again from the point of view of point processes.

## 4.3 The tail empirical measure

The following describes a one-dimensional result, but after converting regular variation to vague convergence as in Section 3.6, the result is really dimensionless. Considering the possibility of doing inference with multidimensional data suggests a broader point of view that is fruitful even in one dimension.

Reviewing the equivalences in Theorem 3.6 (p. 62) suggests that instead of estimating the parameter $\alpha$, we could estimate the measure $\nu_\alpha$ on $(0, \infty]$, which would yield the required information.

Suppose $\{X_j, j \geq 1\}$ is a sequence of random variables with common one-dimensional marginal distribution $F$, which has regularly varying tail probabilities

$$\bar{F}(x) := 1 - F(x) = \mathbb{P}[X_1 > x] = x^{-\alpha} L(x), \quad \alpha > 0. \tag{4.10}$$

For convenience, assume that the variables are nonnegative. A useful scaling quantity is the *quantile function* $b(t)$ defined by

$$b(t) = \left(\frac{1}{1-F}\right)^{\leftarrow}(t) = F^{\leftarrow}\left(1 - \frac{1}{t}\right). \tag{4.11}$$

The *tail empirical measure* is defined as a random element of $M_+(0, \infty]$, the space of nonnegative Radon measures on $(0, \infty]$, by

$$\nu_n := \frac{1}{k} \sum_{i=1}^{n} \epsilon_{X_i / b(\frac{n}{k})}. \tag{4.12}$$

The new feature here is the presence of $k$, which represents the number of upper-order statistics that we think or guess are relevant for estimating tail probabilities. We emphasize that the notation $\nu_n$ suppresses the dependence on $k$ but that the $k$ is critical. The tail empirical measure is used in a variety of inference contexts, but note that, as defined, its statistical use needs to overcome the fact that in a data-driven context, where $F$ is unknown, $b(\cdot)$ is also unknown.

When $\{X_n\}$ are iid, $\nu_n$ approximates $\nu_\alpha$.

**Theorem 4.1.** *Suppose that $\{X_j, j \geq 1\}$ are iid, nonnegative random variables whose common distribution has a regularly varying tail* (4.10), *which implies (see Theorem* 3.6) *that*

$$\frac{n}{k}\mathbb{P}\left[\frac{X_1}{b(n/k)} \in \cdot\right] \xrightarrow{v} \nu_\alpha(\cdot) \tag{4.13}$$

*in $M_+(0,\infty]$ as $n \to \infty$ and $k = k(n) \to \infty$ with $n/k \to \infty$. Then in $M_+(0,\infty]$,*

$$\nu_n \Rightarrow \nu_\alpha, \tag{4.14}$$

*where*

$$\nu_\alpha(x,\infty] = x^{-\alpha}, \quad x > 0, \quad \alpha > 0.$$

*Remark* 4.1. More general versions of this result are possible but await further probability developments in the next chapter. The reason for the odd form of the asymptotics $(n \to \infty, k \to \infty, n/k \to \infty)$ is that we will estimate $b(n/k)$ by $X_{(k)}$, and the condition on $k = k(n)$ forces $X_{(k)} \xrightarrow{P} \infty$. (See Problem 4.3 (p. 115).)

*Proof.* We use some of the methods outlined in Section 3.4.1 (p. 53). It suffices to show for a sequence $h_j \in C_K^+(0,\infty]$ that in $\mathbb{R}^\infty$,

$$(\nu_n(h_j), j \geq 1) \Rightarrow (\nu_\alpha(h_j), j \geq 1) \quad (n \to \infty).$$

Convergence in $\mathbb{R}^\infty$ reduces to convergence in $\mathbb{R}^d$ for any $d$, so it suffices to show that

$$(\nu_n(h_j), 1 \leq j \leq d) \Rightarrow (\nu_\alpha(h_j), 1 \leq j \leq d) \quad (n \to \infty).$$

To show this, we can show the joint Laplace transforms converge so we assume $\lambda_j > 0$, $j = 1, \dots, d$, and show that

$$\mathbf{E}e^{-\sum_{j=1}^d \lambda_j \nu_n(h_j)} \to \mathbf{E}e^{-\sum_{j=1}^d \lambda_j \nu_\alpha(h_j)}.$$

However,

$$\sum_{j=1}^d \lambda_j \nu_n(h_j) = \nu_n\left(\sum_{j=1}^d \lambda_j h_j\right);$$

similarly for $\nu_\alpha$ substituted for $\nu_n$. Since $\sum_{j=1}^d \lambda_j h_j \in C_K^+(0,\infty]$, it suffices to show for any $h \in C_K^+(0,\infty]$ that

$$\mathbf{E}e^{-\nu_n(h)} \to \mathbf{E}e^{-\nu_\alpha(h)}, \tag{4.15}$$

which is a reduction of the original task to a one-dimensional chore. The left side of (4.15) is

$$\begin{aligned}
\mathbf{E}e^{-\frac{1}{k}\sum_{j=1}^{n} h(X_j/b(n/k))} &= \left(\mathbf{E}e^{-\frac{1}{k}h(X_1/b(n/k))}\right)^n \\
&= \left(1 - \int_{(0,\infty]} \left(1 - e^{-\frac{1}{k}h(x)}\right) \mathbb{P}\left[\frac{X_1}{b(n/k)} \in dx\right]\right)^n \\
&= \left(1 - \frac{\int_{(0,\infty]} \left(1 - e^{-\frac{1}{k}h(x)}\right) n\mathbb{P}\left[\frac{X_1}{b(n/k)} \in dx\right]}{n}\right)^n,
\end{aligned}$$

and this converges to $e^{-\nu_\alpha(h)}$ since

$$\int_{(0,\infty]} \left(1 - e^{-\frac{1}{k}h(x)}\right) n\mathbb{P}\left[\frac{X_1}{b(n/k)} \in dx\right] \approx \int_{(0,\infty]} h(x)\frac{n}{k}\mathbb{P}\left[\frac{X_1}{b(n/k)} \in dx\right] \to \nu_\alpha(h),$$

where the approximate equivalence in the previous line can be justified by writing upper and lower bounds resulting from expanding the term $(1 - e^{-\frac{1}{k}h(x)})$ to get upper and lower bounds. Gory details are provided later in Theorem 5.3 (p. 138). See especially the material following (5.18) (p. 140). □

## 4.4 The Hill estimator

Recall the definition of the Hill estimator $H_{k,n}$ given in (4.3) (p. 74) for estimating $1/\alpha$. Suppose at a minimum that $\{X_n\}$ is a sequence of random variables having the same marginal distribution function $F$ and where $\bar{F} := 1 - F$ is regularly varying at $\infty$ and satisfies (4.10). The quantile function (4.11) is $b(t)$. The random measure $\nu_n$ given in (4.12) is a random element of $M_+(0, \infty]$ and is *assumed* to be a vaguely consistent estimator of the measure $\nu_\alpha \in M_+(0, \infty]$, provided $n \to \infty$ and $k/n \to 0$. However, because $b(\cdot)$ is unknown, $b(n/k)$ will be estimated by a consistent estimator, $\hat{b}(n/k)$, to be specified. We set

$$\hat{\nu}_n =: \frac{1}{k}\sum_{i=1}^{n} \epsilon_{X_i/\hat{b}(n/k)}. \tag{4.16}$$

We know from Theorem 4.1 that (4.14) is satisfied if $\{X_j\}$ is iid with common distribution $F$, where $1 - F \in RV_{-\alpha}$, satisfying (4.10). We emphasize in this section that the standing assumption is consistency of the tail empirical measure, a point of view promoted in [252].

### 4.4.1 Random measures and the consistency of the Hill estimator

Consistency of the tail empirical measure given in (4.14) implies consistency of the Hill estimator for $1/\alpha$.

**Theorem 4.2.** *If* (4.14) *holds, then as* $n \to \infty$, $k \to \infty$, *and* $k/n \to 0$,

$$H_{k,n} \xrightarrow{P} \frac{1}{\alpha}.$$

*Proof.* The proof proceeds by a series of steps.

STEP 1. Consistency of the empirical measure given in (4.14) implies

$$\frac{X_{(k)}}{b(n/k)} \xrightarrow{P} 1 \tag{4.17}$$

as $n \to \infty$, $k \to \infty$ and $k/n \to 0$. This allows us to consider $X_{(k)}$ as a consistent estimator of $b(n/k)$.

To see this, write

$$\begin{aligned}\mathbb{P}\left[\left|\frac{X_{(k)}}{b\left(\frac{n}{k}\right)} - 1\right| > \varepsilon\right] &= \mathbb{P}\left[X_{(k)} > (1+\varepsilon)b\left(\frac{n}{k}\right)\right] + \mathbb{P}\left[X_{(k)} < (1-\varepsilon)b\left(\frac{n}{k}\right)\right] \\ &\leq \mathbb{P}\left[\frac{1}{k}\sum_{i=1}^{n} \epsilon_{X_i/b(\frac{n}{k})}(1+\varepsilon, \infty] \geq 1\right) \\ &\quad + \mathbb{P}\left[\frac{1}{k}\sum_{i=1}^{n} \epsilon_{X_i/b(\frac{n}{k})}[1-\varepsilon, \infty] < 1\right].\end{aligned}$$

But (4.14) implies that

$$\frac{1}{k}\sum_{i=1}^{k} \epsilon_{X_i/b(\frac{n}{k})}(1+\varepsilon, \infty] \xrightarrow{P} (1+\varepsilon)^{-\alpha} < 1,$$

and

$$\frac{1}{k}\sum_{i=1}^{k} \epsilon_{X_i/b(\frac{n}{k})}[1-\varepsilon, \infty] \xrightarrow{P} (1-\varepsilon)^{-\alpha} > 1,$$

and therefore (4.17) follows. □

*Bonus.* In fact, more is true. We have that (4.14) implies

$$\frac{X_{(\lceil kt \rceil)}}{b(n/k)} \xrightarrow{P} t^{-1/\alpha} \quad \text{in } D(0, \infty], \tag{4.18}$$

where $\lceil kt \rceil$ is the smallest integer greater than or equal to $kt$. We prove this more muscular version (4.18) as follows: The map from $M_+(0, \infty] \mapsto D[0, \infty)$ defined by

$$\mu \mapsto \mu(t^{-1}, \infty], \quad t \geq 0,$$

is continuous at measures $\mu$ such that $\mu(t, \infty]$ is continuous, strictly decreasing in $t$. So we have from (4.14) and the continuous mapping theorem, Theorem 3.1 (p. 42), that

$$\nu_n(t^{-1}, \infty] \xrightarrow{P} t^{\alpha}, \quad t \geq 0, \tag{4.19}$$

in $D[0, \infty)$. From inversion and Proposition 3.2 (p. 58), we get that inverses also converge in probability

$$(\nu_n((\cdot)^{-1}, \infty])^{\leftarrow}(t) \xrightarrow{P} t^{1/\alpha}, \quad t \geq 0, \tag{4.20}$$

as functions in $D_l[0, \infty)$, where $D_l[0, \infty)$ are the real, left-continuous functions on $[0, \infty)$ with finite right limits on $(0, \infty)$. We now unpack the inverse and see what we get:

$$\begin{aligned}
(\nu_n((\cdot)^{-1}, \infty])^{\leftarrow}(t) &= \inf\{s : \nu_n(s^{-1}, \infty] \geq t\} \\
&= \inf\left\{s : \sum_{i=1}^{n} \epsilon_{X_i/b(n/k)}(s^{-1}, \infty] \geq kt\right\} \\
&= \inf\left\{y^{-1} : \sum_{i=1}^{n} \epsilon_{X_i/b(n/k)}(y, \infty] \geq kt\right\} \\
&= \left(\sup\left\{y : \sum_{i=1}^{n} \epsilon_{X_i/b(n/k)}(y, \infty] \geq kt\right\}\right)^{-1} \\
&= \left(\frac{X_{(\lceil kt \rceil)}}{b(n/k)}\right)^{-1}.
\end{aligned}$$

So

$$\left(\frac{X_{(\lceil kt \rceil)}}{b(n/k)}\right)^{-1} \Rightarrow t^{1/\alpha}$$

in $D[0, \infty)$, and therefore we conclude that

$$\frac{X_{(\lceil kt \rceil)}}{b(n/k)} \xrightarrow{P} t^{-1/\alpha}$$

in $D_l(0, \infty]$. □

Henceforth, set

$$\hat{b}(n/k) = X_{(k)}.$$

STEP 2. The following results from (4.14): In $M_+(0, \infty]$,

$$\hat{\nu}_n \xrightarrow{P} \nu_\alpha, \tag{4.21}$$

as $n \to \infty$, $k \to \infty$, and $k/n \to 0$. This is proved by a scaling argument. Define the operator

$$T : M_+((0, \infty]) \times (0, \infty) \mapsto M_+((0, \infty])$$

by

$$T(\mu, x)(A) = \mu(xA).$$

From (4.14) and Proposition 3.1 (p. 57), we get joint weak convergence

$$\left(\nu_n, \frac{X_{(k)}}{b\left(\frac{n}{k}\right)}\right) \Rightarrow (\nu_\alpha, 1) \tag{4.22}$$

in $M_+(0, \infty] \times (0, \infty)$. Since

$$\hat{\nu}_n(\cdot) = \nu_n\left(\frac{X_{(k)}}{b\left(\frac{n}{k}\right)}\cdot\right) = T\left(\nu_n, \frac{X_{(k)}}{b\left(\frac{n}{k}\right)}\right),$$

the conclusion will follow by the continuous mapping theorem, provided we prove the continuity of the operator $T$ at $(\nu_\alpha, 1)$. If you are anxious to get on with the story, skip to Step 3.

In fact, we prove the continuity of the operator at $(\nu_\alpha, x)$, where $x > 0$. Towards this goal, let $\mu_n \xrightarrow{v} \nu_\alpha$ and $x_n \to x$, where $\mu_n \in M_+(0, \infty]$, and $x_n, x \in (0, \infty)$. It suffices to show for any $f \in C_K^+(0, \infty]$ that

$$\int_{(0,\infty]} f(t)\mu_n(x_n dt) = \int_{(0,\infty]} f(y/x_n)\mu_n(dy) \to \int_{(0,\infty]} f(y/x)\nu_\alpha(dy). \tag{4.23}$$

Write

$$\begin{aligned}
&\left|\int_{(0,\infty]} f(y/x_n)\mu_n(dy) - \int_{(0,\infty]} f(y/x)\nu_\alpha(dy)\right| \\
&\quad \le \left|\int_{(0,\infty]} f(y/x_n)\mu_n(dy) - \int_{(0,\infty]} f(y/x)\mu_n(dy)\right| \\
&\quad\quad + \left|\int_{(0,\infty]} f(y/x)\mu_n(dy) - \int_{(0,\infty]} f(y/x)\nu_\alpha(dy)\right|
\end{aligned}$$

$$\leq \int_{(0,\infty]} |f(y/x_n) - f(y/x)|\mu_n(dy) + o(1),$$

where the second difference goes to 0 because $f(\frac{\cdot}{x}) \in C_K^+(0,\infty]$. To see that the first difference can be made small, note the supports of $f(\frac{\cdot}{x})$ and $f(\frac{\cdot}{x_n})$ for large $n$ are contained in $[\delta_0, \infty]$ for some $\delta_0$. Since $f$ is continuous with compact support, $f$ is uniformly continuous on $(0, \infty]$. To get an idea what this means, metrize $(0, \infty]$ by the metric $(s, t \in (0, \infty])$

$$d(s,t) = |s^{-1} - t^{-1}|;$$

then uniform continuity means

$$\sup_{d(u,v)<\delta} |f(u) - f(v)| \overset{\delta \downarrow 0}{\to} 0.$$

Then

$$d(y/x_n, y/x) = y^{-1}|x_n - x| < \delta$$

if $y > \delta_0$ and $n$ is large, and therefore for any $\epsilon > 0$, we can make

$$\sup_{y \geq \delta_0} |f(y/x_n) - f(y/x)| < \epsilon.$$

Since $\mu_n(\delta_0, \infty]$ is bounded, this completes the proof of continuity of the scaling map. □

STEP 3. Integrate the tails of the measures against $x^{-1}dx$. The integral functional is continuous on $[1, M]$ for any $M$, and so it is only on $[M, \infty]$ that care must be exercised. By the second converging together theorem, Theorem 3.5 (p. 56), we must show that

$$\lim_{M\to\infty} \limsup_{n\to\infty} \mathbb{P}\left[\int_M^\infty \hat{\nu}_n(x,\infty]x^{-1}dx > \delta\right] = 0. \tag{4.24}$$

Recall $\hat{b}(n/k) = X_{(k)}$. Decompose the probability as

$$\begin{aligned}
&\mathbb{P}\left[\int_M^\infty \hat{\nu}_n(x,\infty]x^{-1}dx > \delta\right] \\
&\quad \leq \mathbb{P}\left[\int_M^\infty \hat{\nu}_n(x,\infty]x^{-1}dx > \delta, \frac{\hat{b}(n/k)}{b(n/k)} \in (1-\eta, 1+\eta)\right] \\
&\quad + \mathbb{P}\left[\int_M^\infty \hat{\nu}_n(x,\infty]x^{-1}dx > \delta, \frac{\hat{b}(n/k)}{b(n/k)} \notin (1-\eta, 1+\eta)\right] \\
&= \mathrm{I} + \mathrm{II}.
\end{aligned}$$

Note that

$$\text{II} \le \mathbb{P}\left[\left|\frac{\hat{b}(n/k)}{b(n/k)} - 1\right| \ge \eta\right] \to 0$$

by (4.17). We have that I is bounded above by

$$\mathbb{P}\left[\int_M^\infty \nu_n((1-\eta)x, \infty]x^{-1}dx > \delta\right] = \mathbb{P}\left[\int_{M(1-\eta)}^\infty \nu_n(x, \infty]x^{-1}dx > \delta\right],$$

and the above probability has a bound from Markov's inequality

$$\begin{aligned}
\delta^{-1} E&\left(\int_{M(1-\eta)}^\infty \nu_n(x, \infty]x^{-1}dx\right) \\
&= \delta^{-1}\int_{M(1-\eta)}^\infty \frac{n}{k}\mathbb{P}[X_1 > b(n/k)x]x^{-1}dx \\
&\stackrel{n\to\infty}{\to} \delta^{-1}\int_{M(1-\eta)}^\infty x^{-\alpha-1}dx = (\text{const})M^{-\alpha},
\end{aligned}$$

where we applied Karamata's theorem, Theorem 2.1 (p. 25). This bound goes to 0 as $M \to \infty$, as required. □

STEP 4. We have proved that

$$\int_1^\infty \hat{\nu}_n(x, \infty]x^{-1}dx \xrightarrow{P} \int_1^\infty \nu_\alpha(x, \infty]x^{-1}dx = 1/\alpha.$$

So $\int_1^\infty \hat{\nu}_n(x, \infty]x^{-1}dx$ is a consistent estimator of $1/\alpha$, and we just need to see that this is indeed the Hill estimator, as defined in (4.3). This is done as follows:

$$\begin{aligned}
\int_1^\infty \hat{\nu}_n(x, \infty]x^{-1}dx &= \int_1^\infty \frac{1}{k}\sum_{i=1}^n \epsilon_{X_i/\hat{b}(n/k)}(x, \infty]x^{-1}dx \\
&= \frac{1}{k}\sum_{i=1}^n \int_1^{X_i/\hat{b}(n/k)\vee 1} x^{-1}dx,
\end{aligned}$$

which is equivalent to $H_{k,n}$ defined in (4.3). □

### 4.4.2 The Hill estimator in practice

In practice, the Hill estimator is used as follows: We make the *Hill plot* of $\alpha$,

$$\{(k, H_{k,n}^{-1}), 1 \le k \le n\},$$

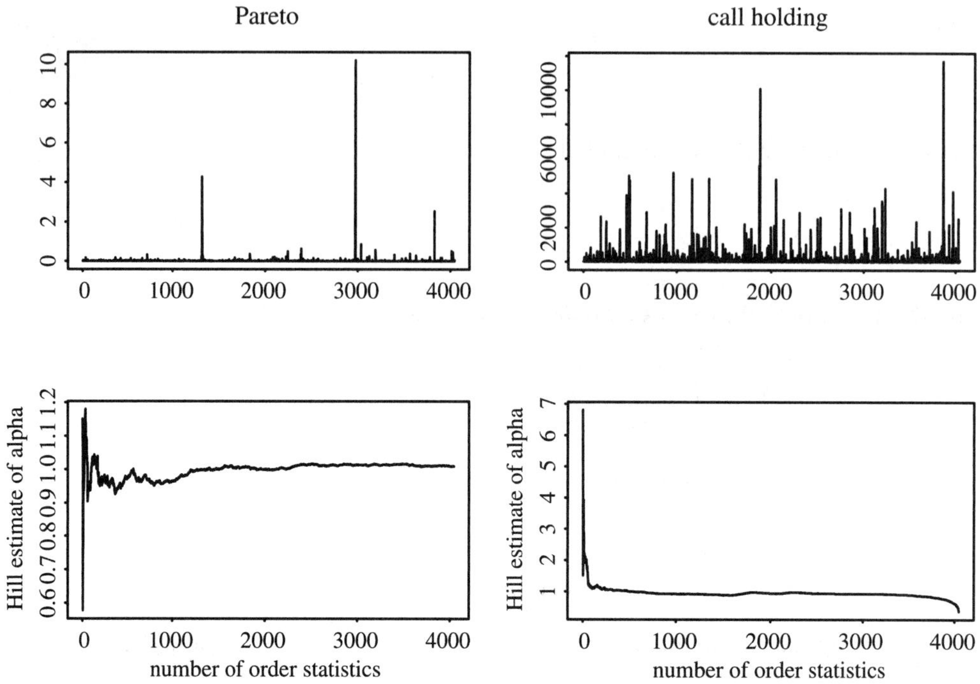


**Fig. 4.1.** Time-series and Hill plots for Pareto (left) and call-holding (right) data.

and hope the graph looks stable so you can pick out a value of $\alpha$.

Sometimes this works beautifully and sometimes the plots are not very revealing. Consider Figure 4.1, which shows two cases where the procedure works gratifyingly well. The top row are time-series plots. The top left plot is 4045 simulated observations from a Pareto distribution with $\alpha = 1$, and the top right plot is 4045 telephone call-holding times indexed according to the time of initiation of the call. The range of the Pareto data is (1.0001, 10206.477), and the range of the call-holding data is (2288, 11714735). The bottom two plots are Hill plots $\{(k, H_{k,n}^{-1}), 1 \le k \le 4045\}$, the bottom left plot being for the Pareto sample and the bottom right plot for the call-holding times. After settling down, both Hill plots are gratifyingly stable and are in a tight neighborhood. The Hill plot for the Pareto seems to estimate $\alpha = 1$ correctly, and the estimate in the call-holding example seems to be between .9 and 1. (So in this case, not only does the variance not exist but the mean appears to be infinite as well.) The Hill plots could be modified to include a confidence interval based on the asymptotic normality of the Hill estimator. McNeil's Hillplot function does just this. See the comments on p. 363.

The Hill plot is not always so revealing. Consider Figure 4.2, one of many Hill Horror Plots. The left plot is for a simulation of size 10,000 from a symmetric $\alpha$-stable

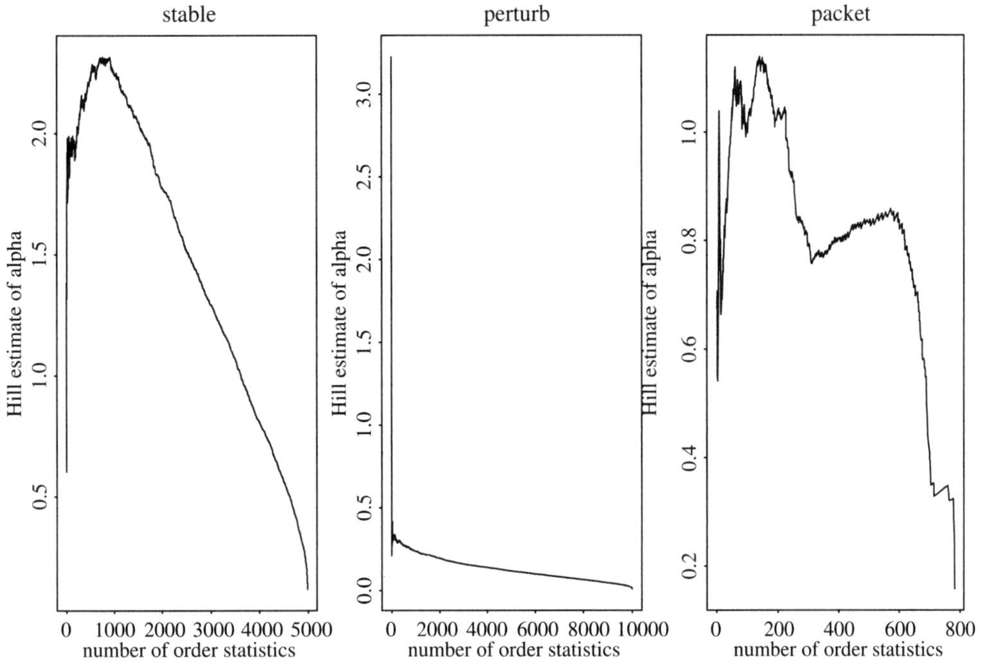


**Fig. 4.2.** A Hill Horror Plot.

distribution with $\alpha = 1.7$. One would have to be paranormal to discern the correct answer of 1.7 from the plot. The middle plot is for a simulated iid sample of size 10,000, called *perturb*, from the distribution tail

$$1 - F(x) \sim x^{-1}(\log x)^{10}, \quad x \to \infty,$$

so that $\alpha = 1$. The plot exhibits extreme bias and comes nowhere close to indicating the correct answer of 1. The problem, of course, is that the Hill estimator is designed for the Pareto distribution and thus does not know how to interpret information correctly from the slowly varying factor $(\log x)^{10}$. It merely readjusts its estimate of $\alpha$ based on this factor rather than identifying the logarithmic perturbation. The third plot is 783 real data called *packet*, representing interarrival times of packets to a server in a network. The problem here is that the graph is volatile and it is not easy to decide what the estimate should be. The sample size may just be too small.

A summary of difficulties when using the Hill estimator include the following:

1. One must get a point estimate from a graph. What value of $k$ should one use?

2. The graph may exhibit considerable volatility and/or the true answer may be hidden in the graph.

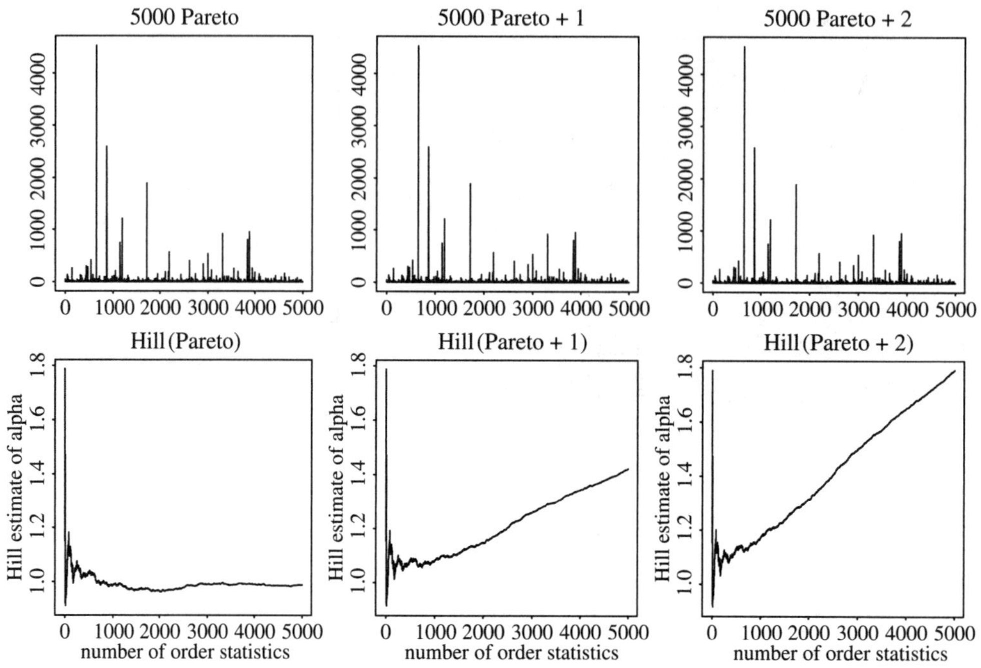


**Fig. 4.3.** Lack of location invariance.

3. The Hill estimator has optimality properties only when the underlying distribution is close to Pareto. If the distribution is far from Pareto, there may be outrageous error.

4. The Hill estimator is not location invariant. A shift in location does not theoretically affect the tail index but may throw the Hill estimate way off.

The lack of location invariance means the Hill estimator can be surprisingly sensitive to changes in location. Figure 4.3 illustrates this. The top plots are time-series plots of 5000 iid Pareto observations where the true $\alpha = 1$. The two right plots on top have the Pareto observations shifted by 1 and then 2. The bottom two plots are the corresponding Hill plots. Shifting by larger and larger amounts soon produces a completely useless plot.

For point 1, several previous studies advocate choosing $k$ to minimize the asymptotic mean squared error of Hill's estimator [155, 235]. In certain cases, the asymptotic form of this optimal $k$ can be expressed, but such a form requires one to know the distribution rather explicitly, and it is not always clear how to obtain finite sample information from an asymptotic formula. There are adaptive methods and bootstrap techniques [66, 108, 145] that try to overcome these problems; it remains to be seen if they will enter the applied research community's toolbox. Bootstrapping heavy-tailed systems presents special problems [9, 10, 79, 88, 133, 189, 313]. See also Section 6.4.

For point 2, there are simple smoothing techniques that can help to overcome the volatility of the plot; plotting on a different scale can sometimes overcome the difficulty associated with the stable example. These techniques are outlined next in Section 4.4.3.

### 4.4.3 Variants of the Hill plot

Some simple techniques of smoothing and rescaling of the Hill plot sometimes are revealing.

#### The smooHill plot

The Hill plot often exhibits extreme volatility, which makes finding a stable regime in the plot more guesswork than science. To counteract this, Resnick and Stărică [252] developed a smoothing technique yielding the smooHill plot: Pick an integer $r$ (usually 2 or 3) and define

$$\text{smooH}_{k,n} = \frac{1}{(r-1)k} \sum_{j=k+1}^{rk} H_{j,n}. \tag{4.25}$$

This is also a consistent estimate of $1/\alpha$; see [252]. To see this is relatively straightforward (modulo some details). We modify the proof of the consistency of the Hill estimator given in Theorem 4.2 (p. 81). In place of (4.22), we use (4.18) coupled with Theorem 4.1 (p. 79) via Proposition 3.1 (p. 57) to get

$$\left( \frac{1}{k} \sum_{i=1}^{n} \epsilon_{X_i/b(n/k)}, \frac{X_{(\lceil kt \rceil)}}{b(n/k)} \right) \Rightarrow (\nu_\alpha, t^{-1/\alpha}) \tag{4.26}$$

in $M_+(0, \infty] \times D(0, \infty]$. Follow the pattern of Theorem 4.2: Compose the two components in (4.26), evaluate the resulting measures on $(x, \infty]$, integrate this function against $\frac{dx}{x}$ (that this is continuous in the right topology needs a verification), and we get

$$\frac{1}{k} \sum_{i=1}^{n} \log \left( \frac{X_i}{X_{(\lceil kt \rceil)}} \wedge 1 \right) = \frac{\lceil kt \rceil}{k} H_{\lceil kt \rceil, n} \Rightarrow \frac{t}{\alpha}$$

in $D[0, \infty)$. Dividing both sides by $t$ leads to

$$H_{\lceil kt \rceil, n} \Rightarrow \frac{1}{\alpha}$$

in $D(0, \infty)$. Therefore, for any integer $r$,

$$\frac{1}{r-1}\int_1^r H_{\lceil kt\rceil,n}dt = \frac{1}{k(r-1)}\int_k^{kr} H_{\lceil s\rceil,n}ds$$
$$= \frac{1}{k(r-1)}\sum_{j=k+1}^{kr} H_{j,n}$$
$$\Rightarrow \frac{1}{\alpha}.$$

The stochastic process $(H_{\lceil kt\rceil,n}, t > 0)$ was named the *Hill process*, studied first in [211] and used in [252].

### Changing the scale, Alt plotting

As an alternative to the Hill plot, it is sometimes useful to display the information provided by the Hill or smooHill estimation as

$$\{(\theta, H^{-1}_{\lceil n^\theta\rceil,n}), 0 \le \theta \le 1\},$$

where we write $\lceil y\rceil$ for the smallest integer greater than or equal to $y \ge 0$. We call this plot the *alternative Hill plot*, abbreviated altHill. The alternative display is sometimes revealing since the initial order statistics are shown more clearly and cover a bigger portion of the displayed space. Unless the distribution is Pareto, the altHill plot spends more of the display space in a small neighborhood of $\alpha$ than in the conventional Hill plot [110].

Figure 4.4 compares several Hill plots for 5000 observations from a stable distribution with $\alpha = 1.7$. Plotting on the usual scale is not revealing and the alt plot is more informative.

A Hill plot was given (p. 5) for file lengths downloaded in BU web sessions in November 1994 in a particular lab under study. The Danish fire insurance data were introduced on p. 13. In Figure 4.5, we have a Hill, an altHill, and a smooHill plot of the Danish data. The altHill plot is not advantageous, probably because the data are well modeled by Pareto.

## 4.5 Alternative estimators I: The Pickands estimator

There are a myriad of other estimators for $\alpha$. We particularly mention the moment estimator of [85–87, 90, 251], the Pickands estimator [85, 112, 235, 236], and the QQ estimator [17, 191]. Here we focus on the Pickands and QQ estimators. Interestingly, a study by de Haan and Peng [93, 235] shows that from the point of view of asymptotic variance, no one estimator dominates the others in the group studied. So it is difficult to

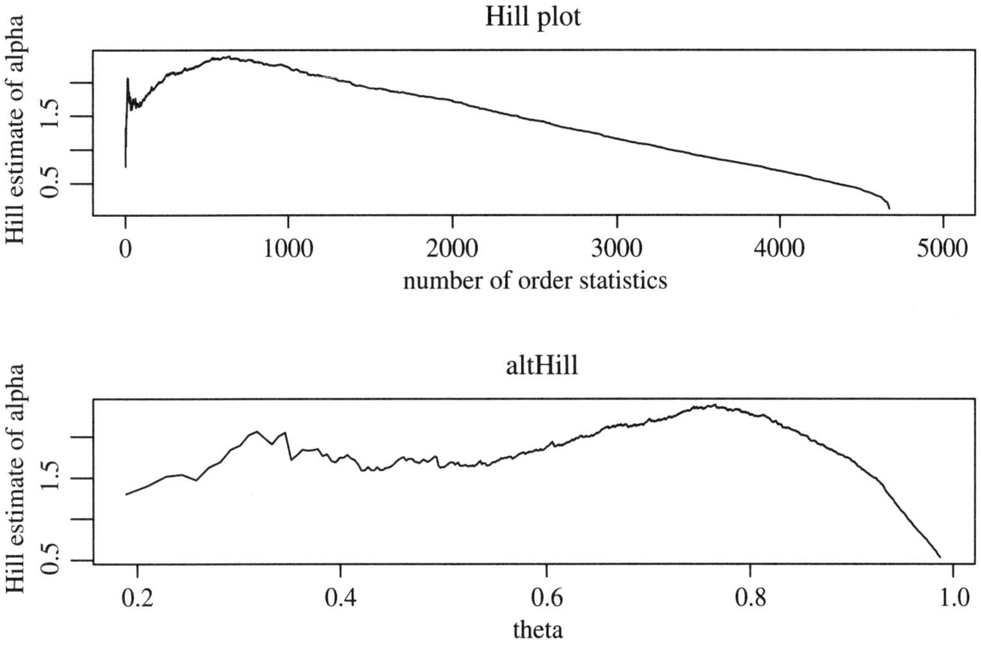


**Fig. 4.4.** Hill and altHill plot for Stable, $\alpha = 1.7$.

imagine one estimator being preferred in all contexts; a sensible practice is not to restrict analysis to one procedure but rather to check that several procedures point toward the same conclusion.

### 4.5.1 Extreme-value theory

The *extreme-value distributions* can be defined as a one-parameter family of types

$$G_\gamma(x) = \exp\{-(1+\gamma x)^{-1/\gamma}\}, \quad \gamma \in \mathbb{R}, \quad 1+\gamma x > 0. \tag{4.27}$$

Define

$$\mathbb{E}_\gamma^0 = \{x : 1+\gamma x > 0\}$$

and observe that

$$\mathbb{E}_\gamma^0 = \begin{cases} \left(-\frac{1}{\gamma}, \infty\right) & \text{if } \gamma > 0, \\ (-\infty, \infty) & \text{if } \gamma = 0, \\ \left(-\infty, \frac{1}{|\gamma|}\right) & \text{if } \gamma < 0. \end{cases}$$

The heavy-tailed case corresponds to $\gamma > 0$, and then $\gamma = 1/\alpha$. For $\gamma = 0$, we interpret $-\log G_\gamma(x) = e^{-x}$. See [16, 50, 102, 129, 238, 260].

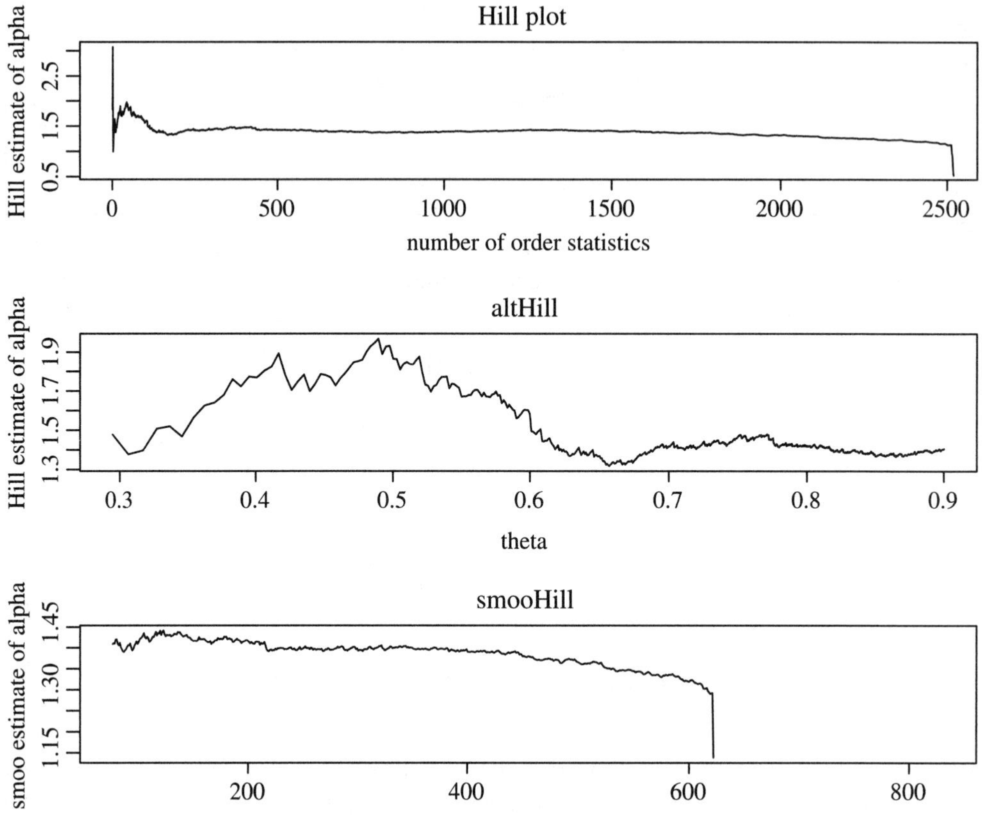


**Fig. 4.5.** Hill, altHill, and smooHill plots for Danish data.

Suppose $\{Z_n, n \geq 1\}$ is iid with common distribution $F$. The distribution $F$ is in the domain of attraction of the extreme-value distribution $G_\gamma$, written $F \in \mathcal{D}(G_\gamma)$, if there exist $a(n) > 0$, $b(n) \in \mathbb{R}$ such that

$$\mathbb{P}\left[\frac{\bigvee_{i=1}^n Z_i - b(n)}{a(n)} \leq x\right] \to G_\gamma(x) \tag{4.28}$$

for $x \in \mathbb{E}_\gamma$ as $n \to \infty$. Equivalently, we have as $n \to \infty$ that

$$n\mathbb{P}[Z_1 > a(n)x + b(n)] \to -\log G_\gamma(x) \tag{4.29}$$

for $x \in \mathbb{E}_\gamma$.

Note that (4.29) is a vague convergence statement about mean measures converging and mimicking the proof of Theorem 4.1 (p. 79) yields

$$\frac{1}{k}\sum_{i=1}^{n} \epsilon_{\frac{Z_i - b(n/k)}{a(n/k)}} \Rightarrow \nu^{(\gamma)} \tag{4.30}$$

on $M_+(\mathbb{E}_\gamma)$, where $\mathbb{E}_\gamma$ is $\mathbb{E}^0_\gamma$ plus the right endpoint of the interval and $\nu^{(\gamma)}(x,\infty] = -\log G_\gamma(x)$. Repeating the procedure that yielded (4.18) gives the equivalent statement

$$\frac{Z_{(\lceil k/y \rceil)} - b(n/k)}{a(n/k)} \to \frac{y^\gamma - 1}{\gamma}, \quad 0 \le y < \infty, \tag{4.31}$$

in $D[0,\infty)$.

Another interpretation of (4.29) is that it gives the limit distribution of *excesses*, where an excess is the exceedance minus the threshold. We may always take the sequence $b_n$ as the quantile function of $F$, the common distribution of $\{Z_n\}$, so that $\mathbb{P}[Z_1 > b_n] \sim 1/n$. Then we have for $x > 0$,

$$\begin{aligned} n\mathbb{P}[Z_1 > a_n x + b_n] &\sim \frac{\mathbb{P}[Z_1 - b_n > a_n x]}{\mathbb{P}[Z_1 > b_n]} \\ &= \mathbb{P}[Z_1 - b_n > a_n x | Z_1 > b_n]. \end{aligned}$$

Referring to (4.5) (p. 76), this is

$$= \mathbb{P}[Z_{\tau_1} - b_n > a_n x] \to -\log G_\gamma(x).$$

For a large threshold, the excess (the exceedance minus the threshold) has a limit distribution whose tail is $-\log G_\gamma(x)$ for values of $x$ such that $0 \le -\log G_\gamma(x) \le 1$. The class of limit distributions is called *generalized Pareto*. The interpretation is that for a large threshold $u$,

$$\mathbb{P}[Z_{\tau_1} - u > x] \approx -\log G_\gamma(\beta x) \tag{4.32}$$

for a scale parameter $\beta$. The POT method assumes the distribution of the excess is exactly the limit distribution and then performs maximum-likelihood estimation on the two parameters $(\gamma, \beta)$ to fit the distribution.

### 4.5.2 The Pickands estimator

The Pickands estimator [85, 112, 235, 236], like the moment estimator [85–87, 90, 251] discussed briefly in Appendix 11.1.3 (p. 369), is a semiparametric estimator of $\gamma$ derived under the sole condition that $F \in \mathcal{D}(G_\gamma)$. The Pickands estimator of $\gamma$ uses differences of quantiles and is based on using three upper-order statistics, $Z_{(k)}$, $Z_{(2k)}$, $Z_{(4k)}$, from a sample of size $n$. The estimator is

$$\hat{\gamma}^{(\text{Pickands})}_{k,n} = \left(\frac{1}{\log 2}\right) \log\left(\frac{Z_{(k)} - Z_{(2k)}}{Z_{(2k)} - Z_{(4k)}}\right). \tag{4.33}$$

Properties of the Pickands estimator include the following:

1. The Pickands estimator is a consistent estimator for $\gamma \in \mathbb{R}$ and does not require the assumption $\gamma > 0$, as does the Hill estimator. The consistency holds as $n \to \infty$, $k \to \infty$, and $n/k \to \infty$.

   We can check consistency easily using (4.31). We have

$$\frac{Z_{(k)} - Z_{(2k)}}{Z_{(2k)} - Z_{(4k)}} = \frac{\frac{(Z_{(k)}-b(n/k))}{a(n/k)} - \frac{(Z_{(2k)}-b(n/k))}{a(n/k)}}{\frac{(Z_{(2k)}-b(n/k))}{a(n/k)} - \frac{(Z_{(4k)}-b(n/k))}{a(n/k)}}$$
$$\xrightarrow{P} \left( \frac{0 - \gamma^{-1}\left(\left(\frac{1}{2}\right)^{\gamma} - 1\right)}{\gamma^{-1}\left(\left(\frac{1}{2}\right)^{\gamma} - 1\right) - \gamma^{-1}\left(\left(\frac{1}{4}\right)^{\gamma} - 1\right)} \right)$$
$$= 2^{\gamma}.$$

   Taking logarithms and dividing by $\log 2$ gives convergence in probability of the estimator to $\gamma$.

2. Usually (under second-order regular variation conditions, which are difficult to check in practice), if $k \to \infty$ and $k/n \to 0$, we have asymptotic normality,

$$\sqrt{k}(\hat{\gamma}_{k,n}^{(\text{Pickands})} - \gamma) \Rightarrow N(0, v(\gamma)),$$

   where

$$v(\gamma) = \frac{\gamma^2(2^{2\gamma+1} + 1)}{(2(2^{\gamma} - 1)\log 2)^2}. \tag{4.34}$$

   More on asymptotic normality later in Chapter 9.1.

3. Unlike the Hill estimator, the Pickands estimator is location invariant. It is also scale invariant.

4. Good plots may require a large sample of the order of several thousand.

5. In terms of asymptotic mean squared error, the Pickands estimator sometimes is preferred over the moment estimator and Hill estimator (where comparable because you know $\gamma > 0$) and sometimes not. See [93, 235].

6. One can make a *Pickands plot* consisting of the points $\{(k, \hat{\gamma}_{k,n}^{(\text{Pickands})}), 1 \le k < n/4\}$. Choice of $k$ and volatility of the plots are issues as they were with the Hill and moments estimators.

7. The Pickands plot often does a good job of warning that a heavy-tail model is inappropriate by indicating $\gamma \le 0$. In circumstances where this is the case, the Hill plot is frequently uninformative.

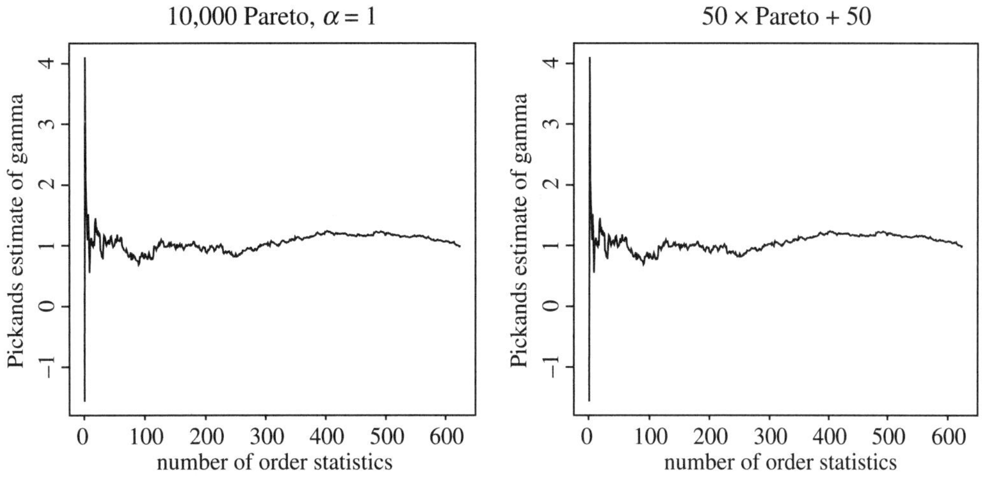


**Fig. 4.6.** Pickands plots of 10,000 simulated Pareto random variables with $\alpha = 1$ (left) and with the same data but multiplied by 50 and shifted by 50.

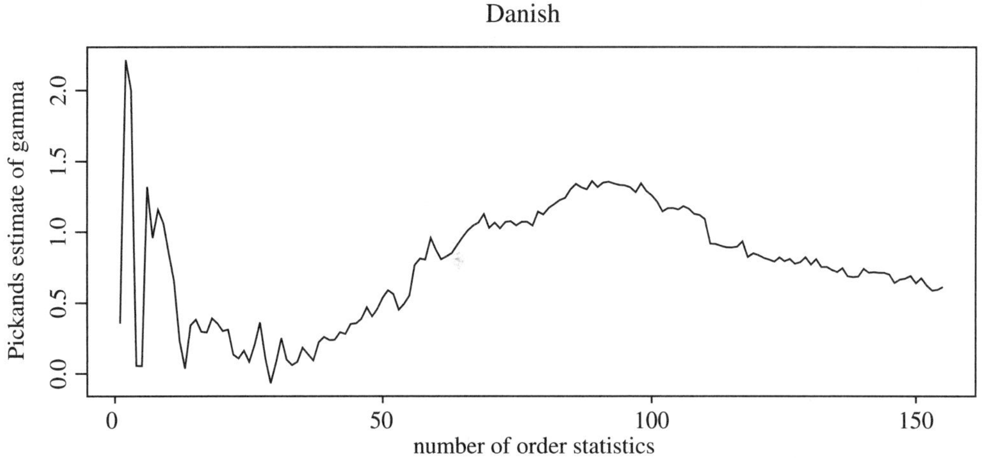


**Fig. 4.7.** Pickands plots of the Danish data (left), where the estimate of $\gamma \approx 0.71$ was obtained from other methods.

Consider Figure 4.6, which is the Pickands estimator applied to 10,000 simulated Pareto random variables with $\alpha = 1$. The Pickands plot on the left picks up the correct value of $\alpha = 1$ quite well. In contrast to the degradation in the Hill plots when the data were shifted (recall Figure 4.3 (p. 88)), the Pickands plot is unaffected.

Earlier (see Figure 4.5 (p. 92)), we found $\alpha \approx 1.4$ for the Danish data. The Pickands plot in Figure 4.7 is not very informative even after accounting for the relation between $\alpha$ and $\gamma$ and taking reciprocals $1/1.4 = 0.71$.

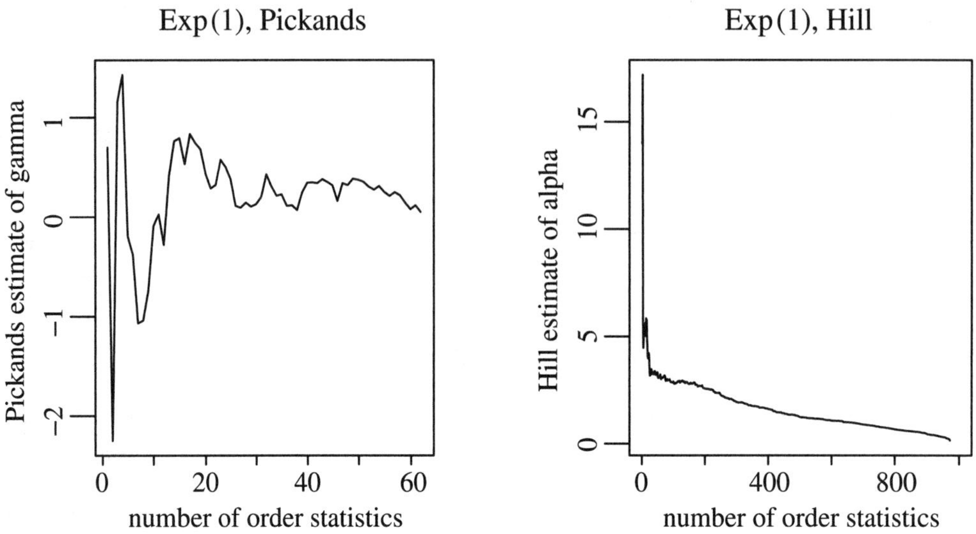


**Fig. 4.8.** Pickands plot (left) for 1000 iid unit exponential variables vs. the Hill plot (right).

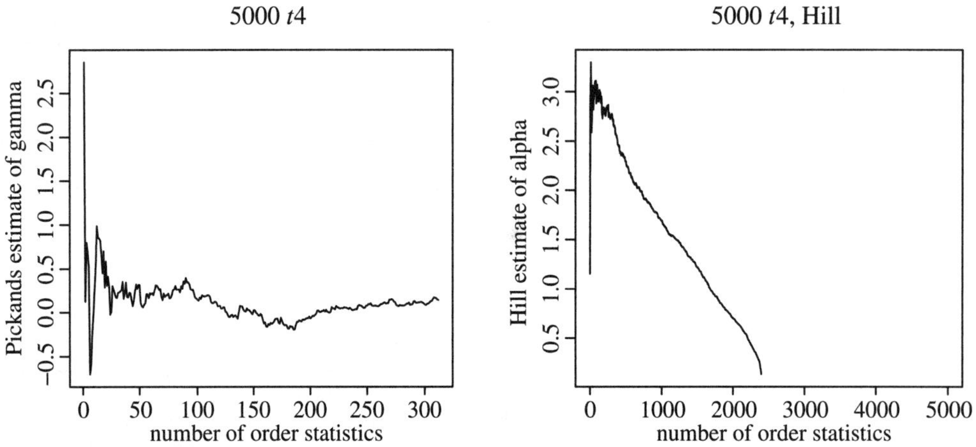


**Fig. 4.9.** Pickands plot (left) for 5000 iid variables from a $t$-density with 4 degrees of freedom vs. the Hill plot (right).

To illustrate why the Pickands plot is useful in deciding on the appropriateness of a heavy-tail model, consider Figure 4.8. On the left is the Pickands plot for 1000 unit exponential variates, which does a reasonable job of identifying a value of $\gamma$ near 0. The Hill plot on the right is not informative.

The last set of plots in Figure 4.9 compares the Pickands plot and the Hill plot for 5000 realizations of the $t$-density with 4 degrees of freedom. Here $\alpha = 4$, so $\gamma = 0.25$. The Pickands plot seems reasonable, but the Hill plot is uninformative.

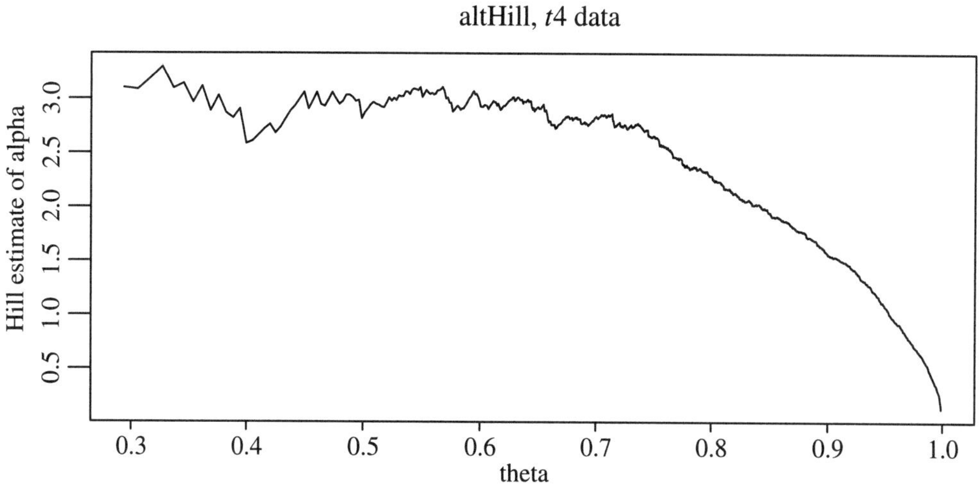


**Fig. 4.10.** altHill plot of 5000 $t4$ variates.

However, the altHill plot is more stable but not particularly close to the true value.

## 4.6 Alternative estimators II: QQ plotting and the QQ estimator

QQ plotting and similar techniques are diagnostic and exploratory methods to graphically assess the goodness-of-fit of a model for data. Suppose we have data $x_1, \dots, x_n$ that could plausibly be a random sample from some distribution $F(x)$; that is, we believe there are iid random variables $X_1, \dots, X_n$ with common distribution $F(x)$ and $x_1, \dots, x_n$ is a realization of $X_1, \dots, X_n$. If we are interested in obtaining such things as high quantile estimates, as is done, for example, in *value-at-risk* estimates, we must find $F(x)$ which provides a good fit in the tail.

### 4.6.1 Quantile-quantile or QQ plots: Preliminaries

Suppose that we have a provisional or null hypothesis that the true distribution function producing the data is $F(x)$. The QQ plot provides a somewhat informal but convenient way to test this hypothesis. The method is empirical process based, and frequently it is convenient to have notation for order statistics indexed from smallest to largest as well as the reverse already introduced. Recall that for a sample $X_1, \dots, X_n$, we set

$$X_{(1)} \geq X_{(2)} \geq \cdots \geq X_{(n)}$$

for the order statistics indexed largest to smallest. For the indexing from smallest to largest, we write

$$X_{1:n} \le X_{2:n} \le \cdots \le X_{n:n},$$

so that $X_{(i)} = X_{n-i+1:n}$. For the empirical distribution of the sample $X_1, \dots, X_n$, we write

$$\hat{F}_n(x) = \frac{1}{n}\sum_{i=1}^{n} 1_{[X_i \le x]} = \frac{1}{n}\sum_{i=1}^{n} \epsilon_{X_i}((-\infty, x]).$$

### 4.6.2 QQ plots: The method

Plot the theoretical quantiles of $F$ vs. the sample quantiles. If the null hypothesis is true, then the result should fall roughly on the straight line $\{(x, x) : x > 0\}$. More precisely, we plot

$$\begin{aligned}&\left\{\left(F^{\leftarrow}\left(\frac{i}{n+1}\right), \hat{F}_n^{\leftarrow}\left(\frac{i}{n+1}\right)\right), 1 \le i \le n\right\}\\ &\quad = \left\{\left(F^{\leftarrow}\left(\frac{i}{n+1}\right), X_{i:n}\right), 1 \le i \le n\right\}.\end{aligned} \tag{4.35}$$

If the plot *looks roughly linear*, there is no evidence against the null hypothesis.

*The rationale*: We know the empirical distribution $\hat{F}_n(x) \approx F(x)$ and hope that $\hat{F}_n^{\leftarrow}(q) \approx F^{\leftarrow}(q)$.

Some potential problems with this procedure:

1. For certain common distributions, for example, the normal or gamma, the distribution is not in a closed, convenient form, and hence the inverse or quantile function $F^{\leftarrow}(q)$ is not in closed form either. Statistics packages typically provide routines to do QQ plots for common densities such as the normal.

2. When does the phrase "roughly linear" become obvious and clear? When in doubt, a common technique for trying to assess variability is as follows: Make your QQ plot. Then simulate 100 data sets from the null distribution $F$. Make each simulation run the size of the original sample. Then superimpose (in a different color) on your QQ plot the 100 QQ plots corresponding to the 100 simulated data sets. The 100 QQ plots of the simulated data sets will form a band around the QQ plot of the real data, and if the real QQ plot does not stick out of the band, you are within acceptable variability.

*Remark* 4.2. Note that

$$\hat{F}_n^{\leftarrow}\left(\frac{i}{n+1}\right) = X_{i:n}, \quad 1 \le i \le n.$$

The reason for this is that $\hat{F}_n^{\leftarrow}(q)$ is left-continuous and we know that

$$\hat{F}_n^{\leftarrow}(q) = X_{i:n} \quad \text{for } \frac{i-1}{n} < q \leq \frac{i}{n}.$$

Since $\hat{F}_n^{\leftarrow}(\frac{i}{n}) = X_{i:n}$, we need only check that

$$\frac{i}{n} - \frac{i}{n+1} < \frac{1}{n}, \tag{4.36}$$

since $\hat{F}_n^{\leftarrow}(\cdot)$ is constant on $(\frac{i-1}{n}, \frac{i}{n}]$.

A modest point: Why are we plotting the points in (4.35) and not the points

$$\left\{\left(F^{\leftarrow}\left(\frac{i}{n}\right), \hat{F}_n^{\leftarrow}\left(\frac{i}{n}\right)\right), 1 \leq i \leq n\right\}?$$

The reason is partly historical, stemming from the following argument: Suppose $U_1, \dots, U_n$ are iid with common $U(0, 1)$ distribution. Sort the random variables to get the order statistics

$$U_{1:n} \leq U_{2:n} \leq \cdots \leq U_{n:n}.$$

Set $U_{0:n} = 0$ and $U_{n+1,n} = 1$. Call the differences

$$U_{i+1:n} - U_{i:n}, \quad i = 0, 1, \dots, n,$$

the *spacings*; by symmetry the spacings should be identically distributed and hence have the same mean. Since

$$\sum_{i=0}^{n}(U_{i+1:n} - U_{i:n}) = 1,$$

we have

$$1 = \boldsymbol{E}\left(\sum_{i=0}^{n}(U_{i+1:n} - U_{i:n})\right) = (n+1)\boldsymbol{E}(U_{i+1:n} - U_{i:n}),$$

and hence the expected spacing is

$$\boldsymbol{E}(U_{i+1:n} - U_{i:n}) = \frac{1}{n+1}.$$

Therefore,

$$\boldsymbol{E}(U_{i:n}) = \frac{i}{n+1}$$

since it is a sum of $i$ consecutive spacings. We hope that

$$\left\{\left(\frac{i}{n+1}, U_{i:n}\right), 1 \le i \le n\right\}$$

will be roughly linear and fall on the line $\{(x, x), 0 \le x \le 1\}$. (This hope is dependent on $i/(n+1) = \boldsymbol{E}(U_{i:n}) \approx U_{i:n}$.)

Recall that if $X_1, \ldots, X_n$ are iid with distribution $F(x)$, then $X_i \stackrel{d}{=} F^{\leftarrow}(U_i)$, which means that

$$\left\{\left(F^{\leftarrow}\left(\frac{i}{n+1}\right), F^{\leftarrow}(U_{i:n})\right), 1 \le i \le n\right\}$$

should be roughly linear and so should

$$\left\{\left(F^{\leftarrow}\left(\frac{i}{n+1}\right), X_{i:n}\right), 1 \le i \le n\right\}.$$

### 4.6.3 QQ plots and location-scale families

Suppose we (null) hypothesize that $X_1, \ldots, X_n$ are iid from the location-scale family

$$F_{\mu,\sigma}(x) = F_{0,1}\left(\frac{x-\mu}{\sigma}\right), \tag{4.37}$$

where $F_{0,1}$ is specified. An example is that $F_{0,1}$ is the standard normal. Invert (4.37) to get

$$F^{\leftarrow}_{\mu,\sigma}(q) = \mu + \sigma F^{\leftarrow}_{0,1}(q).$$

The analysis using QQ plots can be adapted to provide estimates of $\mu$ and $\sigma$. Here is how this is done: Since we assume that $F_{0,1}(x)$ is known, we can plot

$$\left\{\left(F^{\leftarrow}_{0,1}\left(\frac{i}{n+1}\right), X_{i:n}\right), 1 \le i \le n\right\}. \tag{4.38}$$

If the null hypothesis is true,

$$\left\{\left(\mu + \sigma F^{\leftarrow}_{0,1}\left(\frac{i}{n+1}\right), X_{i:n}\right), 1 \le i \le n\right\}$$

should be on the line with angle 45 degrees $\{(x, x), x > 0\}$. Thus

$$\mu + \sigma F^{\leftarrow}_{0,1}\left(\frac{i}{n+1}\right) \approx X_{i:n},$$

and therefore

$$F^{\leftarrow}_{0,1}\left(\frac{i}{n+1}\right) \approx \frac{X_{i:n} - \mu}{\sigma}.$$

So the points plotted in (4.38) should be on the line $\{(\frac{z-\mu}{\sigma}, z), z > 0\}$ or, equivalently, the line $\{(x, \sigma x + \mu), x > 0\}$, and the slope of the fitted (by, for instance, least squares) line is an estimator of $\sigma$ and the intercept is an estimate of $\mu$.

### 4.6.4 Adaptation to the heavy-tailed case: Are the data heavy tailed?

Suppose that the null hypothesis is that for some $x_l > 0$ and random variable $X$, the distribution of $X$ satisfies for some $x_l > 0$ and $\alpha > 0$,

$$\mathbb{P}[X > x] = \left(\frac{x}{x_l}\right)^{-\alpha}, \quad x > x_l. \tag{4.39}$$

Comparing this with (4.1) (p. 74) and referring to (4.5) (p. 76), we see that (4.1) implies that exceedances above threshold $x_l$ have Pareto distribution (4.39) with left endpoint $x_l$. Thus this assumption (4.39) is consistent with the POT method, where the approximate distribution of the large values relative to a threshold is replaced by the limiting Pareto distribution.

Assumption (4.39) means that $X/x_l$ is Pareto with left endpoint 1 and shape parameter $\alpha$, and for $y > 0$,

$$\mathbb{P}\left[\alpha \log \frac{X}{x_l} > y\right] = \mathbb{P}\left[\frac{X}{x_l} > e^{y/\alpha}\right] = e^{-y}.$$

So $\alpha \log \frac{X}{x_l}$ is exponential with parameter 1. Therefore,

$$\begin{aligned} P[\log X > y] &= P\left[\frac{\log X - \log x_l}{\alpha^{-1}} > \frac{y - \log x_l}{\alpha^{-1}}\right] \\ &= P\left[\alpha \log \frac{X}{x_l} > \frac{y - \log x_l}{\alpha^{-1}}\right] \\ &= e^{-(y - \log x_l)/\alpha^{-1})}. \end{aligned}$$

If $W_1(x) = 1 - e^{-x}, x > 0$, then

$$P[\log X > y] = \bar{W}_1\left(\frac{y - \log x_l}{\alpha^{-1}}\right),$$

which is a location-scale family with location parameter $\mu = \log x_l$ and scale parameter $\sigma = \alpha^{-1}$.

What are the quantiles of $W_1$? Solve

$$W_1(x) = 1 - e^{-x} = q,$$

to get

$$W_1^{\leftarrow}(q) = -\log(1 - q).$$

We conclude that we should plot

$$\left\{\left(-\log\left(1-\frac{i}{n+1}\right), \log X_{i:n}\right), 1 \le i \le n\right\}, \tag{4.40}$$

and if the null hypothesis (4.39) is correct, or at least approximately correct, the plot should be roughly linear with slope $\alpha^{-1}$ and intercept $\log x_l$.

*Example* 4.1 (*Internet response data*). This is an Internet measurement study that measured the number of bytes per request transferred from a web server to a browser in response to a request from the browser. The study was conducted around 1997 at the University of North Carolina Computer Science Department under the guidance of Donald Smith. The data were presented with the question, *Are the data heavy tailed*? Some typical diagnostics like the Hill estimator fail miserably for these data but the QQ method works pretty well. In Figure 4.11, we give the time-series plot of the data on the left and the QQ plot of the log transformed data matched against exponential quantiles on the right. Clearly, we should be looking at exceedances, as not all the data fall on the line.

The slope of the fitted line to the QQ plot of the exceedances gives an estimate of $\alpha^{-1}$ (see the discussion in Sections 11.1.2 (p. 366) and 4.6.6 (p. 106)); the estimate is sensitive to the choice of exceedance threshold. Instead of looking at exceedance thresholds, we can choose a number of upper-order statistics and only use those to fit the line. This is demonstrated in the three plots in Figure 4.12. The data set has 131,943 data points. We show the plots obtained by choosing 10,000, 20,000, and 50,000 upper-order statistics.

Just for comparison, Figure 4.13 presents the Hill and the altHill plots for these data. The plots are not stunningly easy to interpret, although after comparison with the QQ plots, one is increasingly confident of an estimate of $\alpha \in (1, 2)$. Interestingly, the Pickands plot Figure 4.14 for these data looks decent.

### 4.6.5 Additional remarks and related plots

Here are some additional remarks and notes.

#### Diagnosing deviations from the line in the QQ plot

If the hypothesized distribution $F_{\text{hyp}}(x)$ is far from the true underlying distribution $F_{\text{true}}(x)$, then the QQ plot will simply look awful. If the QQ plot is not very linear, the deviations from linearity can sometimes indicate what the problem might be.

What if points of the QQ plot are below the line $y = x$, for example, at the right end. Then for $q$ near 1, the points $(F_{\text{hyp}}^{\leftarrow}(q), \hat{F}_n^{\leftarrow}(q))$ are below the diagonal line. This means that

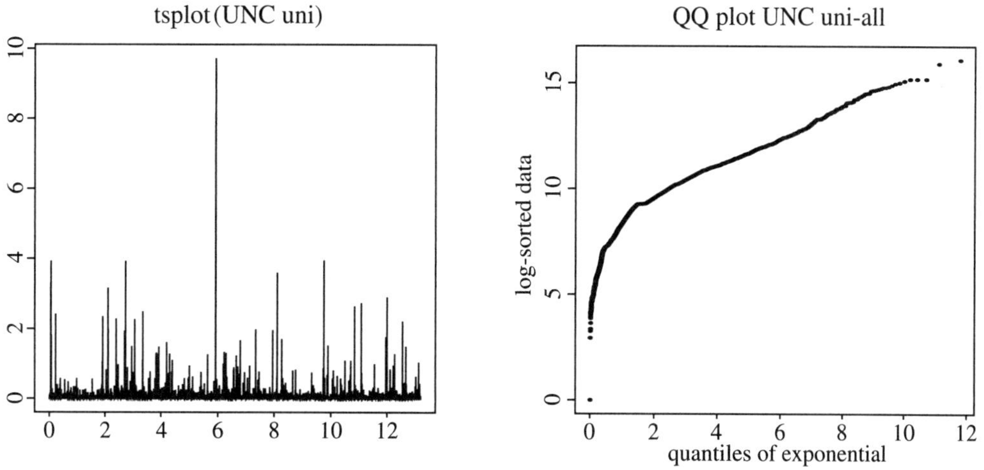


**Fig. 4.11.** Time-series plot of the Internet response data (left); the vertical axis units are millions and the horizontal axis units are 10,000s. QQ plot for all Internet response data (right).

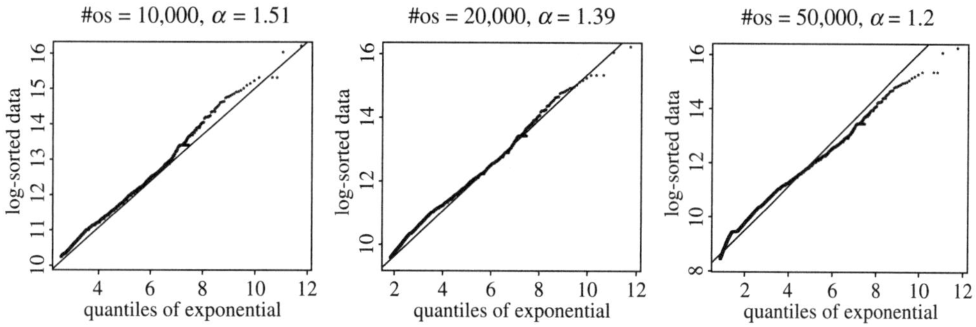


**Fig. 4.12.** Fitting the QQ line to (a) 10K, (b) 20K, and (c) 50K upper-order statistics.

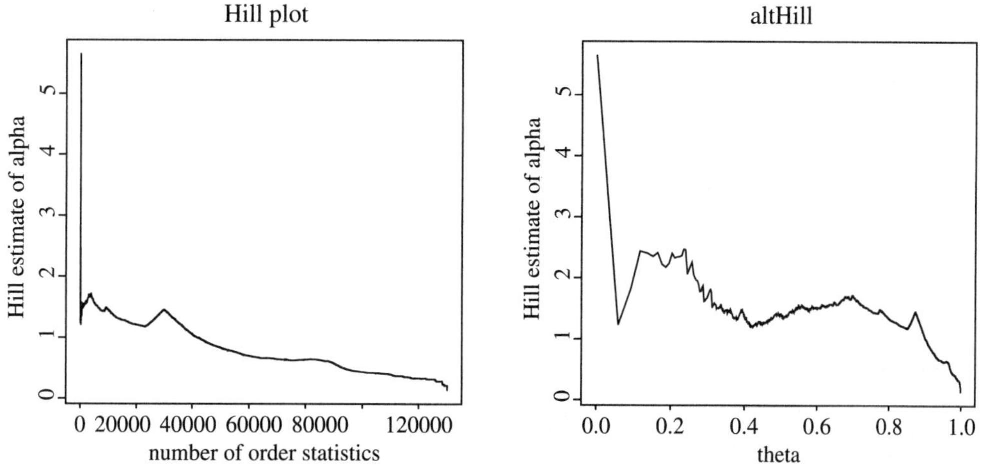


**Fig. 4.13.** Hill and altHill plots for Internet response data.

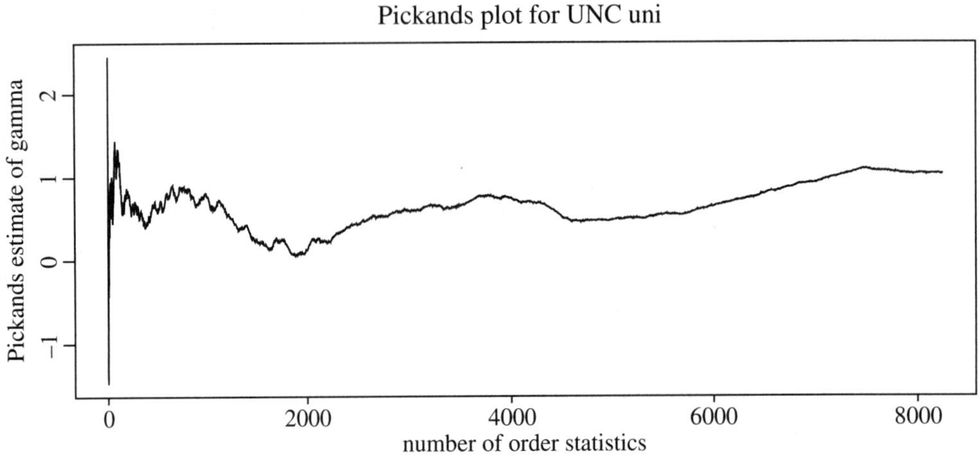


**Fig. 4.14.** Pickands plot for Internet response data.

$$F_{\text{hyp}}^{\leftarrow}(q) > \hat{F}_n^{\leftarrow}(q) \qquad \text{for } q \text{ near } 1,$$

or

$$F_{\text{hyp}}(x) < \hat{F}_n(x) \qquad \text{for large } x,$$

or

$$1 - F_{\text{hyp}}(x) > 1 - \hat{F}_n(x) \quad \text{for large } x.$$

Since, presumably, $1 - \hat{F}_n(x) \approx 1 - F_{\text{true}}(x)$, this means that the hypothesized tail is heavier than the true tail.

### A related plot: The PP plot

This is a plot of the points

$$\left\{\left(\frac{i}{n+1}, F_{\text{hyp}}(X_{i:n})\right), i = 1, \ldots, n\right\}.$$

This is obviously a variant of the QQ plot.

### Another variant: The tail plot for heavy tails

Essentially this is a plot of the tail empirical distribution function in log–log scale. Suppose that

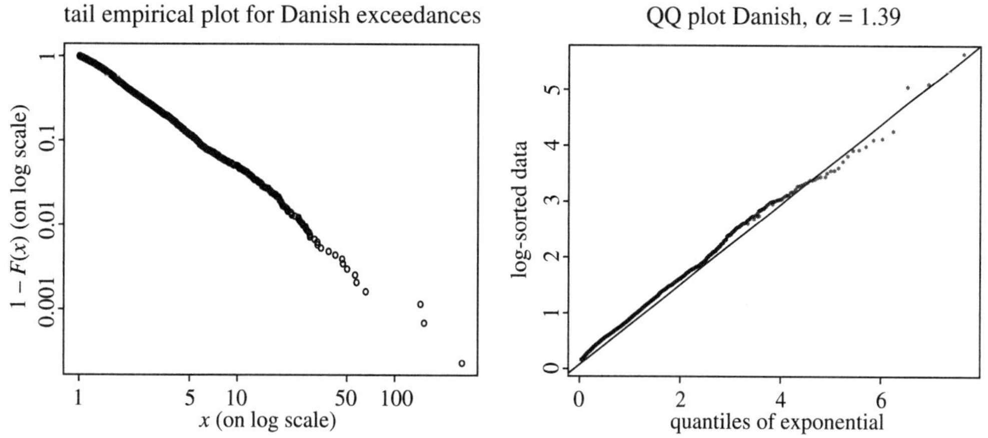


**Fig. 4.15.** Danish data: Tail empirical plot (left) and QQ plot (right).

$$P[X > x] \sim x^{-\alpha}, \quad x > 0, \quad \alpha > 0.$$

Then

$$-\log P[X > x] \sim \alpha \log x,$$

so plotting

$$\{(\log x, -\log P[X > x]), x > 0\}$$

should, at least for large $x$, give a line of slope $\alpha$. Since we are not sure about the form of the distribution of $X$, plot instead

$$\{(\log x, -\log(1 - \hat{F}_n(x)), x > 0\}$$

or

$$\{(\log X_{i:n}, -\log(1 - \hat{F}_n(X_{i:n})), i = 1, \dots, n\}$$

or

$$\left\{\left(\log X_{i:n}, -\log\left(1 - \frac{i}{n}\right)\right), i = 1, \dots, n\right\}.$$

We see this is not much different from the QQ plot. It is customary to plot without the minus sign in the second component and talk piously about plotting in log–log scale.

The McNeil function *emplot* in the Splus add-on WINEVIS or the R package EVIR performs this neatly.

*Example* 4.2 (*Danish data*). As in Section 1.3.3 (p. 13), we consider the 2167 Danish fire claim exceedances. The plot of $\{x, 1 - \hat{F}_n(x), x > 0\}$ looks terrific. For comparison, the QQ plot with the least-squares line is also given in Figure 4.15. The estimate of $\alpha$ is 1.39, so the mean of the fitted distribution will be finite but the second moment infinite.

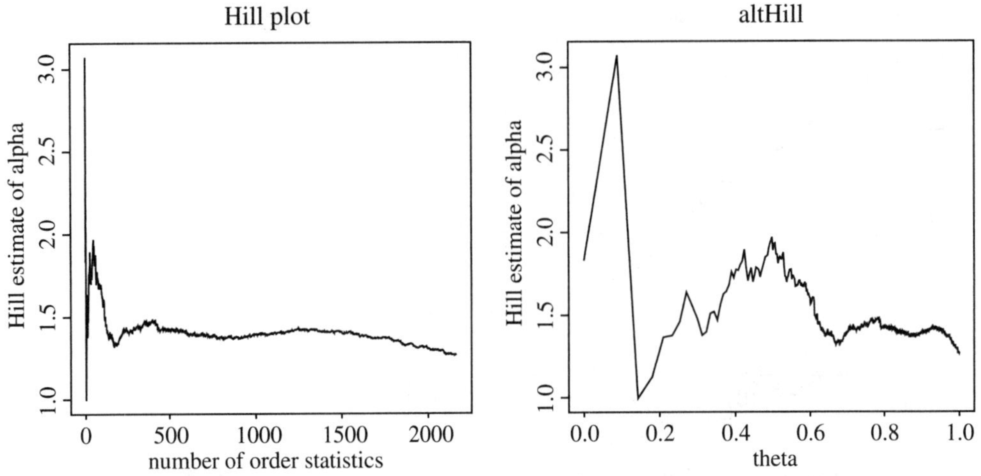


**Fig. 4.16.** Danish data: Hill and altHill plots.

For comparison, Figure 4.16 gives the Hill and altHill plot of the Danish data. Note that the alt plot is not particularly useful here, probably because the data are actually from a distribution that is either Pareto or close to Pareto.

Recall for comparison that the Pickands plot looked poor for the Danish data. See Figure 4.7 (p. 95).

### 4.6.6 The QQ estimator

In this section, we formalize the idea that the slope of the least-squares line fitted to the QQ plot is an estimate of $1/\alpha$. We proceed under the regular variation assumption (4.2), rather than (4.39) or (4.1). This means that we will not put the least-squares line through all the pairs in (4.40), but only through pairs corresponding to $k$ upper-order statistics.

If $\{(x_i, y_i), 1 \le i \le n\}$ are $n$ points in the plane, a standard textbook calculation yields that the slope of the least-squares line through these points is

$$\mathrm{SL}(\{(x_i, y_i), 1 \le i \le n\}) = \frac{\bar{S}_{xy} - \bar{x}\bar{y}}{\bar{S}_{xx} - \bar{x}^2}, \tag{4.41}$$

where we use standard notation:

$$S_{xy} = \sum_{i=1}^{n} x_i y_i, \qquad S_{xx} = \sum_{i=1}^{n} x_i^2,$$

and "bar" indicates mean.

If $X_1, \ldots, X_n$ is a random sample from the Pareto distribution

$$1 - F_\alpha(x) = x^{-\alpha}, \quad \alpha > 0, \quad x \geq 1,$$

then the slope of the least-squares line through the points of (4.40) (p. 102) gives, with

$$x_i = -\log\left(1 - \frac{i}{n+1}\right), \quad y_i = \log X_{i:n},$$

an estimator of $\alpha^{-1}$

$$\begin{aligned}\widehat{\alpha^{-1}} &= \frac{\sum_{i=1}^n -\log\left(\frac{i}{n+1}\right)\left\{n \log X_{n-i+1:n} - \sum_{j=1}^n \log X_{n-j+1:n}\right\}}{n\sum_{i=1}^n \left(-\log\left(\frac{i}{n+1}\right)\right)^2 - \left(\sum_{i=1}^n -\log\left(\frac{i}{n+1}\right)\right)^2} \\ &= \frac{\sum_{i=1}^n -\log\left(\frac{i}{n+1}\right)\left\{n \log X_{(i)} - \sum_{j=1}^n \log X_{(j)}\right\}}{n\sum_{i=1}^n \left(-\log\left(\frac{i}{n+1}\right)\right)^2 - \left(\sum_{i=1}^n -\log\left(\frac{i}{n+1}\right)\right)^2}. \end{aligned} \tag{4.42}$$

We call this estimator the *QQ estimator*.

With only the regular variation assumption (4.2), we modify (4.42) using the POT philosophy, and it is reasonable to define the QQ estimator based on the $k$ upper-order statistics to be

$$\begin{aligned}\widehat{\alpha^{-1}} = \widehat{\alpha^{-1}}_{k,n} &= \mathrm{SL}\left(\left\{\left(-\log\left(1 - \frac{i}{k+1}\right), \log\left(\frac{X_{n-k+i:n}}{X_{n-k:n}}\right)\right), 1 \leq i \leq k\right\}\right) \\ &= \mathrm{SL}\left(\left\{\left(-\log\left(\frac{i}{k+1}\right), \log\left(\frac{X_{(i)}}{X_{(k+1)}}\right)\right), 1 \leq i \leq k\right\}\right). \end{aligned} \tag{4.43}$$

Some modest simplification of (4.43) is possible if we note the following readily checked properties of the SL function: For any real numbers $a$, $b$, we have

$$\mathrm{SL}(\{(x_i, y_i), 1 \leq i \leq n\}) = \mathrm{SL}(\{(x_i + a, y_i + b), 1 \leq i \leq n\}). \tag{4.44}$$

Thus (4.43) simplifies to

$$\begin{aligned}\widehat{\alpha^{-1}} &= \mathrm{SL}\left(\left\{\left(-\log\left(1 - \frac{i}{k+1}\right), \log X_{n-k+i,n}\right), 1 \leq i \leq k\right\}\right) \\ &= \mathrm{SL}\left(\left\{\left(-\log\left(\frac{i}{k+1}\right), \log X_{(i)}\right), 1 \leq i \leq k\right\}\right). \end{aligned} \tag{4.45}$$

We could also drop division by $k+1$ in the first component for the same reason.

In practice, we would make a QQ plot of all the data and choose $k$ based on visual observation of the portion of the graph that looked linear. Then we would compute the slope of the line through the chosen upper $k$-order statistics and the corresponding exponential quantiles. Choosing $k$ is an art as well as a science, and the estimate of $\alpha$ is usually rather sensitive to the choice of $k$. Alternatively, plot $\{(k, \widehat{\alpha^{-1}}_{k,n}), 1 \leq k \leq n\}$ and look for a stable region of the graph as representing the true value of $\alpha^{-1}$. This is analogous to what is done with the Hill estimator of $\alpha^{-1}$,

$$H_{k,n} = \frac{1}{k}\sum_{i=1}^{k} \log\left(\frac{X_{(i)}}{X_{(k+1)}}\right). \tag{4.46}$$

The QQ plot will typically look smoother than the Hill plot.

Choosing $k$ is still the Achilles heel of many of these procedures.

### Consistency of the QQ estimator

Now we prove the weak consistency of the QQ estimator. In view of (4.43), (4.44), and (4.46), we may write the estimator $\widehat{\alpha^{-1}}$ as

$$\begin{aligned}
&= \frac{\frac{1}{k}\sum_{i=1}^{k}\left(-\log\left(1-\frac{i}{k+1}\right)\right)\log\left(\frac{X_{n-k+i,n}}{X_{n-k,n}}\right) - \frac{1}{k}\sum_{i=1}^{k}\left(-\log\left(1-\frac{i}{k+1}\right)\right)H_{k,n}}{\frac{1}{k}\sum_{i=1}^{k}\left(-\log\left(1-\frac{i}{k+1}\right)\right)^2 - \left(\frac{1}{k}\sum_{i=1}^{k}\left(-\log\left(1-\frac{i}{k+1}\right)\right)\right)^2} \\
&= \frac{\frac{1}{k}\sum_{i=1}^{k}\left(-\log\left(\frac{i}{k+1}\right)\right)\log\left(\frac{X_{(i)}}{X_{(k+1)}}\right) - \frac{1}{k}\sum_{i=1}^{k}\left(-\log\left(\frac{i}{k+1}\right)\right)H_{k,n}}{\frac{1}{k}\sum_{i=1}^{k}\left(-\log\left(\frac{i}{k+1}\right)\right)^2 - \left(\frac{1}{k}\sum_{i=1}^{k}\left(-\log\left(\frac{i}{k+1}\right)\right)\right)^2}.
\end{aligned} \tag{4.47}$$

**Theorem 4.3.** *Suppose $X_1, \ldots, X_n$ are a random sample from $F$, a distribution with regularly varying tail satisfying (4.2). Then the QQ estimator $\widehat{\alpha^{-1}}$ given in (4.47) is weakly consistent for $1/\alpha$:*

$$\widehat{\alpha^{-1}} \xrightarrow{P} \alpha^{-1}$$

*as $n \to \infty$, $k = k(n) \to \infty$ in such a way that $k/n \to 0$.*

*Proof.* Write the denominator in (4.47) as

$$\frac{1}{k}S_{xx} - \left(\frac{1}{k}S_x\right)^2,$$

where as $n \to \infty$,

$$\frac{1}{k}S_{xx} = \frac{1}{k}\sum_{i=1}^{k}\left(-\log\left(\frac{i}{k+1}\right)\right)^2 \sim \int_0^1 (-\log x)^2 dx$$
$$= \int_0^\infty y^2 e^{-y} dy = 2 \tag{4.48}$$

and

$$\frac{1}{k}S_{x} = \frac{1}{k}\sum_{i=1}^{k}\left(-\log\left(\frac{i}{k+1}\right)\right) \sim \int_0^1 (-\log x) dx$$
$$= \int_0^\infty y e^{-y} dy = 1. \tag{4.49}$$

Furthermore, as $n \to \infty$, $k \to \infty$, $n/k \to \infty$,

$$\frac{1}{k}\sum_{i=1}^{k}\left(-\log\left(\frac{i}{k+1}\right)\right) H_{k,n} \sim \int_0^1 (-\log x) dx\, H_{k,n} \xrightarrow{P} \alpha^{-1}$$

by the weak consistency of the Hill estimator. So for consistency of the QQ estimator, it suffices to show that

$$A_n := \frac{1}{k}\sum_{i=1}^{k}\left(-\log\left(\frac{i}{k+1}\right)\right)\log\left(\frac{X_{(i)}}{X_{(k+1)}}\right) \xrightarrow{P} \frac{2}{\alpha}. \tag{4.50}$$

Recall from (4.18) (p. 82) that

$$\frac{X_{(\lceil kt\rceil)}}{b\left(\frac{n}{k}\right)} \Rightarrow t^{-1/\alpha}$$

in $D(0, \infty]$. Now write

$$A_n = \frac{k+1}{k}\int_0^1 -\log\left(\frac{\lceil (k+1)t\rceil}{k+1}\right)\log\left(\frac{X_{(\lceil (k+1)t\rceil)}}{X_{(k+1)}}\right) dt.$$

We claim this converges in probability to

$$\xrightarrow{P} \int_0^1 -\log t \cdot (\log t^{-1/\alpha}) dt$$
$$= \frac{1}{\alpha}\int_0^1 (-\log t)^2 dt = \frac{2}{\alpha}. \tag{4.51}$$

The convergence near zero in (4.51) is a problem since (4.18) holds in $D(0, \infty]$ and hence does not cover neighborhoods of zero. So we have to use a converging together argument based on Theorem 3.5 (p. 56). Convergence of the integral in (4.51) over the region $(\delta, 1)$ is guaranteed by the fact that (4.18) is tantamount to local uniform convergence *away from zero*, and hence on $(\delta, 1)$ there is uniform convergence. So it suffices to show that

$$\lim_{\delta \downarrow 0} \limsup_{k \to \infty} \mathbb{P}\left[ \int_0^\delta -\log t \cdot \log \left( \frac{X_{(\lceil (k+1)t \rceil)}}{X_{(k+1)}} \right) dt > \eta \right] = 0. \tag{4.52}$$

We do this by using Potter's inequalities and Rényi's representation of order statistics (see Problem 4.1 (p. 114)). Recall that Potter's inequalities (2.31) take the following form: Since $1/(1-F)$ is regularly varying with index $\alpha$, the inverse $b = (1/(1-F))^{\leftarrow}$ is regularly varying with index $1/\alpha$, and for $\epsilon > 0$, there exists $t_0 = t_0(\epsilon)$ such that if $y \geq 1$ and $t \geq t_0$,

$$(1-\epsilon) y^{\alpha^{-1}-\epsilon} \leq \frac{b(ty)}{b(t)} \leq (1+\epsilon) y^{\alpha^{-1}+\epsilon}. \tag{4.53}$$

We now rephrase this in terms of the function

$$R = -\log(1-F) = \log b^{\leftarrow}.$$

Then $b = R^{\leftarrow} \circ \log$; taking logarithms in (4.53) and then converting from a multiplicative to an additive form yields that

$$\log(1-\epsilon) + (\alpha^{-1}-\epsilon) y \leq \log R^{\leftarrow}(s+y) - \log R^{\leftarrow}(s) \leq \log(1+\epsilon) + (\alpha^{-1}+\epsilon) y \tag{4.54}$$

for $s \geq \log t_0$ and $y \geq 0$.

The reason for introducing the $R$ function is that if $E_1, E_2, \ldots, E_n$ are iid unit exponentially distributed random variables, then

$$(X_1, X_2, \ldots, X_n) \stackrel{d}{=} (R^{\leftarrow}(E_j);\, j = 1, \ldots, n).$$

The Rényi representation gives for the spacings of exponential order statistics,

$$(E_{1,n}, E_{2,n} - E_{1,n}, \ldots, E_{n,n} - E_{n-1,n}) \stackrel{d}{=} \left( \frac{E_n}{n}, \frac{E_{n-1}}{n-1}, \ldots, \frac{E_n}{1} \right)$$

$$(E_{(1)} - E_{(2)}, E_{(2)} - E_{(3)}, \ldots, E_{(n-1)} - E_{(n)}, E_{(n)}) \stackrel{d}{=} \left( \frac{E_1}{1}, \frac{E_2}{2}, \ldots, \frac{E_n}{n} \right). \tag{4.55}$$

Now we have

$$
\begin{aligned}
&\mathbb{P}\left[\int_0^\delta -\log t \cdot \log\left(\frac{X_{(\lceil (k+1)t\rceil)}}{X_{(k+1)}}\right) dt > \eta\right] \\
&= \mathbb{P}\left[\int_0^\delta -\log t \cdot \log\left(\frac{R^{\leftarrow}(E_{(\lceil (k+1)t\rceil)} - E_{(k+1)} + E_{(k+1)})}{R^{\leftarrow}(E_{(k+1)})}\right) dt > \eta\right] \\
&\le \mathbb{P}\left[\int_0^\delta -\log t \cdot \log\left(\frac{R^{\leftarrow}(E_{(\lceil (k+1)t\rceil)} - E_{(k+1)} + E_{(k+1)})}{R^{\leftarrow}(E_{(k+1)})}\right) dt > \eta,\right. \\
&\qquad \left. e^{E_{(k+1)}} > t_0\right] + o_k(1)
\end{aligned}
$$

since $E_{(k)} \stackrel{P}{\to} \infty$ by Problem 4.3 (p. 115). Ignore the term $o_k(1)$ since it goes to zero with $k$. Apply (4.54) and we get the upper bound

$$
\le \mathbb{P}\left[\int_0^\delta -\log t \cdot [(1+\epsilon) + (\alpha^{-1}+\epsilon)(E_{(\lceil (k+1)t\rceil)} - E_{(k+1)})]dt > \eta\right]
$$

and for some small $\eta' > 0$ this is bounded by

$$
\le \mathbb{P}\left[\int_0^\delta -\log t \cdot (E_{(\lceil (k+1)t\rceil)} - E_{(k+1)})]dt > \eta'\right].
$$

Apply Markov's inequality to get the upper bound

$$
\begin{aligned}
&\le \frac{1}{\eta'}\int_0^\delta -\log t \cdot \mathbf{E}(E_{(\lceil (k+1)t\rceil)} - E_{(k+1)})dt \\
&= \frac{1}{\eta'}\int_0^\delta -\log t \cdot \sum_{l=\lceil (k+1)t\rceil}^{k+1} \frac{1}{l}dt;
\end{aligned}
$$

as $k \to \infty$, this is asymptotic to $\frac{1}{\eta'}\int_0^\delta -\log t(-\log t)dt$, and as $\delta \downarrow 0$, this is asymptotic to $\frac{1}{\eta'}\delta(\log\delta)^2 \to 0$. □

## 4.7 How to compute value-at-risk

Review the definitions and discussion in Section 1.3.2 (p. 9), the material on peaks over threshold and exceedances in Section 4.2 (p. 74), and the discussion of the generalized Pareto class in Section 4.5.1 (p. 91).

Computing VaR, requires a good estimate of the tail of the loss distribution. The POT method suggests a solution. Suppose $X$ is a random variable with distribution

$F \in \mathcal{D}(G_\gamma)$. Interest in the heavy-tailed case suggests restricting interest to $\gamma > 0$. For a threshold $u$ and $x > u$, write

$$\begin{aligned}\bar{F}(x) &= \mathbb{P}[X > x] = \mathbb{P}[X > x|X > u]\mathbb{P}[X > u] \\ &= \bar{F}(u)\bar{F}^{[u]}(x)\end{aligned}$$

(where $F^{[u]}(x)$ is the notation in (4.5) (p. 76) for the exceedance distribution)

$$\approx \bar{F}(u)\left(1 + \frac{\gamma}{\beta}(x-u)\right)^{-1/\gamma}. \tag{4.56}$$

Here we replaced $\bar{F}^{[u]}(x)$ by its two-parameter generalized Pareto approximation discussed in (4.32) (p. 93).

In practice, our estimate of the tail probabilities will require $\bar{F}(u)$ to be replaced by the empirical tail probability $\widehat{\bar{F}}(u)$, the fraction of the observed sample exceeding $u$ and the parameters $(\gamma, \beta)$ will be replaced by maximum-likelihood estimators $(\hat{\gamma}, \hat{\beta})$ based on the subsample of excesses relative to $u$.

From (4.56), we get an estimate of the $q$th-order quantile by setting the expression in (4.56) equal to $1 - q$ yielding

$$u + \frac{\hat{\beta}}{\hat{\gamma}}\left(\left(\frac{1-q}{\widehat{\bar{F}}(u)}\right)^{-\hat{\gamma}} - 1\right). \tag{4.57}$$

How do we apply this to VaR? Refer to (1.4) (p. 11) to get with $T = 1$,

$$\frac{L_1}{V_0} = 1 - e^{R_1}, \tag{4.58}$$

expressing the one period loss $L_1$ in terms of $V_0$, the initial asset value, and the one period return. The quantity $L_1/V_0$ is the relative loss after one period assuming $V_0$ is known. Observing a sequence of one period returns from a stationary process amounts to observing observations of relative losses. The tail of the relative loss distribution can be estimated along with its quantiles; this coupled with observed asset values in the prior time period allow for computation of VaR.

*Example* 4.3 (*MSFT*). We consider 2363 daily closing values of Microsoft's stock from January 11, 1993 to March 4, 2003. The time-series plots of the closing values and the returns are given in Figure 4.17. Corresponding to (4.58), we compute relative losses by the transformation $x \mapsto 1 - e^x$. Summary statistics for the nonnegative values are given in Table 4.1.

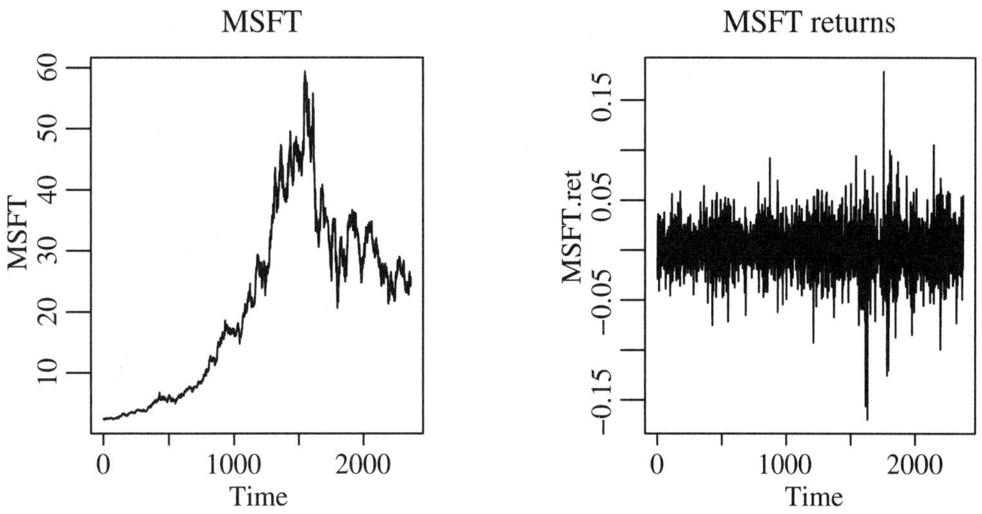


**Fig. 4.17.** Time-series plots of daily closing values of MSFT from January 11, 1993 to March 4, 2003 (left) and the corresponding return series (right).

| Minimum | First quarter | Median | Mean | Third quarter | Maximum |
|---|---|---|---|---|---|
| 0.000379 | 0.006536 | 0.014000 | 0.017670 | 0.024350 | 0.156100 |

**Table 4.1.** Summary statistics for the nonnegative values of relative loss.

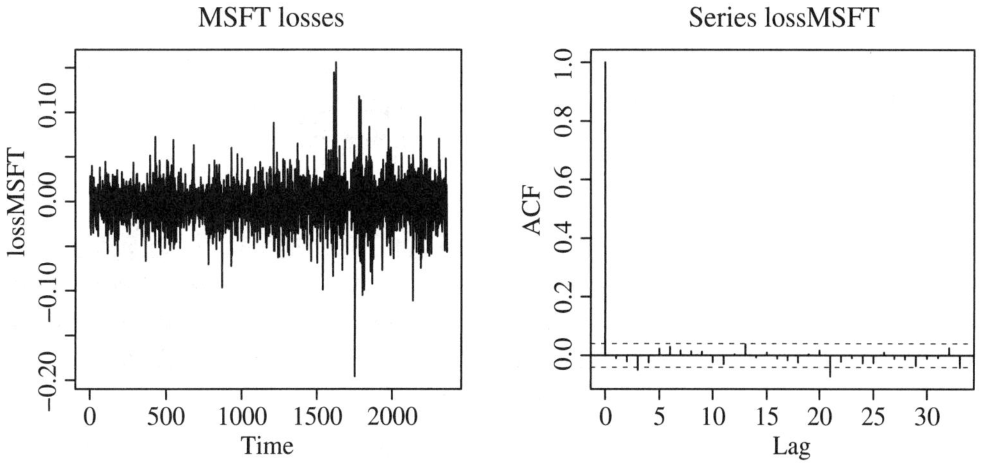


**Fig. 4.18.** Time-series plots of daily relative losses of MSFT (left) and the sample autocorrelation plot of the relative losses (right).

We also check for dependence of the relative losses by computing the sample autocorrelation plot, and this shows surprisingly little correlation. See Figure 4.18.

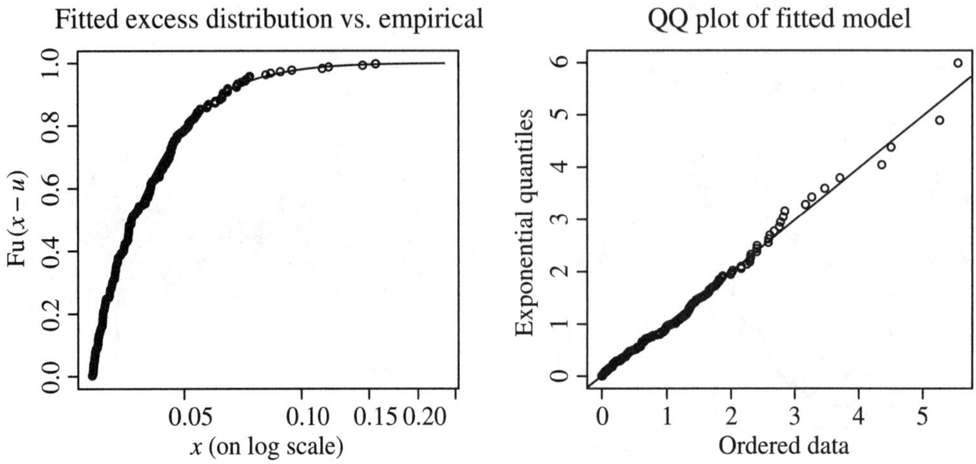


**Fig. 4.19.** Fitted excess distribution and the empirical distribution (left) and QQ plot for fitted model against exponential quantiles.

| $p$ | Quantile |
|---|---|
| 0.900 | 0.0355 |
| 0.990 | 0.0743 |
| 0.999 | 0.1411 |

**Table 4.2.** Quantiles of the relative loss distribution.

Now we take one of the extreme value software packages out for a test ride. We use EVIR (see the discussion in the appendix in Chapter 11 (p. 363)). We fit a generalized Pareto distribution (see (4.32) (p. 93)) by maximum likelihood to the nonnegative excesses of relative losses based on the threshold 0.0289 using the 200 largest-order statistics, yielding $\hat{\alpha} = 1/\hat{\gamma} = 4.242$. The fitted distribution function for relative losses corresponding to (4.56) is shown on the left side in Figure 4.19, and the QQ plot of the fitted model is shown on the right. Neither diagnostic reveals a problem with the fit. Table 4.2 exhibits extreme quantiles of the relative loss distribution. If $V_0 = 1$ and the number of shares stays constant, these quantiles would represent the VaR values.

## 4.8 Problems

**4.1 (Rényi representation).** Suppose $E_1, \dots, E_n$ are iid exponentially distributed random variables with parameter $\lambda > 0$, so that

$$P[E_1 \leq x] = 1 - e^{-\lambda x}, \quad x > 0.$$

Let

$$E_{1,n} \leq E_{2,n} \leq \cdots \leq E_{n,n}$$

be the order statistics. Prove that the $n$ spacings

$$E_{1,n}, E_{2,n} - E_{1,n}, \ldots, E_{n,n} - E_{n-1,n}$$

are independent exponentially distributed random variables, where $E_{k+1,n} - E_{k,n}$ has parameter $(n-k)\lambda$.

Intuitively, this results from the forgetfulness property of the exponential distribution. See [90, 135, 239, 240, 262].

**4.2.** Suppose that $X_1, \ldots, X_n$ are iid from a common continuous distribution $F(x)$ and that

$$X_{(1)} \geq X_{(2)} \geq \cdots \geq X_{(n)}$$

are the order statistics. Prove that conditionally on $X_{(k+1)}$, the family $X_{(1)}, X_{(2)}, \ldots, X_{(k)}$ are distributed as the order statistics from a sample of size $k$ from the conditional distribution of

$$X | X > X_{(k+1)},$$

where $X_{(k+1)}$ is treated as a constant in the conditioning. (You might want to use Problem 4.1.) Furthermore, show that $(X_{(n)}, X_{(n-1)}, \ldots, X_{(1)})$ is a continuous state-space Markov chain.

**4.3.** For the order statistics $E_{(1)}, \ldots, E_{(n)}$ of an iid sample of size $n$ from the unit exponential distribution, show that

$$E_{(k)} \stackrel{P}{\to} \infty$$

iff $k \to \infty$, $n \to \infty$, $n/k \to \infty$.

**4.4 (Découpage de Lévy).** Suppose $\{X_n, n \geq 1\}$ are iid random elements of the metric space $S$ with Borel $\sigma$-field $\mathcal{S}$. Fix a set $B \in \mathcal{S}$ such that $\mathbb{P}[X_1 \in B] > 0$. Let $\tau_0^{\pm} = 0$, and $\tau_i^+ = \inf\{j > \tau_{i-1}^+ : X_j \in B\}$ for $i \geq 1$. The family $\{\tau_j^-, j \geq 0\}$ is defined similarly, with $B^c$ playing the role of $B$. Define the counting function $K_n = \sup\{i : \tau_i^+ \leq n\}$. Show that $\{X_{\tau_j^+}\}$, $\{X_{\tau_j^-}\}$, $\{K_n\}$ are independent with

$$\mathbb{P}[X_{\tau_1^+} \in A] = \mathbb{P}[X_1 \in A | X_1 \in B], \quad A \subset B,$$
$$\mathbb{P}[X_{\tau_1^-} \in A] = \mathbb{P}[X_1 \in A | X_1 \in B^c] \quad A \subset B^c.$$

Furthermore, $\{K_n\}$ is a renewal counting function, $\mathbf{E}(K_n) = n\mathbb{P}[X_1 \in B]$, and $\{X_{\tau_j^{\pm}}\}$ is iid.

**4.5 (For the probabilistically adventurous).** For QQ estimation in the heavy-tailed case, consider the $n$ bivariate pairs consisting of theoretical quantiles paired with sample quantiles for the log-transformed data. Does this set (or a subset of this set corresponding to the $k$ upper quantiles), considered as a random closed subset of the first quadrant, converge in any sense to a limit random set? This limit set is presumably a line and so is not random. Convergence could be in the space of random closed subsets of $\mathbb{R}_+^2$ metrized by, say, the Hausdorff metric.

Is the functional

$$\mathrm{LS} : \{\text{closed sets}\} \mapsto \mathbb{R}_+$$

defined by

$$\mathrm{LS}(F) = \text{slope of the LS line through the closed set } F$$

continuous in the space of closed subsets of $\mathbb{R}_+^2$, or continuous, at least, at any useful elements of the domain?

# Part III

# Probability

# 5

# The Poisson Process

There are many fascinating and useful connections between heavy tails and the Poisson process, some of which we begin to describe here. Many heavy-tailed models are constructed from Poisson processes, which are the most tractable models of point systems. Some of these contructions give paradigms in the theory and some are elegant abstractions of applied systems.

I learned much about this subject from [33–37, 43, 180, 230].

## 5.1 The Poisson process as a random measure

Throughout this discussion, it is enough to assume the state spaces of our random measures and point processes are *nice*; see Section 10.2 (p. 360) if you need a more precise definition.

### 5.1.1 Definition and first properties

Let $N : (\Omega, \mathcal{A}) \mapsto (M_p(\mathbb{E}), \mathcal{M}_p(\mathbb{E}))$ be a point process with state space $\mathbb{E}$, where $\mathcal{M}_p(\mathbb{E})$ is the Borel $\sigma$-algebra of subsets of $M_p(\mathbb{E})$ generated by open sets. The Borel subsets of $\mathbb{E}$ are denoted by $\mathcal{E}$. (If necessary, review Section 3.3.5 (p. 51).)

**Definition 5.1.** $N$ is a *Poisson process with mean measure* $\mu$ or, synonomously, a *Poisson random measure* (PRM($\mu$)), if we have the following:

1. For $A \in \mathcal{E}$,

$$P[N(A) = k] = \begin{cases} \dfrac{e^{-\mu(A)}(\mu(A))^k}{k!} & \text{if } \mu(A) < \infty, \\ 0 & \text{if } \mu(A) = \infty. \end{cases}$$

2. If $A_1, \ldots, A_k$ are disjoint subsets of $\mathbb{E}$ in $\mathcal{E}$, then $N(A_1), \ldots, N(A_k)$ are independent random variables.

So $N$ is Poisson if the random number of points in a set $A$ is Poisson distributed with parameter $\mu(A)$ and the number of points in disjoint regions are independent random variables.

Property 2 is called *complete randomness*. When $\mathbb{E} = \mathbb{R}$, it is called the *independent increments* property since for any $t_1 < t_2 < \cdots < t_k$, $(N((t_i, t_{i+1}]), 1 = 1, \ldots, k-1)$ are independent random variables. When the mean measure is a multiple of Lebesgue measure (that is, length when $\mathbb{E} = [0, \infty)$ or $\mathbb{R}$, area when $\mathbb{E} = \mathbb{R}^2$, volume when $\mathbb{E} = \mathbb{R}^3$, etc.), we call the process *homogeneous*. Thus in the homogeneous case, there is a parameter $\lambda > 0$ such that for any $A$, we have $N(A)$ Poisson distributed with mean $EN(A) = \lambda\, \mathbb{LEB}(A)$, where $\mathbb{LEB}(A)$ is the Lebesgue measure of $A$. When $\mathbb{E} = [0, \infty)$, the parameter $\lambda$ is called the *rate* of the (homogeneous) Poisson process. When $\mathbb{E} = [0, \infty)$, epochs of a pure renewal process in $(0, \infty)$ whose interarrival density is exponential is a homogeneous Poisson process. It can be surprisingly tricky to prove this. We just state the result. See [107, 260, 262].

**Proposition 5.1.** *Let $\{E_j, j \geq 1\}$ be iid random variables with a standard exponential distribution. Define $\Gamma_n = \sum_{i=1}^{n} E_i$ to be the renewal epochs of the renewal process, and set $N = \sum_{n=1}^{\infty} \epsilon_{\Gamma_n}$. Then $N$ is a homogeneous Poisson process on $[0, \infty)$ with unit rate $\lambda = 1$; that is, $N$ satisfies Definition* 5.1, *and the mean measure is* $\mathbb{LEB}(\cdot)$.

### 5.1.2 Point transformations

Useful results are connected with a circle of ideas about what happens to a Poisson process under various types of transformations. The first result, although very elementary, is enormously useful in understanding inhomogeneity. To prepare for this result, suppose $\sum_n \epsilon_{X_n}$ is a Poisson process with state space $\mathbb{E}$ and mean measure $\mu$. Suppose $T$ is some transformation with domain $\mathbb{E}$ and range $\mathbb{E}'$, where $\mathbb{E}'$ is another nice space; that is,

$$T : \mathbb{E} \mapsto \mathbb{E}'.$$

The function $T$ defines a set mapping of subsets of $\mathbb{E}'$ to subsets of $\mathbb{E}$, defined for $A' \subset \mathbb{E}'$ by

$$T^{-1}(A') = \{e \in E : T(e) \in A'\}.$$

Thus $T^{-1}(A')$ is the preimage of $A'$ under $T$; that is, it is the set of points of $\mathbb{E}$ that $T$ maps into $A'$.

As an example, suppose $\mathbb{E} = (0, \infty)$, $\mathbb{E}' = (-\infty, \infty)$, $T(x) = \log x$. If $a < b$ and $A' = (a, b)$, then we have

$$\begin{aligned} T^{-1}((a,b)) &= \{x > 0 : T(x) \in (a,b)\} \\ &= \{x > 0 : \log x \in (a,b)\} \\ &= \{x > 0 : x \in (e^a, e^b)\}. \end{aligned}$$

Given the measures $N$, $\mu$ defined on subsets of $\mathbb{E}$, we may use $T$ to define induced measures $N'$, $\mu'$ on subsets of $\mathbb{E}'$. For $A' \subset E'$, define

$$N'(A') = N(T^{-1}(A')), \qquad \mu'(A') = \mu(T^{-1}(A')).$$

So to get the measure of $A'$, we map $A'$ back into $\mathbb{E}$ and take the measure of the preimage under $T$. Also, if $N$ has points $\{X_n\}$, then $N'$ has points $\{X'_n\} = \{T(X_n)\}$.

The next result asserts that if $N$ is a Poisson process with mean measure $\mu$ and with points $\{X_n\}$ living in the state space $\mathbb{E}$, then $N' = N(T^{-1}(\cdot))$ is a Poisson process with mean measure $\mu'$ and with points $\{T(X_n)\}$ living in the state space $\mathbb{E}'$.

**Proposition 5.2.** *Suppose*

$$T : \mathbb{E} \mapsto \mathbb{E}'$$

*is a measurable mapping of one nice space $\mathbb{E}$ into another $\mathbb{E}'$ such that if $K' \in \mathcal{K}(\mathbb{E}')$ is compact in $\mathbb{E}'$, then so is $T^{-1}K' := \{e \in E : Te \in K'\} \in \mathcal{K}(\mathbb{E})$. If $N$ is* PRM$(\mu)$ *on $\mathbb{E}$, then $N' := N \circ T^{-1}$ is* PRM$(\mu')$ *on $\mathbb{E}'$, where $\mu' := \mu \circ T^{-1}$.*

Remember that if $N$ has the representation

$$N = \sum_n \epsilon_{X_n},$$

then

$$N' = \sum_n \epsilon_{T(X_n)},$$

and the result says that if you transform the points of a Poisson process, you still have a Poisson process.

*Proof.* We have

$$\mathbb{P}[N'(B') = k] = \mathbb{P}[N(T^{-1}(B')) = k] = e^{-\mu(T^{-1}(B))}(\mu(T^{-1}(B)))^k/k!,$$

so $N'$ has Poisson distributions. It is easy to check the independence property since if $B'_1, \ldots, B'_m$ are disjoint, then so are $T^{-1}(B'_1), \ldots, T^{-1}(B_m)$, whence

$$(N'(B'_1), \ldots, N'(B'_m)) = (N(T^{-1}(B'_1), \ldots, N(T^{-1}(B'_m))$$

are independent. Thus requirements 1 and 2 in the definition of a Poisson process (Definition 5.1) are satisfied. □

*Example* 5.1. Consider three easy examples. For each, let $N = \sum_{n=1}^{\infty} \epsilon_{\Gamma_n}$ be a homogeneous Poisson process with rate $\lambda = 1$ on the state space $\mathbb{E} = [0, \infty)$. The mean measure $\mu$ is Lebesgue measure so that $\mu(A) = \mathbb{LEB}(A)$ and, in particular, $\mu([0, t]) = t$.

1. If $T(x) = x^2$, then $\sum_n \epsilon_{\Gamma_n^2}$ is PRM and the mean measure $\mu'$ is given by

$$\mu'[0, t] = \mu\{x : T(x) \le t\} = \mu\left[0, \sqrt{t}\right] = \sqrt{t}.$$

   Note that $\mu'$ has a density

$$g(t) = \frac{d}{dt}\sqrt{t} = \frac{1}{2}t^{-1/2}.$$

2. If $T : \mathbb{E} \mapsto \mathbb{E} \times \mathbb{E}$ via $T(x) = (x, x^2)$, then $\sum_n \epsilon_{T(\Gamma_n)} = \sum_n \epsilon_{(\Gamma_n, \Gamma_n^2)}$ is Poisson on $\mathbb{E} \times \mathbb{E}$. The mean measure concentrates on the graph $\{(x, x^2) : x \ge 0\}$.

3. Given a homogeneous Poisson process $\sum_n \epsilon_{\Gamma_n}$ on $[0, \infty)$, $\sum_n \epsilon_{\Gamma_n^{-1}}$ is Poisson on $(0, \infty]$ with mean measure $\mu'$ given by $(x > 0)$

$$\mu'(x, \infty] = \mu\{t \ge 0 : t^{-1} \ge x\} = \mu[0, x^{-1}) = x^{-1}.$$

   The topology on $\mathbb{E}'$ induced by the map $x \mapsto x^{-1}$ makes the bounded sets of $\mathbb{E}'$ the sets bounded away from 0; that is, the bounded sets are neighborhoods of $\infty$. $\mu'$ has a density

$$g(t) = -\frac{d}{dt}t^{-1} = t^{-2}.$$

   As we will see in Sections 5.5.2 (p. 154) and 5.6 (p. 160), Poisson processes with this mean measure $\mu'$ are particularly important in the theory of stable processes and in extreme-value theory.

### 5.1.3 Augmentation or marking

Given a Poisson process, under certain circumstances it is possible to enlarge the dimension of the points and retain the Poisson structure. One way to do this was given in item 2 of Example 5.1 of the previous section, but the enlargement of dimension was illusory since the points concentrated on a graph $\{(x, x^2) : x > 0\}$. The result presented here allows independent components to be added to the points of the Poisson process. This proves very useful in a variety of applications. We present here the simplest statement of this result. A more sophisticated version will be presented after a discussion of the Laplace functional.

**Proposition 5.3.** *Suppose $\{X_n\}$ are random elements of a nice space $\mathbb{E}_1$ such that*

$$\sum_n \epsilon_{X_n}$$

*is* PRM($\mu$). *Suppose $\{J_n\}$ are iid random elements of a second nice space $\mathbb{E}_2$ with common probability distribution $F$, and suppose the Poisson process and the sequence $\{J_n\}$ are defined on the same probability space and are independent. Then the point process on $\mathbb{E}_1 \times \mathbb{E}_2$,*

$$\sum_n \epsilon_{(X_n, J_n)},$$

*is* PRM *with mean measure $\mu \times F$.*

So if $A_i \subset \mathbb{E}_i$, $i = 1, 2$, are Borel sets, then

$$\mu \times F(A_1 \times A_2) = \mu \times F(\{(e_1, e_2) : e_1 \in A_1, e_2 \in A_2\}) = \mu(A_1)F(A_2).$$

Often this procedure is described by saying we give to point $X_n$ the *mark* $J_n$. Think about a picture where the points of the original Poisson process $\{X_n\}$ appear on the horizontal axis and the marked points appear in the $\mathbb{E}_1 \times \mathbb{E}_2$ plane.

The proof is deferred. For now, note the mean measure is correct since for a rectangle set of the form $A_1 \times A_2 = \{(e_1, e_2) : e_1 \in A_1 \subset \mathbb{E}_1, e_2 \in A_2 \subset \mathbb{E}_2\}$, we have

$$\begin{aligned} E \sum_n \epsilon_{(X_n, J_n)}(A_1 \times A_2) &= \sum_n P[(X_n, J_n) \in A_1 \times A_2] \\ &= \sum_n P[X_n \in A_1]P[J_n \in A_2] \end{aligned}$$

since $\{J_n\}$ is independent of the Poisson process. Since $\{J_n\}$ are iid random variables this is the same as

$$\begin{aligned} &= \sum_n P[X_n \in A_1]P[J_1 \in A_2] \\ &= E\left(\sum_n \epsilon_{X_n}(A_1)\right) P[J_1 \in A_2] \\ &= \mu(A_1)P[J_1 \in A_2]. \end{aligned}$$

## 5.2 Models for data transmission

The infinite-node Poisson model is a simple (probably too simple) model that explains long-range dependence in measured Internet traffic. The simple explanation is based on properties of a Poisson process.

### 5.2.1 Background

The story begins around 1993 with the publication of what is now known as the Bellcore study [118, 203, 305]. Traditional queueing models had thrived on assumptions of exponentially bounded tails, Poisson inputs, and lots of independence. Collected network data studied at what was then Bellcore (now Telcordia) exhibited properties that were inconsistent with traditional queueing models. These anomalies were also found in World Wide Web downloads in the Boston University study [51–56, 63]. The unusual properties found in the data traces included:

- self-similarity and long-range dependence (LRD) of various transmission rates:
  - packet counts per unit time,
  - www bits/time.
- heavy tails of quantities such as
  - file sizes,
  - transmission rates,
  - transmission durations,
  - CPU job completion times,
  - call lengths.

The Bellcore study in the early 1990s resulted in a paradigm shift worthy of a sociological study to understand the frenzy to jump on and off various bandwagons, but after some resistance to the presence of long-range dependence, there was widespread acceptance of the statement that *packet counts per unit time exhibit* self similarity *and* long-range dependence. Research goals then shifted from detection of the phenomena to greater understanding of the causes. The challenges were the following:

- Explain the origins and effects of long-range dependence and self-similarity.
- Understand some connections between self-similarity, long-range dependence, and heavy tails. Use these connections to find an explanation for the perceived long-range dependence in traffic measurements.
- Begin to understand the effect of network protocols and architecture on traffic. This is an ambitious goal, since the simplest models, such as the featured infinite-source Poisson model, pretend protocols, and controls are absent.

### 5.2.2 Probability models

Attempts to explain long-range dependence and self-similarity in traffic rates centered around the paradigm *heavy-tailed file sizes cause long-range dependence in network traffic*. Specific models must be used to explain this and the two most effective and simple models were the following:

- *Superposition of on/off processes* [158, 159, 176, 222, 225, 233, 285, 288, 305]: This is described as follows: Imagine a source/destination pair. The source sends at unit rate for a random length of time to the destination and then is silent or inactive for a random period. Then the source sends again and when finished is silent. And so on. So the transmission schedule of the source follows an alternating renewal or on/off structure. Now imagine the traffic generated by many source/destination pairs being superimposed, which yields the overall traffic.

- *The infinite-node Poisson model* [153, 160, 175, 177, 222, 234, 242, 254]: This is sometimes referred to as the $M/G/\infty$ input model. Imagine infinitely many potential users connected to a single server that processes work at constant rate $r$. At a Poisson time point, some user begins transmitting work to the server at constant rate which, for specificity, we take to be rate 1. The length of the transmission is random with heavy-tailed distribution. The length of the transmission may be considered to be the size of the file needing transmission.

Both models have their adherents and the two models are asymptotically equivalent in a manner nobody (to date) has made fully transparent. We will focus on the infinite-source Poisson model.

Some good news about the model:

- It is somewhat flexible and certainly simple.

- Since each node transmits at unit rate, the overall transmission rate at time $t$ is simply the number of active users $M(t)$ at $t$. From classical $M/G/\infty$ queueing theory, we know that $M(t)$ is a Poisson random variable with mean $\lambda\mu_{\text{on}}$, where $\lambda$ is the rate parameter of the Poisson process and $\mu_{\text{on}}$ is the mean file size or mean transmission length. This is reviewed in Section 5.2.5 (p. 130).

- The length of each transmission is random and heavy tailed.

- The model offers a very simple explanation of long-range dependence being caused by heavy-tailed file sizes.

- The model predicts traffic aggregated over users and accumulated over time $[0, T]$ is approximated by either a Gaussian process (fractional Brownian motion, or FBM) or a heavy-tailed stable Lévy motion [222]. Thus the two approximations are very different in character, but at least both are self-similar.

Some less good news about the model:

- The model does not fit collected data traces all that well.
  - The constant transmission rate assumption is clearly wrong. Each of us knows from personal experience that downloads and uploads do not proceed at constant rate.
  - Not all times of transmissions are Poisson. Identifying Poisson time points in the data can be problematic. Some are machine triggered and these will certainly not be Poisson. While network engineers rightly believe in the *invariant* that behavior associated with humans acting independently can be modeled as a Poisson process, it is highly unlikely that, for example, subsidiary downloads triggered by going to the CNN website (imagine the calls to DoubleClick's ads) would follow a Poisson pattern.
- There is no hope that this simple model can successfully match fine time scale behavior observed below, say, 100 milliseconds.
- The model does not take into account admission and congestion controls such as TCP (transmission control protocol). How can one incorporate a complex object like a control mechanism into an informative probability model?

### 5.2.3 Long-range dependence

There is no universal agreement about how to define long-range dependence, but probably most people associate the term with slow decay of the correlation function as a function of the lag between time points. For us, the most functional definition is this: A stationary $L_2$ sequence $\{\xi_n, n \geq 1\}$ possesses long-range dependence (LRD) if

$$\mathrm{Cov}(\xi_n, \xi_{n+h}) \sim h^{-\beta} L(h), \quad h \to \infty, \tag{5.1}$$

for $0 < \beta < 1$ and $L(\cdot)$ slowly varying [18]. Set $\gamma(h) = \mathrm{Cov}(\xi_n, \xi_{n+h})$ and $\rho(h) = \gamma(h)/\gamma(0)$ for the covariance and correlation functions of the stationary process $\{\xi_n\}$. For other authors, long-range dependence is sometimes taken to mean that covariances are not summable: $\sum_h |\gamma(h)| = \infty$, whereas short-range dependence means that $\sum_h |\gamma(h)| < \infty$. Traditional time-series models, such as ARMA models [31], have covariances that go to zero geometrically fast as a function of the lag $h$. Long-range dependence, like the property of heavy tails, has acquired a mystical, almost religious, significance and generated controversy. Researchers argue over whether it exists, whether it matters if it exists or not, or whether analysts have been fooled into mistaking some other phenomena like shifting levels, undetected trend [193], or nonstationarity for long-range dependence. Discussions about this have been going on since (at least) the mid-1970s in hydrology [20, 27, 28, 32, 270–272], finance [224, 226], and data network modeling [39, 40, 117, 141, 164, 202, 232]. Think of it as one more modeling decision that needs to be made. Since long-range dependence is an asymptotic property, models that possess long-range dependence presumably have different asymptotic properties than those models in which long-range dependence is absent, although even this is sometimes disputed.

### Simple minded detection of long-range dependence using the sample acf plot

Sophisticated methods for detecting long-range dependence exist. However, the most common, ubiquitous, quick, and dirty method to detect long-range dependence (assuming that you are convinced the data comes from a stationary process) is to graph the sample autocorrelation function (acf) $\{\hat{\rho}(h), h = 1, 2, \ldots, N\}$, where $N$ is a large number but not a significant proportion of the whole sample size. The sample acf at lag $h$ corresponding to observations $x_1, \ldots, x_n$ is defined as

$$\hat{\rho}(h) = \frac{\sum_{i=1}^{n-h}(x_i - \bar{x})(x_{i+h} - \bar{x})}{\sum_{i=1}^{n-h}(x_i - \bar{x})^2}.$$

The plot should not decline rapidly. Classical time-series data that one encounters in ARMA (Box–Jenkins) modeling exercises has a sample acf that is essentially zero after a few lags, and acf plots of financial or teletraffic data are often in stark contrast.

*Example* 5.2 (*Company $X$*). This trace is packet counts per 100 milliseconds = 1/10 second for Financial Company X's wide-area network link, including USA–UK traffic. It consists of 288,009 observations corresponding to 8 hours of collection from 9am–5pm. Figure 5.1 shows the time-series plot of a segment.

Figure 5.2 shows the acf plot for 2000 lags. There is little hurry for the plot to approach zero. (Don't try to model this with ARMA.)

### 5.2.4 The infinite-node Poisson model

Understanding the connection between heavy tails and long-range dependence requires a context. For the simplest explanations, one can choose either the superposition of on/off processes or the infinite-node Poisson model, and our preference is for the latter.

In this model, there are potentially an infinite number of sources capable of sending work to the server. Imagine that transmission sources turn on and initiate *sessions*

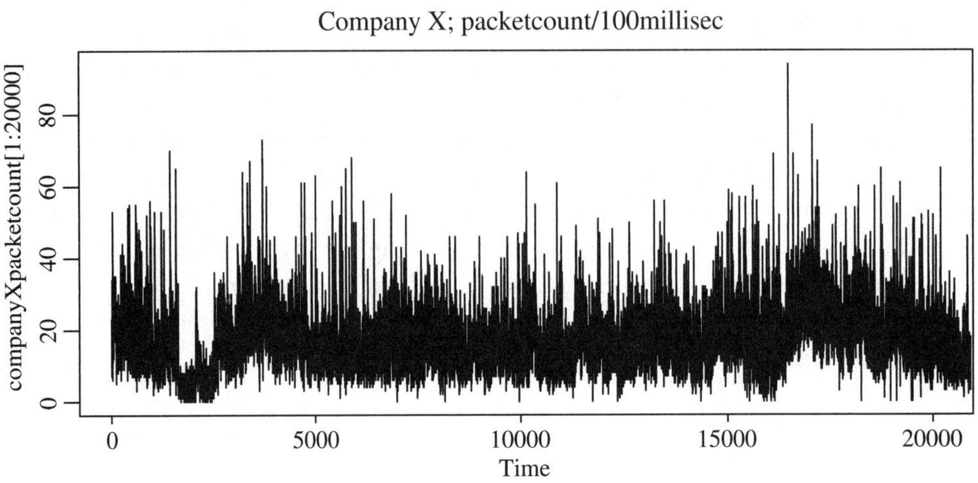


**Fig. 5.1.** Time-series plot for Company X data giving first 50,000 observations.

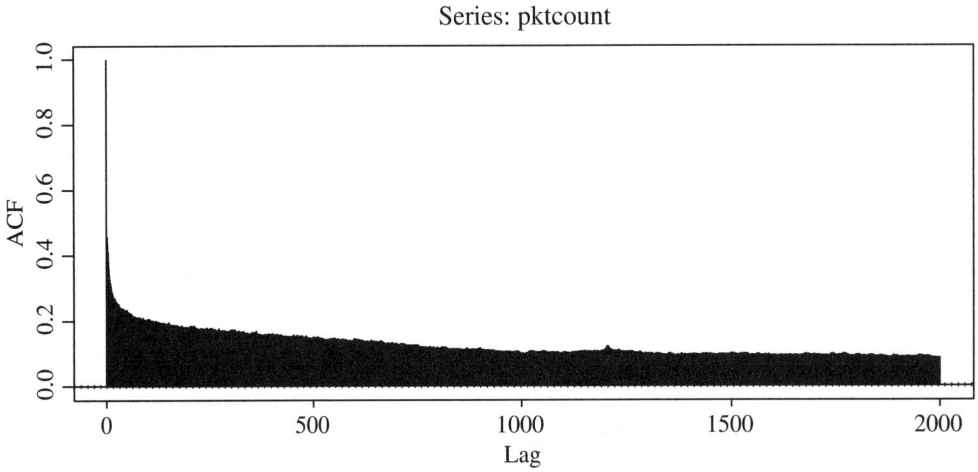


**Fig. 5.2.** Sample autocorrelation plot for Company X data for 2000 lags.

or *connections* at homogeneous Poisson time points $\{\Gamma_k\}$ with rate $\lambda$. The lengths of sessions $\{L_n\}$ are iid nonnegative random variables with common distribution $F_{\text{on}}$, and during a session, work is transmitted to the server at constant rate. As a normalization, we suppose the transmission rate is 1. Assume that

$$1 - F_{\text{on}}(t) := \bar{F}_{\text{on}}(t) = t^{-\alpha} L(t), \quad t \to \infty. \tag{5.2}$$

In practice, empirical estimates of $\alpha$ usually range between 1 and 2 [204, 305]. (However, there are studies of file sizes [7, 242] that report measurements of $\alpha < 1$.) The assumption of a fixed unit transmission rate is obviously an idealization. The model can be modified for greater realism by assuming that either

(i) transmission rates are random and possibly dependent on the size of the file to be transmitted or on the transmission duration [38, 68, 215]

or

(ii) it is assumed that cumulative input from a source follows a random process [194, 216].

For the sake of simplicity and tractability, the fixed unit transmission rate will be assumed.

Note that in the case $1 < \alpha < 2$, the second moment of $F_{\text{on}}$ is infinite, but

$$\mu_{\text{on}} = E(L_1) = \int_0^\infty \bar{F}_{\text{on}}(t)dt < \infty.$$

The processes of primary interest for describing this system are the following:

$$\begin{aligned} M(t) &= \text{number of sessions in progress at } t \\ &= \text{number of busy servers in the M/G/}\infty \text{ model} \\ &= \sum_{k=1}^{\infty} 1_{[\Gamma_k \le t < \Gamma_k + L_k]} \end{aligned} \tag{5.3}$$

and

$$A(t) = \int_0^t M(s)ds = \text{cumulative input in } [0, t], \tag{5.4}$$

$$\begin{aligned} r &= \text{release rate or the rate at which the server} \\ &\quad \text{works off the offered load.} \end{aligned} \tag{5.5}$$

Note that expressing $A(t)$ as an integral gives $M(t)$ the interpretation of "instantaneous input rate at time $t$." So realizations of $M(t)$ correspond to data traces of "packet counts per unit time." So we seek within the model an explanation of why $\{M(t)\}$ possesses long-range dependence.

Stability requires us to assume that the long-term input rate should be less than the output rate, so we require that

$$\lambda\mu_{\text{on}} < r.$$

This means the content or buffer level process $\{X(t), t \ge 0\}$ satisfies

$$dX(t) = M(t)dt - r1_{[X(t)>0]}dt,$$

is regenerative with finite mean regeneration times, and achieves a stationary distribution.

### 5.2.5 Connection between heavy tails and long-range dependence

The common explanation for long-range dependence in the total transmission rate by the system is that *high variability causes long-range dependence*, where we understand high variability means heavy tails. The long-range dependence resulting from the heavy-tailed distribution $F_{\text{on}}$ can be easily seen for the infinite-node Poisson model.

Assume that $1 < \alpha < 2$. To make our argument transparent, we consider the following background. For each $t$, $M(t)$ is a Poisson random variable. Why? When $1 < \alpha < 2$, $M(\cdot)$ has a stationary version on $\mathbb{R}$, the whole real line. Assume that

$$\sum_k \epsilon_{\Gamma_k} = \text{PRM}(\lambda\, \mathbb{LEB})$$

is a homogeneous Poisson random measure on $\mathbb{R}$ with rate $\lambda$. Then using augmentation,

$$\xi := \sum_k \epsilon_{(\Gamma_k, L_k)} = \text{PRM}(\lambda\, \mathbb{LEB} \times F_{\text{on}}) \tag{5.6}$$

is a two-dimensional Poisson random measure on $\mathbb{R} \times [0, \infty)$ with mean measure $\lambda dt \times F_{\text{on}}(dx)$, and

$$\begin{aligned} M(t) &= \sum_k 1_{[\Gamma_k \le t < \Gamma_k + L_k]} \\ &= \xi(\{(s, l) : s \le t < s + l\} = \xi(B) \end{aligned}$$

is Poisson because it is the two-dimensional Poisson process $\xi$ evaluated on the region $B$. See the gorgeous Figure 5.3. Note that $B$ is the region in the $(s, l)$-plane to the left of the vertical line through $t$ and above the $-45$ degree line through $(t, 0)$. The mean of $\xi(B)$ is

$$\begin{aligned} E(\xi(\{(s, l) : s \le t < s + l\}) &= \iint_{\{(s,l):s \le t < s+l\}} \lambda ds\, F_{\text{on}}(dl) \\ &= \int_{s=-\infty}^{t} \bar{F}_{\text{on}}(t - s)\lambda ds = \lambda \mu_{\text{on}}. \end{aligned} \tag{5.7}$$

Understanding the relation between $\{M(t)\}$ and the random measure $\xi$ allows us to easily compute the covariance function. Refer to Figure 5.4. Recall that $M(t)$ corresponds to points to the left of the vertical through $(t, 0)$ and above the $-45$-degree line through $(t, 0)$ with a similar interpretation for $M(t + \tau)$. The process $\{M(t), t \in \mathbb{R}\}$ is stationary with covariance function

$$\text{Cov}(M(t), M(t + \tau)) = \text{Cov}(\xi(A_1) + \xi(A_2), \xi(A_2) + \xi(A_3)),$$

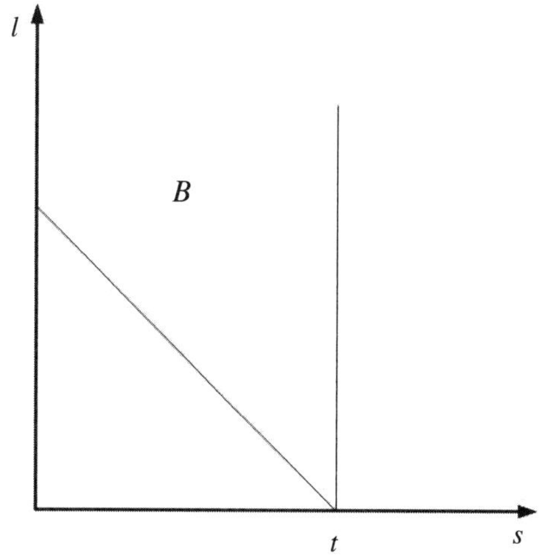


**Fig. 5.3.** The region $B$.

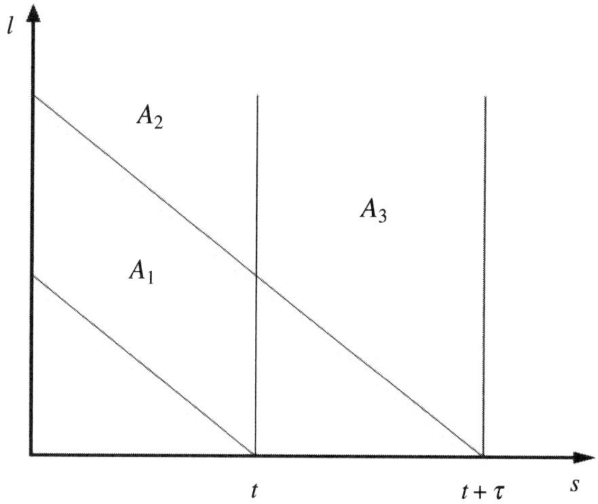


**Fig. 5.4.** The regions $A_1$, $A_2$, $A_3$.

and because $\xi(A_1)$ and $\xi(A_3)$ are independent, the previous expression reduces to

$$= \mathrm{Cov}(\xi(A_2), \xi(A_2)) = \mathrm{Var}(\xi(A_2)).$$

For a Poisson random variable, the mean and the variance are equal, and therefore the above is

$$= E(\xi(A_2)) = \int_{u=-\infty}^{t} \lambda du \bar{F}_{\text{on}}(t + \tau - u)$$

$$= \lambda \int_{\tau}^{\infty} \bar{F}_{\text{on}}(v)dv \sim c\tau^{-(\alpha-1)}L(\tau).$$

Note that we used Karamata's theorem to evaluate the integral of the regularly varying tail.

To summarize, we find that

$$\begin{aligned}
\text{Cov}(M(t), M(t+\tau)) &= \lambda \int_{\tau}^{\infty} \bar{F}_{\text{on}}(v)dv \\
&= (\text{const})\tau^{-(\alpha-1)}L(\tau) \\
&= (\text{const})\tau \bar{F}_{\text{on}}(\tau), \quad \tau \to \infty. \qquad (5.8)
\end{aligned}$$

The slow decay of the covariance as a function of the lag $\tau$ characterizes *long-range dependence*.

## 5.3 The Laplace functional

The Laplace functional is a convenient transform technique that is helpful for manipulating distributions of point processes and random measures. When applied to Poisson processes and empirical measures, algebraic manipulations become familiar to ones used with either characteristic functions or Laplace transforms applied to sums of iid random variables.

### 5.3.1 Definition and first properties

For a nonnegative, bounded measurable function $f : \mathbb{E} \mapsto \mathbb{R}_+$ and for $\mu \in M_+(\mathbb{E})$, we use the notation

$$\mu(f) = \int_{x \in E} f(x)\mu(dx).$$

For $m = \sum_i \epsilon_{x_i} \in M_p(\mathbb{E})$,

$$m(f) = \sum_i f(x_i).$$

A guiding principle is that integrals of measures with respect to arbitrary test functions contain as much information as evaluating the measures on arbitrary sets.

**Definition 5.2 (Laplace functional).** Suppose $\mathcal{B}_+$ are the nonnegative, bounded, measurable functions from $\mathbb{E} \mapsto \mathbb{R}_+$ and let

$$M : (\Omega, \mathcal{A}, \mathbb{P}) \mapsto (M_+(\mathbb{E}), \mathcal{M}_+(\mathbb{E}))$$

be a random measure (that is, a random element of $M_+(\mathbb{E})$). The *Laplace functional* of the random measure $M$ is the nonnegative function on $\mathcal{B}_+$ given by

$$\Psi_M(f) = \mathbf{E}\exp\{-M(f)\} = \int_\Omega \exp\{-M(\omega, f)\}d\mathbb{P}(\omega)$$
$$= \int_{M_+(\mathbb{E})} \exp\{-\mu(f)\}\mathbb{P}\circ M^{-1}(d\mu).$$

Note that if $P$ is a probability measure on $\mathcal{M}_+(\mathbb{E})$, its Laplace functional is

$$\int_{M_+(\mathbb{E})} \exp\{-\mu(f)\}P(d\mu), \quad f \in \mathcal{B}_+. \tag{5.9}$$

**Proposition 5.4.** *If $M$ is a random measure on $\mathbb{E}$, the Laplace functional $\Psi_M(f)$, $f \in C_K^+(\mathbb{E})$, of $M$ uniquely determines the distribution of $M$.*

*Proof.* The distribution of $M$ is the measure $\mathbb{P}\circ M^{-1}$ on $\mathcal{M}_+(\mathbb{E})$, the Borel $\sigma$-algebra generated by the open subsets of $M_+(\mathbb{E})$. Recall (p. 51) that $\mathcal{G}(M_+(\mathbb{E}))$, the class of open subsets of $M_+(\mathbb{E})$, is generated by the class $\mathcal{C}$ of basis sets, given in (3.13), which has a typical member,

$$\{\mu \in M_+(\mathbb{E}) : \mu(f_i) \in (a_i, b_i), i = 1, \dots, d\} \tag{5.10}$$

for $f_i \in C_K^+(\mathbb{E})$, $i = 1, \dots, d$. The class $\mathcal{C}$ has the property that it is closed under finite intersections and hence is a $\Pi$-system generating the Borel $\sigma$-algebra; so it suffices (see, for example, [24, 264]) by Dynkin's $\pi$-$\lambda$ theorem to show the Laplace functional uniquely determines probabilities on $\mathcal{C}$. Since

$$\mathbb{P}\circ M^{-1}\{\mu \in M_+(\mathbb{E}) : \mu(f_i) \in (a_i, b_i), i = 1, \dots, d\}$$
$$= \mathbb{P}[M(f_i) \in (a_i, b_i), i = 1, \dots, d],$$

it suffices to show $\Psi_M(\cdot)$ determines the joint distribution of

$$(M(f_1), \dots, M(f_d)).$$

This joint distribution is determined by its Laplace transform

$$\mathbf{E}\exp\left\{-\sum_{i=1}^d \lambda_i M(f_i)\right\}, \quad \lambda_i > 0, \quad i = 1, \dots, d.$$

However, this is equal to

$$\mathbf{E}\exp\left\{-M(\sum_{i=1}^d \lambda_i f_i)\right\} = \Psi_M\left(\sum_{i=1}^d \lambda_i f_i\right).$$

Since $\sum_{i=1}^d \lambda_i f_i \in C_K^+(\mathbb{E})$, knowledge of $\{\Psi_M(f); f \in C_K^+(\mathbb{E})\}$ is determining. □

*Example* 5.3. Consider the following easy examples, which will have subsequent importance:

1. For $\mu_0 \in M_+(\mathbb{E})$, define the probability measure $P$ on $\mathcal{M}_+(\mathbb{E})$ by
$$P = \epsilon_{\mu_0}.$$
This probability measure concentrates all mass at one point and corresponds to the random measure which is identically $\mu_0$. According to (5.9), the Laplace functional at $f \in \mathcal{B}_+$ is
$$\int_{M_+(\mathbb{E})} \exp\{-\mu(f)\}P(d\mu) = e^{-\mu_0(f)}.$$
2. *Empirical measure*: Suppose $X_1, \dots, X_n$ are iid random elements in $\mathbb{E}$ and define the random point measure
$$M = \sum_{i=1}^{n} \epsilon_{X_i}.$$
Its Laplace functional is
$$\begin{aligned} \mathbf{E}\exp\{-M(f)\} &= \mathbf{E}e^{-\sum_{i=1}^{d} f(X_i)} = \left(\mathbf{E}e^{-f(X_1)}\right)^n \\ &= \left(1 - \int_{\mathbb{E}} (1 - e^{-f(e)})\mathbb{P}[X_1 \in de]\right)^n. \end{aligned}$$
3. *Poissonized empirical measure*: Suppose $\{X_i, i \geq 1\}$ are iid random elements of $\mathbb{E}$ and $\tau$ is a Poisson random variable with parameter $\lambda$ that is independent of $\{X_i\}$. Define
$$M = \sum_{i=1}^{\tau} \epsilon_{X_i},$$
and on $[\tau = 0]$ we understand $M \equiv 0$. The Laplace functional is obtained by conditioning on $\tau$:
$$\begin{aligned} \mathbf{E}\exp\{-M(f)\} &= \mathbf{E}e^{-\sum_{i=1}^{\tau} f(X_i)} = \mathbf{E}((\mathbf{E}e^{-f(X_1)})^{\tau}) \\ &= \exp\left\{-\int_{\mathbb{E}} (1 - e^{-f(e)})\lambda\mathbb{P}[X_1 \in de]\right\}. \end{aligned} \tag{5.11}$$

### 5.3.2 The Laplace functional of the Poisson process

Recall the definition of the Poisson process given in Definition 5.1 and parts 1 and 2 of the definition given on p. 120.

The next result shows that the Poisson process can be identified by the characteristic form of its Laplace functional.

**Theorem 5.1 (Laplace functional of PRM).** *The distribution of* PRM$(\mu)$ *is uniquely determined by* 1 *and* 2 *in Definition* 5.1. *Furthermore, the point process $N$ is* PRM$(\mu)$ *iff its Laplace functional is of the form*

$$\Psi_N(f) = \exp\left\{-\int_{\mathbb{E}} (1 - e^{-f(x)})\mu(dx)\right\}, \quad f \in \mathcal{B}_+. \tag{5.12}$$

*Proof.* We first show 1 and 2 imply (5.12).

STEP 1. If $f = \lambda 1_A$, where $\lambda > 0$, then because $N(f) = \lambda N(A)$ and $N(A)$ is Poisson with parameter $\mu(A)$, we get

$$\begin{aligned}\Psi_N(f) &= \mathbf{E}e^{-\lambda N(A)} = \exp\{(e^{-\lambda} - 1)\mu(A)\} \\ &= \exp\left\{-\int_{\mathbb{E}} (1 - e^{-f(x)})\mu(dx)\right\},\end{aligned}$$

which is the correct form given in (5.12).

STEP 2. Next, suppose $f$ has a somewhat more complex form

$$f = \sum_{i=1}^{k} \lambda_i 1_{A_i},$$

where $\lambda_i \geq 0$, $A_i \in \mathcal{E}$, $1 \leq i \leq k$, and $A_1, \ldots, A_k$ are disjoint. Then

$$\begin{aligned}\Psi_N(f) &= \mathbf{E}\exp\left\{-\sum_{i=1}^{k} \lambda_i N(A_i)\right\} \\ &= \prod_{i=1}^{k} \mathbf{E}\exp\{-\lambda_i N(A_i)\} && \text{from independence} \\ &= \prod_{i=1}^{k} \exp\left\{-\int_{\mathbb{E}} (1 - e^{-\lambda_i 1_{A_i}(x)})\mu(dx)\right\} && \text{from the previous Step 1} \\ &= \exp\left\{\int_{\mathbb{E}} \sum_{i=1}^{k} (1 - e^{-\lambda_i 1_{A_i}(x)})\mu(dx)\right\} \\ &= \exp\left\{\int_{\mathbb{E}} \left(1 - e^{-\sum_{i=1}^{k} \lambda_i 1_{A_i}(x)}\right)\mu(dx)\right\} \\ &= \exp\left\{\int_{\mathbb{E}} (1 - e^{-f(x)})\mu(dx)\right\},\end{aligned}$$

which again verifies (5.12).

STEP 3. Now the last step is to take general $f \in \mathcal{B}_+$ and verify (5.12) for such $f$. We may approximate $f$ from below by simple $f_n$ of the form just considered in Step 2. We may take, for instance,

$$f_n(x) = \sum_{i=1}^{n2^n} \frac{i-1}{2^n} 1_{[\frac{i-1}{2^n}, \frac{i}{2^n})}(f(x)) + n 1_{[n,\infty)}(f(x))$$

so that

$$0 \le f_n(x) \uparrow f(x).$$

By monotone convergence $N(f_n) \uparrow N(f)$, and since $e^{-f} \le 1$, we get by dominated convergence that

$$\Psi_N(f) = \lim_{n\to\infty} \Psi_N(f_n).$$

We have from the previous step that

$$\Psi_N(f_n) = \exp\left\{-\int_{\mathbb{E}} (1 - e^{-f_n(x)})\mu(dx)\right\}.$$

Since

$$1 - e^{-f_n} \uparrow 1 - e^{-f},$$

we conclude by monotone convergence that

$$\int_{\mathbb{E}} (1 - e^{-f_n(x)})\mu(dx) \uparrow \int_{\mathbb{E}} (1 - e^{-f(x)})\mu(dx),$$

and thus we conclude that (5.12) holds for any $f \in \mathcal{B}_+$. Since the distribution of $N$ is uniquely determined by $\Psi_N$, we have shown that 1 and 2 in Definition 5.1 determine the distribution of $N$.

Conversely, if the Laplace functional of $N$ is given by (5.12), then $N(A)$ must be Poisson distributed with parameter $\mu(A)$ for any $A \in \mathcal{E}$, which is readily checked by substituting $f = \lambda 1_A$ in (5.12) to get a Laplace transform of a Poisson distribution. Furthermore, if $A_1, \dots, A_k$ are disjoint sets in $\mathcal{E}$ and $\lambda_1, \dots, \lambda_k$ are positive and $f = \sum_{i=1}^k \lambda_i 1_{A_i}$, then substituting in (5.12) gives

$$\begin{aligned}
\mathbf{E} e^{-\sum_{i=1}^k \lambda_i N(A_i)} &= \exp\left\{-\int_{\mathbb{E}} \left(1 - e^{-\sum_{i=1}^k \lambda_i 1_{A_i}}\right) d\mu\right\} \\
&= \exp\left\{-\int_{\mathbb{E}} \sum_{i=1}^k (1 - e^{-\lambda_i 1_{A_i}}) d\mu\right\} \\
&= \prod_{i=1}^k \exp\{-(1 - e^{-\lambda_i})\mu(A_i)\}
\end{aligned}$$

$$= \prod_{i=1}^{k} \mathbf{E}e^{-\lambda_i N(A_i)},$$

and so the joint Laplace transform of $(N(A_i), 1 \le i \le k)$ factors into a product of Laplace transforms, which shows independence. □

## 5.4 See the Laplace functional flex its muscles!

This section discusses why the Laplace functional is such a useful theoretical tool.

### 5.4.1 The Laplace functional and weak convergence

We can test for weak convergence of a sequence of random measures in $M_+(\mathbb{E})$ by showing for $f \in C_K^+(\mathbb{E})$ that the Laplace functionals of the random measures converge. To do this, we will rely on the criterion 3 of Section 3.4.1 (p. 54). For more detail, see [260, Section 3.5] or [180, 230].

**Theorem 5.2 (convergence criterion).** *Let $\{\eta_n, n \ge 0\}$ be random elements of $M_+(\mathbb{E})$. Then*

$$\eta_n \Rightarrow \eta_0 \quad \textit{in } M_p(\mathbb{E}),$$

*iff*

$$\Psi_{\eta_n}(f) = \mathbf{E}e^{-\eta_n(f)} \to \mathbf{E}e^{-\eta_0(f)} = \Psi_{\eta_0}(f) \quad \forall f \in C_K^+(\mathbb{E}). \tag{5.13}$$

So weak convergence is characterized by convergence of Laplace functionals on $C_K^+(\mathbb{E})$.

*Proof.* Suppose $\eta_n \Rightarrow \eta_0$ in $M_+(\mathbb{E})$. The map $M_+(\mathbb{E}) \mapsto [0, \infty)$ defined by $\mu \mapsto \mu(f)$ is continuous, so the continuous mapping theorem gives $\eta_n(f) \Rightarrow \eta_0(f)$ in $\mathbb{R}$. Thus

$$e^{-\eta_n(f)} \Rightarrow e^{-\eta_0(f)},$$

and by Lebesgue's dominated convergence theorem,

$$\mathbf{E}e^{-\eta_n(f)} \to \mathbf{E}e^{-\eta_0(f)},$$

as required. This was straightforward.

Conversely, suppose (5.13) holds. According to criterion 3 of Section 3.4.1 (p. 54), we have to prove for any family $\{h_j\} \subset C_K^+(\mathbb{E})$ that

$$(\eta_n(h_j), j \ge 1) \Rightarrow (\eta_0(h_j), j \ge 1)$$

in $\mathbb{R}^\infty$, and for this it suffices to prove for any integer $d$ that

$$(\eta_n(h_j), 1 \leq j \leq d) \Rightarrow (\eta_0(h_j), 1 \leq j \leq d),$$

in $\mathbb{R}^d$. So it suffices to show multivariate Laplace transforms converge. Set $\lambda_i > 0$ for $i = 1, \ldots, d$, and we have

$$\mathbf{E}\exp\left\{-\sum_{i=1}^{d}\lambda_i\eta_n(h_i)\right\} = \mathbf{E}\exp\left\{-\eta_n\left(\sum_{i=1}^{d}\lambda_i h_i\right)\right\}$$

and because $\sum_{i=1}^{d}\lambda_i h_j \in C_K^+(\mathbb{E})$ and (5.13) holds, we get

$$\begin{aligned}\to \mathbf{E}\exp\left\{-\eta_0\left(\sum_{i=1}^{d}\lambda_i h_i\right)\right\}\\ = \mathbf{E}\exp\left\{-\sum_{i=1}^{d}\lambda_i\eta_0(h_i)\right\},\end{aligned}$$

as required. □

### Convergence of empirical measures

We will give extensive applications of Theorem 5.2 in the next chapter. For now, we indicate why the criterion provides a natural way to consider weak convergence of empirical measures.

We now give two convergence results. One gives necessary and sufficient conditions for empirical measures to converge to a Poisson random measure limit, and the other discusses convergence to a constant limit measure. The first is the basis for manipulating iid random elements with regularly varying tails by means of the Poisson transform, and the second is the basis for consistency of estimates of heavy-tailed parameters, which has already been considered in Theorem 4.1 (p. 79).

**Theorem 5.3 (basic convergence).** *Suppose that for each $n \geq 1$, we have that $\{X_{n,j}, j \geq 1\}$ is a sequence of iid random elements of $(\mathbb{E}, \mathcal{E})$. Let $\xi$ be* PRM$(\mu)$ *on $M_p(\mathbb{E})$, that is, the Poisson random measure with mean measure $\mu$.*

(i) *We have*

$$\sum_{j=1}^{n}\epsilon_{X_{n,j}} \Rightarrow \xi = \text{PRM}(\mu) \tag{5.14}$$

*on $M_p(\mathbb{E})$ iff*

$$n\mathbb{P}[X_{n,1} \in \cdot] = \mathbf{E}\left(\sum_{j=1}^{n} \epsilon_{X_{n,j}}(\cdot)\right) \xrightarrow{v} \mu \tag{5.15}$$

*in* $M_+(\mathbb{E})$.

(ii) *Suppose additionally that* $0 < a_n \uparrow \infty$. *Then for a measure* $\mu \in M_+(\mathbb{E})$, *we have*

$$\frac{1}{a_n}\sum_{j=1}^{n} \epsilon_{X_{n,j}} \Rightarrow \mu \tag{5.16}$$

*on* $M_+(\mathbb{E})$ *iff*

$$\frac{n}{a_n}P[X_{n,1} \in \cdot] = \mathbf{E}\left(\frac{1}{a_n}\sum_{j=1}^{n} \epsilon_{X_{n,j}}(\cdot)\right) \xrightarrow{v} \mu \tag{5.17}$$

*in* $M_+(\mathbb{E})$.

*Remark* 5.1. Note that the mean measure of $\sum_{j=1}^{n} \epsilon_{X_{n,j}}$ is $n\mathbb{P}[X_{n,1} \in \cdot]$, and likewise, the mean measure of $\frac{1}{a_n}\sum_{j=1}^{n} \epsilon_{X_{n,j}}$ is $\frac{n}{a_n}P[X_{n,1} \in \cdot]$.

*Proof.*

(i) We compute Laplace functionals of the empirical measures and decide when they converge. As in part 2 of Example 5.3 (p. 134), for $f \in C_K^+(\mathbb{E})$,

$$\begin{aligned}
\mathbf{E}e^{-\sum_{j=1}^{n}\epsilon_{X_{n,j}}(f)} &= \mathbf{E}e^{-\sum_{j=1}^{n} f(X_{n,j})} = (\mathbf{E}e^{-f(X_{n,1})})^n \\
&= \left(1 - \frac{\mathbf{E}(n(1-e^{-f(X_{n,1})}))}{n}\right)^n \\
&= \left(1 - \frac{\int_{\mathbb{E}}(1-e^{-f(x)})nP[X_{n,1} \in dx]}{n}\right)^n,
\end{aligned}$$

and this converges to

$$\exp\left\{\int_{\mathbb{E}}(1-e^{-f(x)})\mu(dx)\right\},$$

the Laplace functional of PRM$(\mu)$, iff

$$\int_{\mathbb{E}}(1-e^{-f(x)})nP[X_{n,1} \in dx] \to \int_{\mathbb{E}}(1-e^{-f(x)})\mu(dx).$$

This last statement is equivalent to vague convergence in (5.15). (See Problem 3.1 (p. 64.))

(ii) Here again we prove the result by showing that Laplace functionals converge. We compute the Laplace functional for the quantity on the left side of (5.16):

$$\mathbf{E}e^{-\frac{1}{a_n}\sum_{i=1}^n \epsilon_{X_{n,1}}(f)} = \left(\mathbf{E}e^{-\frac{1}{a_n}f(X_{n,1})}\right)^n$$
$$= \left(1 - \frac{\int_{\mathbb{E}}\left(1-e^{-\frac{1}{a_n}f(x)}\right)n\mathbb{P}[X_{n,1}\in dx]}{n}\right)^n,$$

and we claim that this converges to $e^{-\mu(f)}$, the Laplace functional of $\mu$, iff

$$\int_{\mathbb{E}}\left(1-e^{-\frac{1}{a_n}f(x)}\right)nP[X_{n,1}\in dx] \to \mu(f). \tag{5.18}$$

We show that (5.18) is equivalent to (5.17) as follows: Suppose (5.17) holds. On one hand,

$$\int_{\mathbb{E}}(1-e^{-f(x)/a_n})n\mathbb{P}[X_{n,1}\in dx] \le \int_{\mathbb{E}} f(x)\frac{n}{a_n}P[X_{n,1}\in dx] \to \mu(f),$$

so

$$\limsup_{n\to\infty}\int_{\mathbb{E}}(1-e^{-f(x)/a_n})nP[X_{n,1}\in dx] \le \mu(f).$$

On the other hand,

$$\int_{\mathbb{E}}(1-e^{-f(x)/a_n})nP[X_{n,1}\in dx]$$
$$\ge \int_{\mathbb{E}} f(x)\frac{n}{a_n}P[X_{n,1}\in dx] - \int_{\mathbb{E}}\frac{f^2(x)}{2a_n}\frac{n}{a_n}P[X_{n,1}\in dx]$$
$$= \mathrm{I} + \mathrm{II}.$$

Now $\mathrm{I} \to \mu(f)$ from (5.17), and since $f^2 \in C_K^+(\mathbb{E})$, we have

$$\mathrm{II} \sim \frac{\mu(f^2)}{2a_n} \to 0$$

since $a_n \uparrow \infty$. So

$$\liminf_{n\to\infty}\int_{\mathbb{E}}(1-e^{-f(x)/a_n})nP[X_{n,1}\in dx] \ge \mu(f),$$

providing the other half of the sandwich.

Conversely, let $f \in C_K^+(\mathbb{E})$, and suppose that $f \le 1$. Assuming that (5.18) is true, we get

$$f/a_n \ge 1 - e^{-f/a_n},$$

leading to

$$\liminf_{n\to\infty} \int_{\mathbb{E}} f(x) \frac{n}{a_n} P[X_{n,1} \in dx] \geq \mu(f)$$

and

$$\frac{f}{a_n} - \frac{f^2}{2a_n^2} \leq 1 - e^{-f/a_n},$$

so

$$\limsup_{n\to\infty} \int_{\mathbb{E}} \left( \frac{f(x)}{a_n} - \frac{f^2(x)}{2a_n^2} \right) n\mathbb{P}[X_{n,1} \in dx] \leq \mu(f).$$

As before, we may show that

$$\int_{\mathbb{E}} \frac{f^2(x)}{2a_n^2} nP[X_{n,1} \in dx] \to 0. \qquad \square$$

### Preservation of weak convergence under mappings of the state space

Consider two *nice* state spaces $\mathbb{E}_1$ and $\mathbb{E}_2$ with a mapping $T : \mathbb{E}_1 \mapsto \mathbb{E}_2$ from one into the other. A measure $\mu$ on $(\mathbb{E}_1, \mathcal{E}_1)$ has an image $\hat{T}(\mu)$ on $(\mathbb{E}_2, \mathcal{E}_2)$ given by the map

$$\hat{T}(\mu) = \mu \circ T^{-1}.$$

If $T$ is a continuous point transformation, is $\hat{T} : M_+(\mathbb{E}_1) \mapsto M_+(\mathbb{E}_2)$ continuous? Note that if $m \in M_p(\mathbb{E})$ is a point measure of the form $\sum_i \epsilon_{x_i}$, then

$$\hat{T}(m) = m \circ T^{-1} = \sum_i \epsilon_{T(x_i)}.$$

Continuity of $T$ does not guarantee continuity of $\hat{T}$ without a condition. We call condition (5.19) the *compactness condition.*

**Proposition 5.5.** *Suppose $T : \mathbb{E}_1 \mapsto \mathbb{E}_2$ is a continuous function such that*

$$T^{-1}(K_2) \in \mathcal{K}(\mathbb{E}_1) \quad \forall K_2 \in \mathcal{K}(\mathbb{E}_2). \tag{5.19}$$

(a) *If $\mu_n \xrightarrow{v} \mu_0$ in $M_+(\mathbb{E}_1)$, then*

$$\hat{T}(\mu_n) = \mu_n \circ T^{-1} \xrightarrow{v} \mu_n \circ T^{-1} = \hat{T}(\mu_0) \tag{5.20}$$

*in $M_+(\mathbb{E}_2)$.*

(b) *Furthermore, if $T$ is continuous and (5.19) holds, if $\{\eta_n(\cdot), n \geq 0\}$ is a family of random measures in $M_+(\mathbb{E}_1)$ such that*

$$\eta_n \Rightarrow \eta_0,$$

*then*

$$\hat{T}(\eta_n) \Rightarrow \hat{T}(\eta_0) \tag{5.21}$$

*in $M_+(\mathbb{E}_2)$.*

*Remark* 5.2. If $T$ is continuous and $\mathbb{E}_1$ is compact, then (5.19) is automatically satisfied. In cases where $\mathbb{E}_1$ is not compact, a commonly employed strategy for constructing proofs is to truncate $\mathbb{E}_1$ to a compact set, apply Proposition 5.5 to the convergence restricted to the compact set, and then use a Slutsky argument to remove the truncation level.

*Proof.*

(a) Suppose $\mu_n \overset{v}{\to} \mu_0$. Let $f_2 \in C_K^+(\mathbb{E}_2)$. We must show that

$$\mu_n \circ T^{-1}(f_2) \to \mu_0 \circ T^{-1}(f_2). \tag{5.22}$$

Unpack the notation:

$$\mu_n \circ T^{-1}(f_2) = \int_{\mathbb{E}_2} f_2(e_2)\mu_n \circ T^{-1}(de_2)$$

and using the change of variable formula or transformation theorem for integrals [264, p. 135], this is

$$= \int_{\mathbb{E}_1} f_2(T(e_1))\mu_n(de_1).$$

What remains is to show that $f_2 \circ T \in C_K^+(\mathbb{E}_1)$. Now $f_2$ and $T$ are both continuous, so $f_2 \circ T$ is continuous. Since $f_2 \in C_K^+(\mathbb{E}_2)$, there exists $K_2 \in \mathcal{K}(\mathbb{E}_2)$ such that $f_2(e_2) = 0$ if $e_2 \notin K_2$. So

$$f(T(e_1)) = 0 \quad \text{if } T(e_1) \notin K_2,$$

that is,

$$f(T(e_1)) = 0 \quad \text{if } e_1 \notin T^{-1}(K_2).$$

From the hypothesis (5.19), $T^{-1}(K_2) \in \mathcal{K}(\mathbb{E}_1)$. So this says that $f_2 \circ T$ is null off a compact set. Thus $f_2 \circ T \in C_K^+(\mathbb{E}_1)$, and since $\mu_n \overset{v}{\to} \mu_0$ in $M_+(\mathbb{E}_1)$, we have

$$\int_{\mathbb{E}_1} f_2(T(e_1))\mu_n(de_1) \to \int_{\mathbb{E}_1} f_2(T(e_1))\mu_0(de_1),$$

which gives (5.22).

(b) For $f_2 \in C_K^+(\mathbb{E}_2)$, it is enough to show that Laplace functionals converge:

$$\mathbf{E}(e^{-\eta_n \circ T^{-1}(f_2)}) \to \mathbf{E}(e^{-\eta_0 \circ T^{-1}(f_2)}). \tag{5.23}$$

Again, we unpack the notation

$$\mathbf{E}(e^{-\eta_n \circ T^{-1}(f_2)}) = \mathbf{E}(e^{-\eta_n(f_2 \circ T)}),$$

and as in (a), the fact that $f_2 \circ T \in C_K^+(\mathbb{E}_1)$ and $\eta_n \Rightarrow \eta_0$ imply convergence on $C_K^+(\mathbb{E}_1)$ of the Laplace functionals. Thus

$$\mathbf{E}(e^{-\eta_n(f_2 \circ T)}) \to \mathbf{E}(e^{-N_0(f_2 \circ T)}) = \mathbf{E}(e^{-N_0 \circ T^{-1}(f_2)}),$$

which is (5.23). □

### 5.4.2 A general construction of the Poisson process

Here is a general scheme for constructing a Poisson process with given mean measure $\mu$.

Start by supposing that $\mu(\mathbb{E}) < \infty$. Define the probability measure $F$,

$$F(dx) = \mu(dx)/\mu(\mathbb{E}),$$

on $\mathcal{E}$. Let $\{X_n, n \geq 1\}$ be iid random elements of $\mathbb{E}$ with common distribution $F$, and let $\tau$ be independent of $\{X_n\}$ with a Poisson distribution with parameter $\mu(\mathbb{E})$. Define

$$N = \begin{cases} \sum_{i=1}^{\tau} \epsilon_{X_i} & \text{if } \tau \geq 1, \\ 0 & \text{if } \tau = 0. \end{cases}$$

Then $N$ is PRM$(\mu)$ since its Laplace functional is given in (5.11) with $\lambda = \mu(\mathbb{E})$, which is of the correct form (5.12).

When the condition $\mu(\mathbb{E}) < \infty$ fails, we make a minor modification in the foregoing construction: Decompose $\mathbb{E}$ into disjoint sets $\mathbb{E}_1, \mathbb{E}_2, \dots$ so that $\mathbb{E} = \cup_i \mathbb{E}_i$, where each $\mathbb{E}_i$ satisfies $\mu(\mathbb{E}_i) < \infty$ for each $i$. Let $\mu_i(dx) = \mu(dx)1_{\mathbb{E}_i}(x)$, let $N_i$ be PRM$(\mu_i)$ on $\mathbb{E}$ (do the construction just outlined), and arrange things so the collection $\{N_i\}$ is independent. Define $N := \sum_i N_i$. $N$ is PRM$(\mu)$ since

$$\begin{aligned} \Psi_N(f) &= \prod_i \Psi_{N_i}(f) \\ &= \prod_i \exp\left\{-\int_{\mathbb{E}_i} (1 - e^{-f(x)})\mu_i(dx)\right\} \end{aligned}$$

$$
\begin{aligned}
&= \exp\left\{-\sum_i \int_{\mathbb{E}} (1-e^{-f(x)})\mu_i(dx)\right\}\\
&= \exp\left\{-\int_{\mathbb{E}} (1-e^{-f(x)})\sum_i \mu_i(dx)\right\}\\
&= \exp\left\{-\int_{\mathbb{E}} (1-e^{-f(x)})\mu(dx)\right\}
\end{aligned}
$$

since $\sum_i \mu_i = \mu$. This completes the construction.

### 5.4.3 Augmentation, location-dependent marking

Given a Poisson process, recall that one can enlarge the dimension of the points by appending independent marks and that this retains the Poisson structure in a product space. In fact, the marks need not be independent of the Poisson points.

**Proposition 5.6.** *Suppose* $\sum_n \epsilon_{X_n}$ *is* PRM$(\mu)$ *on* $M_p(\mathbb{E}_1)$. *Suppose we have a second nice space* $(\mathbb{E}_2, \mathcal{E}_2)$ *and* $K : \mathbb{E}_1 \times \mathcal{E}_2 \mapsto [0,1]$ *is a transition function. This means that* $K(\cdot, A_2)$ *is a measurable function of the first variable for every fixed* $A_2 \in \mathcal{E}_2$*, and for every* $x \in \mathbb{E}_1$*, we have* $K(x,\cdot)$ *is a probability measure on* $\mathcal{E}_2$. *Let* $\{J_i\}$ *be random elements of* $\mathbb{E}_2$ *that are conditionally independent given* $\{X_n\}$*; that is,*

$$\mathbb{P}[J_i \in A_2 | X_i, \{X_j, j \neq i\}, \{J_j, j \neq i\}] = K(X_i, A_2), \tag{5.24}$$

*so that only* $X_i$ *is relevant in the conditioning. Then the point process on* $\mathbb{E}_1 \times \mathbb{E}_2$*,*

$$\sum_n \epsilon_{(X_n, J_n)},$$

*is* PRM *with mean measure*

$$\mu_1(dx, dy) = \mu(dx)K(x, dy).$$

So it is not necessary for $\{J_n\}$ to be independent of $\{X_n\}$. Conditional independence will do. If the distribution of $J_n$ depends on the $\{X_i\}$, it must do so only through $X_n$ and not the other $X$s.

*Proof.* We begin by first proving Proposition 5.3, which is the case in which the marks $\{J_n\}$ are independent of the points so that $K(x,\cdot) = F(\cdot)$, $F$ being the distribution of the $J$s. Assume initially that $\mu$ is finite. From the construction in Section 5.4.2, we may, without loss of generality, assume that the PRM$(\mu)$ is of the form $\sum_{i=1}^{\tau} \epsilon_{Y_i}$, where $\tau$, $\{Y_n\}$, and $\{J_n\}$ are independent, $\tau$ is a Poisson random variable with parameter

$\mu(\mathbb{E}_1)$, and $\{Y_n\}$ are iid with common distribution $\mu(dx)/\mu(\mathbb{E}_1)$. Then it follows that $\{(Y_n, J_n)\}$ is iid in $\mathbb{E}_1 \times \mathbb{E}_2$ with common distribution

$$\frac{\mu(\cdot)}{\mu(\mathbb{E}_1)} \times F.$$

So from the construction of Section 5.4.2,

$$\sum_n \epsilon_{(X_n, J_n)} \stackrel{d}{=} \sum_{i=1}^{\tau} \epsilon_{(Y_i, J_i)}$$

is PRM with mean measure $\mu(\mathbb{E}_1)\mu(\cdot)/\mu(\mathbb{E}_1) \times F = \mu \times F(\cdot)$, as required.

If $\mu$ is not finite, then as in the construction of Section 5.4.2, we patch things together by repeating the argument of the previous paragraph on partition sets of $\mathbb{E}_1$ where $\mu$ is finite. We need an at most countable number of such partition sets that are disjoint and exhaust $\mathbb{E}_1$.

Now for the proof of general case stated in Proposition 5.6. Write

$$K(x, A_2) = \mathbb{P}[J_1 \in A_2 | X_1 = x]$$

for the conditional distribution of $J_1$. It is always possible to realize a distribution as a function of a uniform random variable (see, for example, [23]). That is, there exists a function, say, $g(x, u)$, such that

$$K(x, A_2) = \mathbb{P}[g(x, U_1) \in A_2],$$

where we suppose that $\{U_n\}$ are iid $U(0, 1)$ random variables, independent of $\{X_n\}$.

The impact of this transformation is that

$$\{(X_n, J_n)\} \stackrel{d}{=} \{X_n, g(X_n, U_n)\}.$$

We know from the proof of Proposition 5.3 that

$$\sum_n \epsilon_{(X_n, U_n)}$$

is PRM with mean measure $\mu(dx) \times \mathbb{LEB}(dy)1_{[0,1]}(y)$. Therefore, from Proposition 5.2, we get that

$$\sum_n \epsilon_{(X_n, J_n)} \stackrel{d}{=} \sum_n \epsilon_{(X_n, g(X_n, U_n))}$$

is PRM$(\mu_1)$. To compute $\mu_1$, define $T : \mathbb{E}_1 \times [0, 1] \mapsto \mathbb{E}_1 \times \mathbb{E}_2$ via $T(x, u) = (x, g(x, u))$. Then the mean measure $\mu_1$ is $(A_1 \in \mathcal{E}_1, A_2 \in \mathcal{E}_2)$

$$\begin{aligned}\mu_1(A_1 \times A_2) &= (\mu \times \mathbb{LEB}) \circ T^{-1}(A_1 \times A_2)\\ &= \int_{[x \in A_1]} \mu(dx)\, \mathbb{LEB}\{u \in [0,1] : g(x,u) \in A_2\}\\ &= \int_{[x \in A_1]} \mu(dx) P[g(x, U_1) \in A_2]\\ &= \int_{A_1} K(x, A_2)\mu(dx). \qquad \square\end{aligned}$$

There are alternative proofs using Laplace functionals and induction. See [260, p. 135].

## 5.5 Lévy processes

Poisson processes serve as the building blocks for many heavy-tailed models, and pride of place goes to the Itô construction of Lévy processes discussed in this section. The next section will discuss extremal processes. Crudely speaking, Lévy processes can be considered as summation functionals applied to PRMs, while extremal processes can be considered as maximal functionals applied to PRMs.

I learned much about the Itô construction from [172–174]. Other fine references include [4, 19, 273, 274].

### 5.5.1 Itô's construction of Lévy processes

We work in the space $\mathbb{E} = \mathbb{R}^d \setminus \{\mathbf{0}\}$ with generic element $\boldsymbol{u} = (u^{(1)}, \ldots, u^{(d)})$ and use the Euclidean metric

$$\|\boldsymbol{u}\| = \sqrt{\sum_{i=1}^{d} (u^{(i)})^2}, \quad \boldsymbol{u} \in \mathbb{E}.$$

#### Lévy measure

We begin by assuming that $\nu$ is a measure on $\mathbb{E}$ satisfying the following:

(i) For every $x > 0$,

$$\nu\{\boldsymbol{u} \in \mathbb{E} : \|\boldsymbol{u}\| > x\} < \infty. \tag{5.25}$$

(ii) $\int_{0<\|\boldsymbol{x}\|\le 1} \|\boldsymbol{x}\|^2 \nu(d\boldsymbol{x}) < \infty$.

In fact, combining the two properties allows recasting of (ii) as

$$\int_{0<\|\boldsymbol{x}\|\le c} \|\boldsymbol{x}\|^2 \nu(d\boldsymbol{x}) < \infty \quad \forall c \in (0,\infty). \tag{5.26}$$

Measures on $\mathbb{E}$ satisfying (i) and (ii) are called *Lévy measures*. When $d = 1$, if $\nu(-\infty, 0) = 0$, the resulting process that we will construct will be called *totally skewed to the right*, while if $\nu(0, \infty) = 0$, the resulting process is called *totally skewed to the left*.

Let $N$ be PRM($\mathbb{LEB} \times \nu$) on $[0, \infty) \times \mathbb{E}$. Represent $N$ as

$$N = \sum_k \epsilon_{(t_k, \boldsymbol{j}_k)}.$$

In the simplest cases, we can define a Lévy process

$$\boldsymbol{X}(t) = (X^{(1)}(t), \dots, X^{(d)}(t))$$

by

$$\boldsymbol{X}(t) := \sum_{t_k \le t} \boldsymbol{j}_k, \quad t \ge 0,$$

but in general, we have to be careful to first center the summands to zero expectation in order for the infinite sum to converge.

### Compound Poisson representations

Fix $t$ and let $I \subset \mathbb{E}$ be a set bounded away from $\mathbf{0}$. Define

$$\boldsymbol{S}_I(t) = \sum_{\substack{t_k \le t \\ \boldsymbol{j}_k \in I}} \boldsymbol{j}_k = \iint_{[0,t] \times I} \boldsymbol{u} N(ds, d\boldsymbol{u}).$$

From the construction of a PRM with finite mean measure as a sprinkling of a Poisson number of iid random elements into the space $\mathbb{E}$, we have the restriction of $N$ to $[0, t] \times I$, representable as

$$N\Big|_{[0,t]\times I} \stackrel{d}{=} \sum_1^{\tau} \epsilon_{(T_k, \boldsymbol{J}_k)}, \tag{5.27}$$

where

(i) $\{(T_k, \boldsymbol{J}_k), k \ge 1\}$ are iid pairs,
(ii) $T_k$ and $\boldsymbol{J}_k$ are independent of each other,
(iii) $T_k$ is uniformly distributed on $(0, t)$, and $\boldsymbol{J}_k$ has distribution $\frac{\nu}{\nu(I)}$ restricted to $I$.
(iv) $\tau$ is independent of $\{(T_k, \boldsymbol{J}_k)\}$, and $\tau$ is a Poisson random variable with parameter $t\nu(I)$).

It follows from representation (5.27) that

$$\boldsymbol{S}_I(t) \stackrel{d}{=} \sum_1^{\tau} \boldsymbol{J}_k \tag{5.28}$$

is a compound Poisson random vector and

$$\mathbf{E}(\boldsymbol{S}_I(t)) = \mathbf{E}(\tau)\mathbf{E}(\boldsymbol{J}_1) = t\nu(I)\int_I \boldsymbol{x}\frac{\nu(d\boldsymbol{x})}{\nu(I)},$$

that is,

$$\mathbf{E}(\boldsymbol{S}_I(t)) = t\int_I \boldsymbol{x}\nu(d\boldsymbol{x}), \tag{5.29}$$

which is finite. Furthermore, the characteristic function can be computed: For $\boldsymbol{\zeta} \in \mathbb{R}^d$ and $\boldsymbol{x} \in \mathbb{R}^d$, write $\boldsymbol{\zeta} \cdot \boldsymbol{x} = \sum_{i=1}^d \zeta^{(i)} x^{(i)}$. Then we have from (5.28),

$$\begin{aligned}
\mathbf{E}e^{i\boldsymbol{\zeta}\cdot\boldsymbol{S}_I(t)} &= \mathbf{E}e^{i\sum_{k=1}^{\tau}\boldsymbol{\zeta}\cdot\boldsymbol{J}_k} = \sum_{j=0}^{\infty}(\mathbf{E}e^{i\boldsymbol{\zeta}\cdot\boldsymbol{J}_1})^j P(\tau = j) \\
&= \exp\{t\nu(I)(\mathbf{E}e^{i\boldsymbol{\zeta}\cdot\boldsymbol{J}_1} - 1)\} \\
&= \exp\{t\nu(I)\int_I (e^{i\boldsymbol{\zeta}\cdot\boldsymbol{x}} - 1)\nu(d\boldsymbol{x})/\nu(I)\},
\end{aligned}$$

and so we get

$$\mathbf{E}e^{i\boldsymbol{\zeta}\cdot\boldsymbol{S}_I(t)} = \exp\left\{t\int_I (e^{i\boldsymbol{\zeta}\cdot\boldsymbol{x}} - 1)\nu(d\boldsymbol{x})\right\}. \tag{5.30}$$

### Variance calculations

Suppose, in addition to being bounded away from $\mathbf{0}$, $I$ also satisfies

$$I \subset \{\boldsymbol{x} : \|\boldsymbol{x}\| \le c\}$$

for some $c > 0$. Recalling (5.26) and again using (5.28), we have, for $l = 1, 2, \ldots, d$,

$$\begin{aligned}
\mathrm{Var}(S_I^{(l)}(t)) &= \mathrm{Var}\left(\sum_{i=1}^{\tau} J_i^{(l)}\right) = \mathbf{E}(\tau)\mathbf{E}(J_1^{(l)})^2 \qquad (5.31)\\
&= t\nu(I)\int_I (x^{(l)})^2\nu(d\boldsymbol{x})/\nu(I) \\
&= t\int_I (x^{(l)})^2\nu(d\boldsymbol{x}), \qquad (5.32)
\end{aligned}$$

which results from a standard fact about compound Poisson random variables. The calculation used to verify (5.31) is reviewed in the next lemma, which can be skipped by the impatient or the knowledgeable. Note that

$$\int_I (x^{(l)})^2 \nu(d\boldsymbol{x}) \leq \int_I \|\boldsymbol{x}\|^2 \nu(d\boldsymbol{x}) \leq \int_{\{\boldsymbol{x}:\|\boldsymbol{x}\|\leq c\}} \|\boldsymbol{x}\|^2 \nu(d\boldsymbol{x}) < \infty \tag{5.33}$$

from assumption (5.26).

**Lemma 5.1.** *Suppose $\tau$ is a Poisson distributed random variable with parameter $\lambda$ that is independent of the iid random variables $\{J_k, k \geq 1\}$. Then*

$$\operatorname{Var}\left(\sum_{k=1}^{\tau} J_k\right) = \lambda \mathbf{E}(J_1^2).$$

*Proof.* This is a standard calculation using the formula

$$\operatorname{Var}\left(\sum_{k=1}^{\tau} J_k\right) = \operatorname{Var}\left(\mathbf{E}\left(\sum_{i=1}^{\tau} J_k|\tau\right)\right) + \mathbf{E}\left(\operatorname{Var}\left(\sum_{i=1}^{\tau} J_k|\tau\right)\right)$$

and the fact that for a Poisson distributed $\tau$, the mean and variance are the same. Alternatively, one can compute $\mathbf{E}(\sum_{k=1}^{\tau} J_k)^2$ by opening the square into a double sum over $i, j$, separating the double sum into terms where $i = j$ and $i \neq j$ and then condition on $\tau$. □

### Process definition

Suppose we have a sequence $\varepsilon_n \downarrow 0$ such that $1 = \varepsilon_0 > \varepsilon_1 > \varepsilon_2 > \dots$. Define

$$I_{j+1} := \{\boldsymbol{x} \in \mathbb{E} : \varepsilon_{j+1} < \|\boldsymbol{x}\| \leq \varepsilon_j\}, \quad j = 0, 1, 2, \dots,$$

and the stochastic process

$$\begin{aligned} \boldsymbol{X}_{j+1}(t) &:= \boldsymbol{S}_{I_{j+1}}(t) - \mathbf{E}(\boldsymbol{S}_{I_{j+1}}(t)) \\ &= \iint_{\substack{0\leq s\leq t \\ \boldsymbol{u}\in I_{j+1}}} \boldsymbol{u} N(ds, d\boldsymbol{u}) - t \int_{\boldsymbol{u}\in I_{j+1}} \boldsymbol{u}\nu(d\boldsymbol{u}). \end{aligned} \tag{5.34}$$

Note from (5.32) that for $l = 1, 2, \dots, d$,

$$\operatorname{Var}(X_{j+1}^{(l)}(t)) = t \int_{I_{j+1}} (u^{(l)})^2 \nu(d\boldsymbol{u}).$$

Also, for $l = 1, \ldots, d$,

$$\sum_{j=0}^{\infty} \mathrm{Var}(X_{j+1}^{(l)}(t)) \leq t \int_{\{u:\|u\|\leq 1\}} \|u\|^2 \nu(du) < \infty$$

from (5.33). Recall the Kolmogorov convergence criterion [24], [264, Section 7.3] for sums of independent random variables: If $\{\eta_k\}$ are independent random variables such that $\sum_k \mathrm{Var}(\eta_k) < \infty$, then $\sum_k (\eta_k - \mathbf{E}(\eta_k))$ converges almost surely.

Finally, set

$$X_0(t) = \sum_{t_k \leq t} j_k 1[\|j_k\| > 1] = \iint_{[0,t]\times\{x:\|x\|>1\}} u N(ds, du).$$

Note this is a finite sum since

$$\mathbf{E}(N([0,t] \times \{x : \|x\| > 1\})) = t\nu\{x : \|x\| > 1\} < \infty,$$

and hence $N([0,t] \times \{x : \|x\| > 1\}) < \infty$ almost surely.

We define

$$\begin{aligned} X(t) &= X_0(t) + \sum_{j=0}^{\infty} X_{j+1}(t) && (5.35) \\ &= \iint_{\substack{s\leq t \\ \|u\|>1}} u N(ds, du) + \sum_{j=0}^{\infty} \left[ \iint_{\substack{s\leq t \\ u\in I_{j+1}}} u N(ds, du) - \iint_{\substack{s\leq t \\ u\in I_{j+1}}} u ds \nu(du) \right] \end{aligned}$$

and call $(X(t), t \geq 0)$ a *Lévy process* with Lévy measure $\nu$. Note the series converges because of the Kolmogorov convergence criterion.

The equivalent representation,

$$\begin{aligned} X(t) = &\iint_{\substack{s\leq t \\ \|u\|>1}} u N(ds, du) \\ &+ \lim_{\epsilon\downarrow 0} \left[ \iint_{\substack{s\leq t \\ \|u\|\in(\epsilon,1]}} u N(ds, du) - \iint_{\substack{s\leq t \\ \|u\|\in(\epsilon,1]}} u ds \nu(du) \right], && (5.36) \end{aligned}$$

is sometimes called the *Itô representation* of the Lévy process.

### 5.5.2 Basic properties of Lévy processes

In this section, we survey some basic properties of Lévy processes that come from the construction in a fairly straightforward manner.

### The characteristic function of $X(t)$

To compute the characteristic function of $\boldsymbol{X}(t)$, note that all the summands in (5.35) are independent, and therefore for $\boldsymbol{\zeta} \in \mathbb{R}^d$,

$$\mathbf{E}e^{i\boldsymbol{\zeta}\cdot\boldsymbol{X}(t)} = \mathbf{E}e^{i\boldsymbol{\zeta}\cdot\boldsymbol{X}_0(t)} \prod_{j=0}^{\infty} \mathbf{E}e^{i\boldsymbol{\zeta}\cdot\boldsymbol{X}_{j+1}(t)};$$

using (5.30), we get

$$\begin{aligned}
&= \exp\left\{t \int_{\|\boldsymbol{x}\|>1} (e^{i\boldsymbol{\zeta}\cdot\boldsymbol{x}} - 1)\nu(d\boldsymbol{x})\right\} \\
&\quad \times \prod_{j=0}^{\infty} \exp\left\{t \int_{I_{j+1}} (e^{i\boldsymbol{\zeta}\cdot\boldsymbol{x}} - 1)\nu(d\boldsymbol{x}) - i \int_{I_{j+1}} \boldsymbol{x}\nu(d\boldsymbol{x})\right\} \\
&= \exp\left\{t \int_{\|\boldsymbol{x}\|>1} (e^{i\boldsymbol{\zeta}\cdot\boldsymbol{x}} - 1)\nu(d\boldsymbol{x}) + t \sum_{j=0}^{\infty} \int_{I_{j+1}} (e^{i\boldsymbol{\zeta}\cdot\boldsymbol{x}} - 1 - i\boldsymbol{\zeta}\cdot\boldsymbol{x})\nu(d\boldsymbol{x})\right\} \\
&= (\phi(\boldsymbol{\zeta}))^t,
\end{aligned}$$

where

$$\begin{aligned}
\phi(\boldsymbol{\zeta}) &= \mathbf{E}e^{i\boldsymbol{\zeta}\cdot\boldsymbol{X}(1)} \\
&= \exp\left\{\int_{\|\boldsymbol{x}\|>1} (e^{i\boldsymbol{\zeta}\cdot\boldsymbol{x}} - 1)\nu(d\boldsymbol{x}) + \int_{\|\boldsymbol{x}\|\in(0,1]} (e^{i\boldsymbol{\zeta}\cdot\boldsymbol{x}} - 1 - i\boldsymbol{\zeta}\cdot\boldsymbol{x})\nu(d\boldsymbol{x})\right\}. \quad (5.37)
\end{aligned}$$

### Independent increment property of $X(t)$

The process $\{\boldsymbol{X}(t), t \geq 0\}$ has independent increments that means that for $0 < s \leq t$, $\boldsymbol{X}(t) - \boldsymbol{X}(s)$ is independent of the $\sigma$-algebra generated by $\{\boldsymbol{X}(v), v \leq s\}$.

The reason for the independent increment property is that the Poisson random measure on which $\boldsymbol{X}(\cdot)$ is built has the independence property of complete randomness given in item 2 on p. 120. Denote the $\sigma$-algebra generated by a collection of random elements $\{\xi_t, t \in T\}$ by $\sigma(\xi_t, t \in T)$, and let $\mathcal{E}$ be the $\sigma$-algebra of Borel subsets of $\mathbb{E}$. The independence property of PRM $N$ means that

$$\sigma\{N((s_1, s_2] \times A), 0 \leq s_1 < s_2 \leq s, A \in \mathcal{E}\}$$

and

$$\sigma\{N([t_1, t_2] \times B), s < t_1 \leq t_2 \leq t, B \in \mathcal{E}\}$$

are independent. The variable $X(t) - X(s)$ is measurable with respect to the second $\sigma$-algebra, and $\sigma(X(v), v \leq s)$ is a sub-$\sigma$-algebra of the first.

**Stationary increment property**

The process $\{X(t), t \geq 0\}$ has stationary increments; that is, for $s > 0$,

$$\{X(t+s) - X(s), t \geq 0\} \stackrel{d}{=} \{X(t), t \geq 0\},$$

where $\stackrel{d}{=}$ means equality of the finite-dimensional distributions. Recall that the distribution of a PRM is only dependent on the mean measure. Define $T_t : [0, \infty) \times \mathbb{E} \mapsto [t, \infty) \times \mathbb{E}$ by

$$T_t(s, \boldsymbol{x}) = (s + t, \boldsymbol{x}),$$

and define

$$N_t = N \circ T_t^{-1}.$$

If $N = \sum_k \epsilon_{(t_k, \boldsymbol{j}_k)}$ is PRM($\mathbb{LEB} \times \nu$) in $M_p([0, \infty) \times \mathbb{E})$, then $N_t = \sum_k \epsilon_{(t_k+t, \boldsymbol{j}_k)}$ is PRM($\mathbb{LEB} \times \nu$) in $M_p([t, \infty) \times \mathbb{E})$. This follows from Lebesgue measure being translation invariant. Now

$$\begin{aligned}
X(t+s) - X(s) &= \iint_{\substack{\|u\|>1 \\ \tau\in(s,s+t]}} \boldsymbol{u} N(d\tau, d\boldsymbol{u}) \\
&\quad + \lim_{\varepsilon\downarrow 0} \left[ \iint_{\substack{\|u\|\in(\varepsilon,1] \\ \tau\in(s,s+t]}} \boldsymbol{u} N(d\tau, d\boldsymbol{u}) - t \int_{\varepsilon<\|u\|\leq 1} \boldsymbol{u}\nu(d\boldsymbol{u}) \right] \\
&= \iint_{\substack{\|u\|>1 \\ \tau\in(0,t]}} \boldsymbol{u} N(d\tau + s, d\boldsymbol{u}) \\
&\quad + \lim_{\varepsilon\downarrow 0} \left[ \iint_{\substack{\|u\|\in(\varepsilon,1] \\ \tau\in(0,t]}} \boldsymbol{u} N(d\tau + s, d\boldsymbol{u}) - t \int_{\varepsilon<\|u\|\leq 1} \boldsymbol{u}\nu(d\boldsymbol{u}) \right] \\
&\stackrel{d}{=} \iint_{\substack{\|u\|>1 \\ \tau\in[0,t]}} \boldsymbol{u} N(d\tau, d\boldsymbol{u}) \\
&\quad + \lim_{\varepsilon\downarrow 0} \left[ \iint_{\substack{\|u\|\in(\varepsilon,1] \\ \tau\in(0,t]}} \boldsymbol{u} N(d\tau, d\boldsymbol{u}) - t \int_{\varepsilon<\|u\|\leq 1} \boldsymbol{u}\nu(d\boldsymbol{u}) \right] \\
&= X(t).
\end{aligned}$$

**Stochastic continuity of $X(\cdot)$**

Stochastic continuity means that if $s_n \to t$, then $X(s_n) \stackrel{P}{\to} X(t)$. To show this, it suffices to show that $X(t) - X(s_n) \stackrel{P}{\to} \boldsymbol{0}$. Suppose for simplicity that $t > s_n$. Then because of stationary increments,

$$\mathbf{E}e^{i\boldsymbol{\zeta}\cdot(\boldsymbol{X}(t)-\boldsymbol{X}(s_n))} = \mathbf{E}e^{i\boldsymbol{\zeta}\cdot\boldsymbol{X}(t-s_n)} = (\phi(\boldsymbol{\zeta}))^{t-s_n}$$
$$\to (\phi(\boldsymbol{\zeta}))^0 = 1 \quad (n \to \infty),$$

and hence by the continuity theorem for characteristic functions, $\boldsymbol{X}(t) - \boldsymbol{X}(s_n) \xrightarrow{P} \mathbf{0}$ [24, 264].

### Subordinators

If $\nu$ satisfies the stronger condition

$$\int_{\|\boldsymbol{x}\|\le 1} \|\boldsymbol{x}\|\nu(d\boldsymbol{x}) < \infty \tag{5.38}$$

rather than just $\int_{\|\boldsymbol{x}\|\le 1} \|\boldsymbol{x}\|^2\nu(d\boldsymbol{x}) < \infty$, as required by the definition of a Lévy measure, then

$$\boldsymbol{X}_0(t) + \sum_{j=0}^{\infty} \boldsymbol{S}_{I_{j+1}}(t)$$

converges absolutely almost surely without centering, since by the triangle inequality

$$\sum_{j=0}^{\infty} \mathbf{E}\|\boldsymbol{S}_{I_{j+1}}(t)\| \le \sum_{j=0}^{\infty} \mathbf{E} \sum_{t_k\le t} \|\boldsymbol{j}_k\| 1[\|\boldsymbol{j}_k\| \in I_{j+1}]$$
$$= \sum_{j=0}^{\infty} t \int_{I_{j+1}} \|\boldsymbol{x}\|\nu(d\boldsymbol{x}) = t\int_{(0,1]} \|\boldsymbol{x}\|\nu(d\boldsymbol{x}) < \infty.$$

In this case,

$$\mathbf{E}e^{i\boldsymbol{\zeta}\cdot(\boldsymbol{X}_0(t)+\sum_{j=0}^{\infty}\boldsymbol{S}_{I_{j+1}}(t))} = \exp\left\{t\int_{\|\boldsymbol{x}\|>0} (e^{i\boldsymbol{\zeta}\cdot\boldsymbol{x}} - 1)\nu(d\boldsymbol{x})\right\}.$$

If the dimension $d = 1$ and $\nu(-\infty, 0) = 0$, so that all $j_k$s are positive, then

$$\sum_{j=0}^{\infty} S_{I_{j+1}}(t) + X_0(t)$$

is nondecreasing and is called a *subordinator* or an *increasing Lévy process*.

### Stable Lévy motion

Suppose $d = 1, 0 < \alpha < 2, 0 \le p \le 1$, and $q = 1 - p$, and define the Lévy measure

$$\nu_\alpha(dx) = p\alpha x^{-\alpha-1} dx 1_{(0,\infty)}(x) + q\alpha |x|^{-\alpha-1} dx 1_{(-\infty,0)}. \tag{5.39}$$

The Lévy process with this Lévy measure is called stable Lévy motion and is denoted by $X_\alpha(\cdot)$.

Stable Lévy motion has the self-similarity property that for any $c > 0$,

$$X(c\cdot) \stackrel{d}{=} c^{1/\alpha} X(\cdot), \tag{5.40}$$

where equality in distribution means equality of finite-dimensional distributions.

To verify the self-similarity, we suppose that

$$\sum \epsilon_{(t_k, j_k)} = \text{PRM}(\mathbb{LEB} \times \nu_\alpha).$$

For $c > 0$,

$$\sum \epsilon_{(\frac{t_k}{c}, j_k)} = \text{PRM}(c\, \mathbb{LEB} \times \nu_\alpha),$$

but so is

$$\sum \epsilon_{(t_k, c^{1/\alpha} j_k)} = \text{PRM}(c\, \mathbb{LEB} \times \nu_\alpha),$$

and therefore

$$\sum_k \epsilon_{(\frac{t_k}{c}, j_k)} \stackrel{d}{=} \sum_k \epsilon_{(t_k, c^{1/\alpha} j_k)}.$$

$X(c\cdot)$ is built on the first Poisson process and $c^{1/\alpha} X(\cdot)$ on the second, so that (5.40) follows.

### Symmetric $\alpha$-stable Lévy motion

A special case of Section 5.5.2 is called symmetric $\alpha$-stable motion: If $p = q = 1/2$, then $\nu_\alpha$ given in (5.39) is symmetric, which means that

$$\nu_\alpha(a, b] = \nu_\alpha([-b, -a)), \quad 0 < a < b.$$

The characteristic function is real, and therefore $X(t) \stackrel{d}{=} -X(t)$ and

$$\phi_\alpha(\zeta) = \mathbf{E} e^{i\zeta X_\alpha(t)} = e^{-ct|\zeta|^\alpha}, \quad \zeta \in \mathbb{R}, \tag{5.41}$$

for some $c > 0$.

We verify (5.41). For $\zeta > 0$, we get from (5.37) that twice the log characteristic function of $X_\alpha(t)$ is (remember that $p = q = \frac{1}{2}$)

$$2\left[\int_{|x|>1}(e^{i\zeta x}-1)\nu_\alpha(dx)+\int_{0<|x|\le 1}(e^{i\zeta x}-1-i\zeta x)\nu_\alpha(dx)\right]$$
$$=\int_{x=1}^{\infty}(e^{i\zeta x}-1)\alpha x^{-\alpha-1}dx+\int_{x=-\infty}^{-1}(e^{i\zeta x}-1)\alpha|x|^{-\alpha-1}dx$$
$$+\int_{0<x\le 1}(e^{i\zeta x}-1-i\zeta x)\alpha x^{-\alpha-1}dx$$
$$+\int_{-1\le x<0}(e^{i\zeta x}-1-i\zeta x)\alpha|x|^{-\alpha-1}dx$$
$$=\int_{1}^{\infty}(e^{i\zeta x}+e^{-i\zeta x}-2)\alpha x^{-\alpha-1}dx+\int_{0}^{1}(e^{i\zeta x}+e^{-i\zeta x}-2)\alpha x^{-\alpha-1}dx.$$

Making the change of variable $y=\zeta x$, we get

$$=\zeta^\alpha\int_0^\infty(e^{iy}+e^{-iy}-2)\alpha y^{-\alpha-1}dy$$
$$=\zeta^\alpha\int_0^\infty 2(\cos y-1)\alpha y^{-\alpha-1}dy=-c\zeta^\alpha.$$

### 5.5.3 Basic path properties of Lévy processes

Let $D([0,\infty),\mathbb{R}^d)$ be the space of $\mathbb{R}^d$-valued functions on $[0,\infty)$ that are right continuous and have finite left-hand limits on $(0,\infty)$. Also, recall that two $\mathbb{R}^d$-valued random processes $\boldsymbol{X}(\cdot)$ and $\boldsymbol{Y}(\cdot)$ are *versions* if

$$\mathbb{P}[\boldsymbol{X}(t)=\boldsymbol{Y}(t)]=1 \quad \text{for all } t.$$

This, of course, assumes that $\boldsymbol{X}(\cdot)$ and $\boldsymbol{Y}(\cdot)$ are defined on the same probability space.

**Theorem 5.4.** *If $\boldsymbol{X}(\cdot)$ is a Lévy process in $\mathbb{R}^d$ with Lévy measure $\nu$, there is a version $\boldsymbol{Y}(\cdot)$ with almost all paths in $D([0,\infty),\mathbb{R}^d)$.*

To prove this, it suffices to suppose that $d=1$, work in $D[0,\infty)$, and prove that the infinite series used in the definition of the Lévy process converges almost surely in a stronger sense than previously considered in (5.35) (p. 150), which was only for a fixed time point $t$. The stronger sense is uniform convergence on compact $t$-sets. To review: For functions $f_n:[0,\infty)\mapsto\mathbb{R}$, we say that $f_n$ converges *uniformly on compact sets*, denoted by $f_n\overset{\text{uc}}{\to}f_0$, if for any $k$,

$$\sup_{0\le x\le k}|f_n(x)-f_0(x)|\to 0.$$

This is also called local uniform convergence.

The next result shows $D[0,\infty)$ is closed under local uniform convergence.

**Lemma 5.2.** *Suppose for each $n \geq 1$ that $x_n(\cdot) \in D[0, \infty)$ and that $x_n \stackrel{\text{uc}}{\to} x_0$. Then it follows that $x_0 \in D[0, \infty)$.*

*Proof.* Start by assuming $t_j \downarrow t$. Then for any $m$,

$$|x_0(t_j) - x_0(t)| \leq |x_0(t_j) - x_m(t_j)| + |x_m(t_j) - x_m(t)| + |x_m(t) - x_0(t)|,$$

and for $k > t$, this is eventually bounded above by

$$\leq 2 \sup_{0 \leq s \leq k} |x_0(s) - x_m(s)| + |x_m(t_j) - x_m(t)|.$$

Given $\varepsilon > 0$, choose $m_0$ so large that

$$2 \sup_{0 \leq s \leq k} |x_0(s) - x_{m_0}(s)| \leq \varepsilon/36.$$

This is possible by local uniform convergence. Then having chosen $m_0$, and using the fact that $x_{m_0} \in D[0, \infty)$, we may pick $j_0$ such that for $j \geq j_0$,

$$|x_{m_0}(t_j) - x_{m_0}(t)| \leq \varepsilon/36,$$

which means that for $j \geq j_0$,

$$|x_0(t_j) - x_0(t)| \leq \varepsilon/18.$$

Thus $x_0(\cdot)$ is right continuous.

If $t_j \uparrow t$, a similar argument shows that $\{x(t_j)\}$ is Cauchy and hence convergent. The limit is $x(t-)$ by definition. □

To prove Theorem 5.4, define for $k = 2, 3, \ldots,$

$$T_k(t) = \iint_{\substack{|u|>\frac{1}{k} \\ 0 \leq s \leq t}} uN(ds, du) - t \int_{\frac{1}{k}<|u|\leq 1} u\nu(du). \tag{5.42}$$

The construction of the Lévy process proved that for each $t$, as $k \to \infty$,

$$\lim_{k\to\infty} T_k(t) = X(t)$$

almost surely in $\mathbb{R}$. The following result proves Theorem 5.4.

**Proposition 5.7.** *For a Lévy process $X(\cdot)$ in $\mathbb{R}$ built from the* PRM($\mathbb{LEB} \times \nu$),

$$N = \sum_k \epsilon_{(t_k, j_k)};$$

*the approximation* (5.42) *has the following properties:*

Property 1. *$T_k(\cdot)$ has almost all paths in $D[0,\infty)$.*

Property 2. *There exists $T_\infty(\cdot)$ with almost all paths in $D[0,\infty)$ such that for any positive $K$,*

$$\lim_{k\to\infty} \sup_{0\le s\le K} |T_k(s) - T_\infty(s)| = 0,$$

*almost surely.*

*This $T_\infty(\cdot)$ is the desired version of $X(\cdot)$ in $D[0,\infty)$.*

*Proof of Property* 1. Obviously, $t\int_{\frac{1}{k}<|u|\le 1} u\nu(du)$ is continuous in $t$ and hence in $D[0,\infty)$, so we only have to show for $k$ fixed that

$$\iint_{\substack{|u|>1/k\\ 0\le s\le K}} uN(ds,du) = \sum_{t_k\le t} j_k 1_{[|j_k|>k^{-1}]} =: V_k(t)$$

is a nice function of $t$. For any $n \ge 1$,

$$P[N[0,n]\times\{y : |y| > k^{-1}\} < \infty] = 1,$$

and therefore

$$\Lambda_n = \{\omega : \text{the set } \{(t_k(\omega), j_k(\omega)) : t_k(\omega) \le n, |j_k(\omega)| > k^{-1}\} \text{ is finite}\}$$

has probability 1. For $\omega \in \Lambda_n$,

$$V_k(t,\omega) \in D[0,n],$$

and $\mathbb{P}(\Lambda_n) = 1$. So $\mathbb{P}(\cap_n\Lambda_n) = 1$, and for $\omega \in \cap_n\Lambda_n$,

$$V_k(t,\omega) \in D[0,\infty). \qquad \square$$

*Proof of Property* 2. Recall Kolmogorov's inequality [24]. If $\{\xi_j\}$ are independent random variables, with $\mathrm{Var}(\xi_j) < \infty$, then for any $\varepsilon > 0$ we have

$$\mathbb{P}\left[\sup_{j\le N}\left|\sum_{i=1}^{j}\xi_i - \mathbf{E}\left(\sum_{i=1}^{j}\xi_i\right)\right| > \varepsilon\right] \le \frac{\mathrm{Var}\left(\sum_1^N \xi_i\right)}{\varepsilon^2}. \tag{5.43}$$

We need the following simple variant.

**Lemma 5.3 (continuous version of Kolmogorov's inequality).** *Let $\{Z(t), t\ge 0\}$ be a Lévy process with almost all paths in $D[0,\infty)$ and satisfying $\mathrm{Var}(Z(1)) < \infty$. Then for any $N$ and any $\varepsilon > 0$, we have*

$$\mathbb{P}\left[\sup_{0\le t\le N} |Z(t) - \mathbf{E}(Z(t))| > \varepsilon\right] \le \frac{\mathrm{Var}(Z(N))}{\varepsilon^2}.$$

*Proof.* Think of $Z(\frac{jN}{2^n})$, $j = 1, \ldots, 2^n$, as successive sums of independent random variables. Then we have

$$\mathbb{P}\left[\sup_{0 \le j \le 2^n} \left| Z\left(\frac{jN}{2^n}\right) - \mathbf{E}Z\left(\frac{jN}{2^n}\right) \right| > \varepsilon \right] \le \frac{\mathrm{Var}(Z(N))}{\varepsilon^2}. \tag{5.44}$$

Let $n \to \infty$. The left side of (5.44) converges upward to

$$\mathbb{P}\left[\sup_{0 \le s \le N} |Z(s) - \mathbf{E}(Z(s))| > \varepsilon \right].$$

Note we used that $\mathbf{E}(Z(s))$ is continuous in $s$, which follows from the fact that

$$\mathbf{E}(Z(s)) = s\mathbf{E}(Z(1)),$$

which is a consequence of the stationary, independent increments. We also have

$$\mathrm{Var}(Z(N)) = N\,\mathrm{Var}(Z(1))$$

for the same reason. □

We continue with the proof of Property 2 of Proposition 5.7. We seek a set $\Lambda$ with $\mathbb{P}(\Lambda) = 1$ such that if $\omega \in \Lambda$, then for any $K$, the sequence $\{T_k(\cdot, \omega), k \ge 1\}$ is Cauchy with respect to uniform convergence on $[0, K]$. If we find such a $\Lambda$, we have $\lim_{k\to\infty} T_k(\cdot, \omega)$ exists uniformly on compacta for $\omega \in \Lambda$, which is the desired result.

For $x \in D[0, \infty)$, write $\| x \|_K = \sup_{0 \le s \le K} |x(s)|$.

Pick and fix $K$. To show the Cauchy property, we prove that

$$Y_N = \sup_{\substack{m \ge N \\ n \ge N}} \| T_m - T_n \|_K \overset{\text{a.s.}}{\to} 0, \tag{5.45}$$

as $N \to \infty$. Since $\{Y_N\}$ is nonincreasing, it suffices to show that $Y_N \overset{P}{\to} 0$ [24, 264], so we aim to show that

$$\lim_{N\to\infty} \mathbb{P}\left[\sup_{\substack{m \ge N \\ n \ge N}} \| T_m - T_n \|_K > \varepsilon \right] = 0. \tag{5.46}$$

This is the same as showing that

$$\lim_{N\to\infty} \lim_{M\to\infty} \mathbb{P}\left[\sup_{\substack{M \ge m \ge N \\ M \ge n \ge N}} \| T_m - T_n \|_K > \varepsilon \right] = 0, \tag{5.47}$$

and since

$$\| T_m - T_n \|_K \le \| T_m - T_N \|_K + \| T_N - T_n \|_K,$$

it suffices to show that

$$\lim_{N\to\infty} \lim_{M\to\infty} P\left[ \sup_{M\ge n\ge N} \| T_n - T_N \|_K > 2\varepsilon \right] = 0. \tag{5.48}$$

The triangle inequality implies that

$$\| T_M - T_N \|_K \ge \| T_i - T_N \|_K - \| T_M - T_i \|_K,$$

and hence

$$\bigcup_{i=N+1}^{M} \left[ \bigvee_{j=N}^{i-1} \|T_j - T_N\|_K \le 2\varepsilon,\ \|T_i - T_N\|_K > 2\varepsilon,\ \|T_M - T_i\|_K \le \varepsilon \right] \subset [\|T_M - T_N\|_K > \varepsilon].$$

Note that the union is a disjoint union since we decompose according to the first index where a difference exceeds $2\epsilon$. Therefore,

$$\begin{aligned} &\mathbb{P}[\| T_M - T_N \|_K > \varepsilon] \\ &\ge \sum_{i=N+1}^{M} \mathbb{P}\left[ \bigvee_{j=N}^{i-1} \|T_j - T_N\|_K \le 2\varepsilon,\ \|T_i - T_N\|_K > 2\varepsilon,\ \|T_M - T_i\|_K \le \varepsilon \right] \\ &= \sum_{i=N+1}^{M} \mathbb{P}\left[ \bigvee_{j=N}^{i-1} \|T_j - T_N\|_K \le 2\varepsilon,\ \|T_i - T_N\|_K > 2\varepsilon \right] \mathbb{P}[\|T_M - T_i\|_K \le \varepsilon]. \end{aligned} \tag{5.49}$$

Note that we have used the fact that for $m > N$,

$$T_m(t) - T_N(t) = \int_{\substack{\frac{1}{m} < |u| \le \frac{1}{N} \\ 0 \le s \le t}} u N(ds, du) - t \int_{\frac{1}{m} < |u| \le \frac{1}{N}} u \nu(du),$$

so $T_{N+1} - T_N, \ldots, T_{i-1} - T_N, T_i - T_N$ involve points in the horizontal strip with boundaries at $(\frac{1}{i}, \frac{1}{N}]$, while $T_M - T_i$ uses points in the disjoint strip $(\frac{1}{M}, \frac{1}{i}]$; therefore $T_M - T_i$ is independent of $T_{N+1} - T_N, \ldots, T_i - T_N$.

Finally, we have

$$\begin{aligned} \mathbb{P}[\| T_M - T_i \|_K \le \varepsilon] &= 1 - \mathbb{P}[\| T_M - T_i \|_K > \varepsilon] \\ &\ge 1 - \frac{\mathrm{Var}(T_M(K) - T_i(K))}{\varepsilon^2} \end{aligned}$$

(by Kolmogorov's inequality as given in Lemma 5.3 (p. 157))

$$
\begin{aligned}
&= 1 - K\varepsilon^{-2} \int_{\frac{1}{M}<|u|<\frac{1}{i}} u^2 \nu(du) \\
&\geq 1 - K\varepsilon^{-2} \int_{\frac{1}{M}<|u|\leq\frac{1}{N}} u^2 \nu(du) \\
&\geq 1 - K\varepsilon^{-2} \int_{0<|u|\leq\frac{1}{N}} u^2 \nu(du) \to 1 \quad (N \to \infty)
\end{aligned}
$$

since $\int_{|u|\in(0,1]} u^2\nu(du) < \infty$. Pick $N_0$ so large that for $N \geq N_0$, we have that for $N \leq i \leq M$

$$\mathbb{P}[\| T_M - T_i \|_K \leq \varepsilon] \geq \frac{1}{2}.$$

From (5.49)

$$
\begin{aligned}
&2\mathbb{P}[\| T_M - T_N \|_K > \varepsilon] \\
&\quad \geq \sum_{i=N+1}^{M} \mathbb{P}\left[\bigvee_{j=N}^{i-1} \|T_j - T_N\|_K \leq 2\varepsilon, \|T_i - T_N\|_K > 2\varepsilon\right] \\
&\quad = \mathbb{P}\left[\bigvee_{n=N+1}^{M} \|T_n - T_N\|_K > 2\varepsilon\right],
\end{aligned}
$$

and again applying Kolmogorov's inequality, we get

$$
\begin{aligned}
\mathbb{P}\left[\bigvee_{n=N+1}^{M} \|T_n - T_N\|_K > 2\varepsilon\right] &\leq 2\mathbb{P}[\|T_M - T_N\|_K > \varepsilon] \\
&\leq \frac{2}{\varepsilon^2}\operatorname{Var}(T_M(K) - T_N(K)) = \frac{2K}{\varepsilon^2}\int_{\frac{1}{M}<|u|\leq\frac{1}{N}} u^2\nu(du).
\end{aligned}
$$

This gives (5.48) and (5.45) follows. Let $\Lambda_K$ be the set of probability 1 on which (5.45) holds. Then $\Lambda = \bigcap_{K\in\mathbb{Z}_+} \Lambda_K$ is a set of probability 1 on which $\| T_m - T_n \|_K \to 0$ as $m, n \to \infty$ for any $K$ and therefore on which uniform convergence takes place for any $K$. □

## 5.6 Extremal processes

Extremal processes are another simple class of processes derived from Poisson random measures.

References for this section include [11, 82–84, 119–122, 244–247, 257–260, 276–278, 291, 294–296].

### 5.6.1 Construction

In what follows, vector notation relies on the convention that operations on vectors are performed componentwise. Usage should be self-explanatory; Appendix 10 (p. 359) collates conventions and notations.

For simplicity, take $\mathbb{E} = [0, \infty)^d$. Let $\nu$ be a measure on $\mathbb{E}$ satisfying

$$\nu\{\boldsymbol{x} \in \mathbb{E} : \|\boldsymbol{x}\| > \delta\} < \infty \tag{5.50}$$

for any $\delta > 0$. Suppose that

$$N = \sum_k \epsilon_{(t_k, \boldsymbol{j}_k)}$$

is PRM($\mathbb{LEB} \times \nu$) on $[0, \infty) \times \mathbb{E}$, and define the extremal process generated by $N$ as

$$\boldsymbol{Y}(t) = \bigvee_{t_k \le t} \boldsymbol{j}_k, \quad t > 0. \tag{5.51}$$

Then for $\boldsymbol{x} \ge 0$, $\boldsymbol{x} \ne \boldsymbol{0}$, and any $t \ge 0$,

$$\begin{aligned} \mathbb{P}[\boldsymbol{Y}(t) \le \boldsymbol{x}] &= \mathbb{P}[N((0, t] \times [\boldsymbol{0}, \boldsymbol{x}]^c) = 0] = e^{-t\nu([\boldsymbol{0},\boldsymbol{x}]^c)} \\ &=: F^t(\boldsymbol{x}). \end{aligned} \tag{5.52}$$

Notice that for $\|\boldsymbol{x}\| > 0$, $\nu([\boldsymbol{0}, \boldsymbol{x}]^c) < \infty$ from (5.50), and so $F$ is not identically 0 and converges to 1 as $\wedge_{i=1}^d x^{(i)} \to \infty$.

The distribution $F(\boldsymbol{x})$ constructed this way is *max-infinitely divisible*, which means that for any $t$, $F^t$ is a multivariate distribution function [11, 260]. The measure $\nu$ is called the exponent measure. (Conversely, any max-infinitely divisible distribution has an exponent measure.)

### 5.6.2 Discussion

The extremal process $\boldsymbol{Y}(\cdot)$ given by (5.51) is a stochastically continuous Markov jump process that is constant between jump times. It is nondecreasing in each component, and as constructed in (5.51), the paths are almost surely in $D([0, \infty), \mathbb{R}^d)$.

When $d = 1$ and $F(x)$ given in (5.52) is continuous, much is known about the structure of the process. For instance, we have the following:

1. Jump times of the process form a Poisson process with mean measure having density $x^{-1}dx, x > 0$.
2. The range of the process is also a Poisson process with mean measure having distribution function $V(x) := -\log(-\log F(x))$, so that the mean measure of the interval $(a, b]$ is $V(b) - V(a)$.

3. $Y^{\leftarrow}(x), x > 0$ is a process with independent increments.

For details, see, for example, [260, Chapter 4.3].

One final comment: When $d = 1$ and the exponent measure is $\nu_\alpha$ for some $\alpha > 0$ given by

$$\nu_\alpha(x, \infty] = x^{-\alpha}, \quad x > 0,$$

we have

$$F(x) = \exp\{-x^{-\alpha}\} = \Phi_\alpha(x), \quad x > 0.$$

This is the Fréchet distribution, one of the classical extreme-value distributions given in Section 2.2.1 (p. 23).

## 5.7 Problems

**5.1 (Thinning).** Suppose $\sum_n \epsilon_{X_n}$ is a Poisson process on the state space $\mathbb{E}$ with mean measure $\mu$. Suppose we inspect each point independently of others and decide with probability $p$ to retain the point and with probability $1 - p = q$ to delete the point. Let $N_r$ be the point process of retained points and $N_d$ be the point process of deleted points. Then $N_r$, $N_d$ are independent Poisson processes with mean measures $p\mu$ and $q\mu$, respectively. Analyze this using augmentation by iid Bernoulli random variables having values $\{\pm 1\}$ with probabilities $p, q$.

*Generalize*: Previously, we categorized or marked the points in two ways: retained or deleted. However, we could just as well randomly assign the points to any of $d > 1$ categories, thereby splitting a Poisson input stream into $d$ independent Poisson substreams. The Bernoulli random variables need to be replaced by multinomial random variables with $d$ cells.

**5.2 (The order statistics property).** The construction in Section 5.4.2 proves that PRM($\mu$) exists and also gives information about the distribution of the points: Conditional on there being $n$ points in a region $A$ with $\mu(A) < \infty$, these points are distributed as $n$ iid random elements of $A$ with common distribution $F(dx) = \mu(dx)/\mu(A)$. Show that when $\mathbb{E} = [0, \infty)$, this yields the order statistics property for a homogeneous Poisson process: If $N = \sum_{n=1}^{\infty} \epsilon_{\Gamma_n}$ is a homogeneous Poisson process on $[0, \infty)$ with rate $\lambda$, then conditional on

$$[N((0, t]) = n],$$

the points of $N$ in $[0, t]$ in increasing order are distributed as the order statistics from a sample of size $n$ from the uniform distribution $U(0, t)$ on $[0, t]$; that is,

$$(\Gamma_1, \ldots, \Gamma_n | N[0, t] = n) \stackrel{d}{=} (U_{1:n}, \ldots, U_{n:n}).$$

**5.3 (Weak convergence of Poisson random measures).** Suppose for each $n \geq 0$ that $N_n$ is PRM$(\mu_n)$ on $\mathbb{E}$, where $\mu_n$ is a Radon measure on $\mathbb{E}$. Then as $n \to \infty$,

$$N_n \Rightarrow N_0$$

if and only if

$$\mu_n \stackrel{v}{\to} \mu_0$$

in $M_+(\mathbb{E})$. (Laplace functionals make short work of this.)

**5.4.** Suppose $\nu$ is a measure on $\mathbb{R}$ with

$$Q(x) := \nu(x, \infty) < \infty \quad \forall x \in \mathbb{R}.$$

Define for $y > 0$,

$$Q^{\leftarrow}(y) = (1/Q)^{\leftarrow}(y^{-1}).$$

Suppose $\{\Gamma_n, n \geq\}$ are successive sums of iid unit exponential random variables.

1. Show that
$$\sum_n \epsilon_{Q^{\leftarrow}(\Gamma_n)}$$
is PRM$(\nu)$ on $\mathbb{R}$. What is the distribution of the largest point of the point process?
2. Suppose $\{U_n, n \geq 1\}$ are iid $U(0,1)$ random variables that are independent of $\{\Gamma_n, n \geq\}$. Show that
$$\sum_n \epsilon_{(U_n, Q^{\leftarrow}(\Gamma_n))}$$
is PRM$(\mathbb{LEB} \times \nu)$ on $[0,1] \times \mathbb{R}$. Give a representation of the extremal process generated by this Poisson process.

**5.5.**

(a) Suppose $\sum_{i=1}^{\infty} \epsilon_{y_i} \in M_p(\mathbb{E})$. Let $\{\xi_i, i \geq 1\}$ be iid Poisson random variables, each with mean 1 . Compute the Laplace functional of

$$\sum_i \xi_i \epsilon_{y_i}.$$

This is the point process in which, for each $i$, a Poisson number of points is assigned to location $y_i$

(b) Now suppose $\sum_i \epsilon_{Y_i}$ is PRM($\nu$) on $\mathbb{E}$, and let $\{\xi_i, i \geq 1\}$ be iid Poisson random variables, each with mean 1, and independent of $\{Y_i\}$. Compute the Laplace functional of

$$\sum_i \xi_i \epsilon_{Y_i}.$$

**5.6 (Variant).** As a variant of Problem 5.5, suppose that

$$m = \sum_i \epsilon_{y_i} \in M_p(\mathbb{E})$$

and that $\{\xi_n\}$ are iid nonnegative integer-valued random variables with Laplace transform

$$\phi(\lambda) = \mathbf{E}e^{-\lambda\xi_1}, \quad \lambda > 0.$$

Compute the Laplace functional at $f$ for

$$\sum_i \xi_i \epsilon_{y_i},$$

and express the answer in terms of $\phi$, $f$, and $m$.

**5.7 (Largest-jump functional [247]).** Suppose $\{X(t), t \geq 0\}$ is a one-dimensional Lévy process with Lévy measure $\nu$ and paths in $D[0, \infty)$. Define

$$Y(t) = \sup\{X(s) - X(s-) : s \leq t; X(s) - X(s-) > 0\}$$

to be the largest positive jump of $X(\cdot)$ in $[0, t]$. Show that $Y(\cdot)$ is an extremal process, and compute $\mathbb{P}[Y(t) \leq x]$ for $x > 0$.

**5.8 (Cluster processes).** Let $\{\Gamma_j, j \geq 1\}$ be the points of a homogeneous Poisson process on $\mathbb{R}_+$, and suppose $\{Y_i^{(k)}, i \geq 1, k \geq 1\}$ are iid, nonnegative random variables independent of $\{\Gamma_j, j \geq 1\}$. Set $S_0^{(k)} = 0$ and

$$S_n^{(k)} = \sum_{i=1}^n Y_i^{(k)}, \quad n \geq 1.$$

Finally, let $\{\tau^{(k)}, k \geq 1\}$ be iid nonnegative integer-valued random variables independent of $\{\Gamma_i\}$ and $\{Y_i^{(k)}\}$.

A Poisson cluster process consists of points determined by Poisson "centers" and points sprinkled around a Poisson center according to some rule.

1. Define

$$N_1 = \sum_{k=1}^{\infty} \sum_{n=0}^{\tau^{(k)}} \epsilon_{\Gamma_k + S_n^{(k)}}$$

consisting of Poisson points trailed by a renewal process terminated at a random index. Compute the Laplace functional. (Here we count the Poisson points.)

2. Define the *Neyman–Scott model* (see [231] for background) by

$$N_2 = \sum_{k=1}^{\infty} \sum_{n=1}^{\tau^{(k)}} \epsilon_{\Gamma_k + Y_n^{(k)}}$$

consisting of Poisson points trailed by order statistics. (Here we do not count the Poisson points.) Compute the Laplace functional; presumably the answer comes out in terms of the generating function of the $\tau$s.

# 6

# Multivariate Regular Variation and the Poisson Transform

This chapter discusses the relationship between (multivariate) regular variation and the Poisson process. We begin with a survey of multivariate regular variation as it applies to distributions. The goal is to make the results of Theorem 3.6 applicable to higher dimensions.

For other treatments and additional material, see [13, 90, 220, 308–312].

## 6.1 Multivariate regular variation: Basics

We begin by discussing regular variation of functions and then move to measures.

### 6.1.1 Multivariate regularly varying functions

A subset $C \subset \mathbb{R}^d$ is called a *cone* if whenever $\boldsymbol{x} \in C$, then also $t\boldsymbol{x} \in C$ for any $t > 0$. A function $h : C \mapsto (0, \infty)$ is monotone if it is either nondecreasing in each component or nonincreasing in each component. For $h$ nondecreasing, this is equivalent to saying that whenever $\boldsymbol{x}, \boldsymbol{y} \in C$ and $\boldsymbol{x} \leq \boldsymbol{y}$, we have $h(\boldsymbol{x}) \leq h(\boldsymbol{y})$.

Suppose $h \geq 0$ is a measurable function defined on $C$. Suppose $\mathbf{1} = (1, \ldots, 1) \in C$. Call $h$ *multivariate regularly varying with limit function* $\lambda$, provided $\lambda(\boldsymbol{x}) > 0$ for $\boldsymbol{x} \in C$, and for all $\boldsymbol{x} \in C$, we have

$$\lim_{t\to\infty} \frac{h(t\boldsymbol{x})}{h(t\mathbf{1})} = \lambda(\boldsymbol{x}). \tag{6.1}$$

More properly, this should be called regular variation at $\infty$. Note that $\lambda(\mathbf{1}) = 1$. Fix $\boldsymbol{x} \in C$ and define $U : (0, \infty) \mapsto [0, \infty)$ by $U(t) = h(t\boldsymbol{x})$. For any $s > 0$,

$$\lim_{t\to\infty} \frac{U(ts)}{U(t)} = \lim_{t\to\infty} \frac{h(ts\boldsymbol{x})}{h(t\boldsymbol{x})} = \lim_{t\to\infty} \frac{h(ts\boldsymbol{x})}{h(t\mathbf{1})} \Big/ \frac{h(t\boldsymbol{x})}{h(t\mathbf{1})} = \frac{\lambda(s\boldsymbol{x})}{\lambda(\boldsymbol{x})}.$$

From Proposition 2.3, we have for some $\rho(\boldsymbol{x}) \in \mathbb{R}$ that $U \in \mathrm{RV}_{\rho(\boldsymbol{x})}$ and

$$\frac{\lambda(s\boldsymbol{x})}{\lambda(\boldsymbol{x})} = s^{\rho(\boldsymbol{x})}.$$

In the next paragraph, we verify that $\rho$ does not depend on $\boldsymbol{x}$, and thus we conclude from this simple scaling argument that $\lambda(\cdot)$ is homogeneous:

$$\lambda(s\boldsymbol{x}) = s^{\rho}\lambda(\boldsymbol{x}), \quad s > 0, \quad \boldsymbol{x} \in C. \tag{6.2}$$

Why is $\rho(\boldsymbol{x})$ constant in $\boldsymbol{x}$? For $\boldsymbol{x}, \boldsymbol{y} \in C$, we have for any $s > 0$,

$$\begin{aligned} s^{\rho(\boldsymbol{y})} &= \lim_{t\to\infty} \frac{h(ts\boldsymbol{y})}{h(t\boldsymbol{y})} = \lim_{t\to\infty} \frac{\frac{h(ts\boldsymbol{y})}{h(ts\boldsymbol{x})}}{\frac{h(t\boldsymbol{y})}{h(t\boldsymbol{x})}} \cdot \frac{h(ts\boldsymbol{x})}{h(t\boldsymbol{x})} \\ &= \frac{\frac{\lambda(\boldsymbol{y})}{\lambda(\boldsymbol{x})}}{\frac{\lambda(\boldsymbol{y})}{\lambda(\boldsymbol{x})}} \cdot s^{\rho(\boldsymbol{x})} = s^{\rho(\boldsymbol{x})}. \end{aligned}$$

This is true for any $s > 0$, and hence

$$\rho(\boldsymbol{x}) = \rho(\boldsymbol{y}).$$

Thus reassured, we note that (6.1) could be rephrased as $h$ *is multivariate regularly varying with limit function* $\lambda$ if there exists $V : (0, \infty) \mapsto (0, \infty)$ with $V \in \mathrm{RV}_\rho$ for some $\rho \in \mathbb{R}$ such that

$$\lim_{t\to\infty} \frac{h(t\boldsymbol{x})}{V(t)} = \lambda(\boldsymbol{x}) \quad \forall \boldsymbol{x} \in C. \tag{6.3}$$

### 6.1.2 The polar coordinate transformation

It is frequently convenient when considering multivariate regular variation to transform the state space using a generalized polar coordinate transformation. After the transformation, the homogeneity property (6.2) in Cartesian coordinates becomes a product property in polar coordinates. We soon state equivalences for multivariate regular variation for the distribution of a random vector, where this will be particularly convenient.

A *norm* on $\mathbb{R}^d$ is a mapping $\|\cdot\| : \mathbb{R}^d \mapsto [0, \infty)$ such that

1. $\|\boldsymbol{x}\| \geq 0$ for all $\boldsymbol{x} \in \mathbb{R}^d$, and $\|\boldsymbol{x}\| = 0$ iff $\boldsymbol{x} = \boldsymbol{0}$.
2. $\|c\boldsymbol{x}\| = |c|\|\boldsymbol{x}\|$ for all $\boldsymbol{x} \in \mathbb{R}^d$ and $c \in \mathbb{R}$.
3. The triangle inequality holds: For $\boldsymbol{x}, \boldsymbol{y} \in \mathbb{R}^d$, we have

$$\|\boldsymbol{x} + \boldsymbol{y}\| \leq \|\boldsymbol{x}\| + \|\boldsymbol{y}\|.$$

Examples of norms are the following:

- The usual Euclidean norm:

$$\|\boldsymbol{x}\| = \sqrt{\sum_{i=1}^{d} (x^{(i)})^2}.$$

- The $L_p$-norm:

$$\|\boldsymbol{x}\| = \left(\sum_{i=1}^{d} |x^{(i)}|^p\right)^{1/p}, \quad p \geq 0.$$

- The $L_\infty$-norm:

$$\|\boldsymbol{x}\| = \bigvee_{i=1}^{n} |x^{(i)}|.$$

Given a chosen norm $\|\cdot\|$, the unit sphere is

$$\aleph := \{\boldsymbol{x} : \|\boldsymbol{x}\| = 1\}.$$

Note that for the Euclidean norm, the unit sphere is really a sphere in the conventional sense. If $d = 2$, the "unit sphere" in the $L_1$-norm is a diamond, and the "unit sphere" in the $L_\infty$-norm is a square.

A norm always defines a distance on $\mathbb{R}^d$ by

$$d(\boldsymbol{x}, \boldsymbol{y}) = \|\boldsymbol{x} - \boldsymbol{y}\|.$$

Norms on $\mathbb{R}^d$ are all topologically equivalent in that convergence in one norm implies convergence in another. This follows from the fact that for any two norms $\|\cdot\|_i, i = 1, 2$, some constants $c > 0$ and $C > 0$ exist such that

$$c\|\boldsymbol{x}\|_1 \leq \|\boldsymbol{x}\|_2 \leq C\|\boldsymbol{x}\|_1.$$

Fix a norm. (Theoretically, it does not matter which norm is chosen, but for considering particular examples, some norms are more appropriate and convenient.) We want to define the polar coordinate transform of a vector $\boldsymbol{x} \in \mathbb{R}^d$ as

$$\boldsymbol{x} \mapsto \left(\|\boldsymbol{x}\|, \frac{\boldsymbol{x}}{\|\boldsymbol{x}\|}\right) =: (r, \boldsymbol{a}).$$

This obviously creates difficulties if $\|\boldsymbol{x}\| = 0$, so due to property 1 in the property list for norms, we exclude $\boldsymbol{0} \in \mathbb{R}^d$, and we define the polar coordinate transformation $T : \mathbb{R}^d \setminus \{\boldsymbol{0}\} \mapsto (0, \infty) \times \aleph$ by

$$T(\boldsymbol{x}) = \left( \|\boldsymbol{x}\|, \frac{\boldsymbol{x}}{\|\boldsymbol{x}\|} \right) =: (r, \boldsymbol{a}).$$

This has inverse transformation $T^{\leftarrow} : (0, \infty) \times \aleph \mapsto \mathbb{R}^d \setminus \{\mathbf{0}\}$ given by

$$T^{\leftarrow}(r, \boldsymbol{a}) = r\boldsymbol{a}.$$

Think of $\boldsymbol{a} \in \aleph$ as defining a direction and $r$ as telling how far in direction $\boldsymbol{a}$ to proceed. Both $T$ and $T^{\leftarrow}$ are continuous bijections when we exclude $\mathbf{0}$.

When $d = 2$, it is customary, but not obligatory, to write

$$T(\boldsymbol{x}) = (r \cos \theta, r \sin \theta),$$

where $0 \leq \theta < 2\pi$, rather than the more consistent notation $T(\boldsymbol{x}) = (r, (\cos \theta, \sin \theta))$.

For a random vector $\boldsymbol{X}$ in $\mathbb{R}^d$, we sometimes write

$$T(\boldsymbol{X}) = (R, \boldsymbol{\Theta}).$$

The problem with all this is that for multivariate regular variation of tail probabilities, we have to deal with a punctured space such as $[\mathbf{0}, \boldsymbol{\infty}] \setminus \{\mathbf{0}\}$. (See Section 6.1.3.) The polar coordinate transformation is not defined on the lines through $\boldsymbol{\infty}$, so when discussing multivariate regular variation, some sort of restriction argument is necessary to get around this.

### 6.1.3 The one-point uncompactification

In reformulating the function-theory concept of regularly varying functions into a measure-theory concept, there is continual need to deal with sets that are bounded away from the origin. Such sets need to be regarded as "bounded" in an appropriate topology so sequences of measures of such sets can converge nontrivially. This is necessitated by focusing on tail probabilities or exceedance probabilities, which naturally consider probabilities of sets in a neighborhood of infinity. A convenient way to think about this is by means of the *one-point uncompactification*. We have already seen an example of this in Section 3.6 (p. 62).

Let $(\mathbb{X}, \mathcal{T})$ be a nice topological space; $\mathbb{X}$ is the set and $\mathcal{T}$ is the topology, that is, a collection of subsets of $\mathbb{X}$ designated as *open*, satisfying the following:

(i) Both $\emptyset \in \mathcal{T}$ and $\mathbb{X} \in \mathcal{T}$.

(ii) The collection $\mathcal{T}$ is closed under finite intersections and arbitrary unions.

| $\mathbb{X}$ compact set | $\mathbb{X} \setminus \{x\}$ punctured version | Uses |
|---|---|---|
| $[0, \infty]$ | $[0, \infty] \setminus \{0\} = (0, \infty]$ | extremes<br>positive jump Lévy processes |
| $[-\infty, \infty]$ | $[-\infty, \infty] \setminus \{0\}$ | Lévy processes<br>stable processes |
| $[0, \infty]^d = [\mathbf{0}, \boldsymbol{\infty}]$ | $[\mathbf{0}, \boldsymbol{\infty}] \setminus \{\mathbf{0}\}$ | multivariate exceedances<br>$\mathbb{R}^d$-valued positive jump processes |
| $[-\infty, \infty]^d = [-\boldsymbol{\infty}, \boldsymbol{\infty}]$ | $[-\boldsymbol{\infty}, \boldsymbol{\infty}] \setminus \{\mathbf{0}\}$ | multivariate Lévy processes<br>multivariate stable processes |

**Table 6.1.** Compact spaces, their punctured modifications and their uses.

(For example, $\mathbb{X}$ could be a subset of Euclidean space.) Consider a subset $\mathbb{D} \subset \mathbb{X}$, define

$$\mathbb{X}^{\#} = \mathbb{X} \setminus \mathbb{D} = \mathbb{X} \cap \mathbb{D}^c,$$

and give $\mathbb{X}^{\#}$ the relative topology

$$\mathcal{T}^{\#} = \mathcal{T} \cap \mathbb{D}^c = \mathcal{T} \cap \mathbb{X}^{\#}.$$

So a set is open in $\mathbb{X}^{\#}$ if it is an open subset of $\mathbb{X}$ intersected with $\mathbb{X}^{\#}$.

We need to identify the compact sets of $\mathbb{X}^{\#}$. This is done next.

**Proposition 6.1.** *Suppose, as usual, the compact subsets of* $\mathbb{X}$ *are denoted by* $\mathcal{K}(\mathbb{X})$. *Then*

$$\mathcal{K}(\mathbb{X}^{\#}) = \{K \in \mathcal{K}(\mathbb{X}) : K \cap \mathbb{D} = \emptyset\}$$

*are the compact subsets of* $\mathbb{X}^{\#}$.

The compact sets of $\mathbb{X}^{\#}$ are the original compact sets of $\mathbb{X}$, provided they do not intersect the piece $D$ chopped away from $\mathbb{X}$ to form $\mathbb{X}^{\#}$.

Specialize this to the *one-point uncompactification*: Suppose $\mathbb{X}$ is a compact set and $x \in \mathbb{X}$. Give $\mathbb{X}\setminus\{x\}$ the relative topology consisting of sets in $\mathbb{X}\setminus\{x\}$ of the form $G\setminus\{x\}$, where $G \in \mathcal{G}(\mathbb{X})$, the open subsets of $\mathbb{X}$. The compact sets of $\mathbb{X}\setminus\{x\}$ are those compact subsets $K \subset \mathbb{X}$ such that $x \notin K$. Thus the *one-point uncompactification* describes the compact sets of a compact space punctured by the removal of a point.

Special cases, each of which will be of use, are summarized in Table 6.1, which lists compact spaces, the one-point uncompactified versions, and intended uses.

*Proof of Proposition* 6.1. Begin by assuming that

$$K \in \mathcal{K}(\mathbb{X}), \quad K \cap \mathbb{D} = \emptyset;$$

we show that $K \in \mathcal{K}(\mathbb{X}^{\#})$. Let

$$\{G_\gamma^{\#} = G_\gamma \cap \mathbb{X}^{\#}, \gamma \in \Lambda\}$$

be some arbitrary cover of $K$ by open subsets of $\mathbb{X}^{\#}$, where $G_\gamma \in \mathcal{G}(\mathbb{X})$ and $\Lambda$ is some index set. So

$$K \subset \bigcup_{\gamma \in \Lambda} G_\gamma \cap \mathbb{X}^{\#} \subset \bigcup_{\gamma \in \Lambda} G_\gamma.$$

Since $K \in \mathcal{K}(\mathbb{X})$, there is a finite subcollection indexed by $\Lambda' \subset \Lambda$ such that $K \subset \bigcup_{\gamma \in \Lambda'} G_\gamma$. Since $K \cap \mathbb{D} = \emptyset$,

$$K \subset \bigcup_{\gamma \in \Lambda'} G_\gamma \cap \mathbb{X}^{\#}.$$

Therefore, any cover of $K$ by open subsets of $\mathbb{X}^{\#}$ has a finite subcover, and thus $K$ is compact in $\mathbb{X}^{\#}$. Thus

$$\{K \in \mathcal{K}(\mathbb{X}) : K \cap \mathbb{D} = \emptyset\} \subset \mathcal{K}(\mathbb{X}^{\#}).$$

The converse is quite similar. □

### 6.1.4 Multivariate regular variation of measures

The equivalences in Theorem 3.6 (p. 62) suggest a way to proceed with a definition of multivariate regular variation that is useful for probability and statistics. We assume that $\boldsymbol{Z} \geq \boldsymbol{0}$ is a $d$-dimensional random vector that takes values in the nonnegative quadrant $[\boldsymbol{0}, \boldsymbol{\infty})$. (Extensions to positive and negative components are straightforward and discussed later in Section 6.5.5 (p. 201).) Suppose the distribution of $\boldsymbol{Z}$ is $F$. We could say that $F$ has a regularly varying tail if there exist $b_n \to \infty$ and a limit measure $\nu(\cdot)$ on the Borel subsets of the quadrant such that

$$nF(b_n \cdot) = n\mathbb{P}\left[\frac{\boldsymbol{Z}}{b_n} \in \cdot\right] \xrightarrow{v} \nu(\cdot) \tag{6.4}$$

in $M_+([\boldsymbol{0}, \boldsymbol{\infty}] \setminus \{\boldsymbol{0}\})$. This is the correct analogue of (3.29); it is more fully developed in the next theorem.

To deal with multivariate regular variation of tail probabilities, we work in the punctured space with a one-point uncompactification $\mathbb{E} = [\boldsymbol{0}, \boldsymbol{\infty}] \setminus \{\boldsymbol{0}\}$. Equivalent formulations in terms of polar coordinates then have to deal with the fact that the polar coordinate transformation is not defined on the lines through $\boldsymbol{\infty}$, so some kind of restriction argument is necessary. For a somewhat different treatment, see [13, 15, 227]. Set $\aleph_+ = \aleph \cap \mathbb{E}$. Continue to denote vague convergence of measures by $\xrightarrow{v}$.

**Theorem 6.1 (multivariate regularly varying tail probabilities).** *The following statements are equivalent. (In each, we understand the phrase* Radon measure *to mean a Radon measure that is not identically zero and that is not degenerate at a point. Also, repeated use of the symbols $\nu$, $b(\cdot)$, $\{b_n\}$ from statement to statement does not require these objects to be exactly the same in different statements. See Remark* 6.1 *after Theorem* 6.1.)

1. *There exists a Radon measure $\nu$ on $\mathbb{E}$ such that*

$$\lim_{t\to\infty}\frac{1-F(t\boldsymbol{x})}{1-F(t\boldsymbol{1})}=\lim_{t\to\infty}\frac{\mathbb{P}\left[\frac{\boldsymbol{Z}}{t}\in[\boldsymbol{0},\boldsymbol{x}]^c\right]}{\mathbb{P}\left[\frac{\boldsymbol{Z}}{t}\in[\boldsymbol{0},\boldsymbol{1}]^c\right]}=\nu([\boldsymbol{0},\boldsymbol{x}]^c) \tag{6.5}$$

*for all points $\boldsymbol{x}\in[\boldsymbol{0},\infty)\setminus\{\boldsymbol{0}\}$ which are continuity points of the function $\nu([\boldsymbol{0},\cdot]^c)$.*

2. *There exists a function $b(t)\to\infty$ and a Radon measure $\nu$ on $\mathbb{E}$, called the* limit measure, *such that in $M_+(\mathbb{E})$,*

$$t\mathbb{P}\left[\frac{\boldsymbol{Z}}{b(t)}\in\cdot\right]\xrightarrow{v}\nu,\quad t\to\infty. \tag{6.6}$$

3. *There exists a sequence $b_n\to\infty$ and a Radon measure $\nu$ on $\mathbb{E}$ such that in $M_+(\mathbb{E})$,*

$$n\mathbb{P}\left[\frac{\boldsymbol{Z}}{b_n}\in\cdot\right]\xrightarrow{v}\nu,\quad n\to\infty. \tag{6.7}$$

4. *There exists a probability measure $S(\cdot)$ on $\aleph_+$, called the* angular measure, *and a function $b(t)\to\infty$ such that for $(R,\boldsymbol{\Theta})=(\|\boldsymbol{Z}\|,\frac{\boldsymbol{Z}}{\|\boldsymbol{Z}\|})$, we have*

$$t\mathbb{P}\left[\left(\frac{R}{b(t)},\boldsymbol{\Theta}\right)\in\cdot\right]\xrightarrow{v}c\nu_\alpha\times S \tag{6.8}$$

*in $M_+((0,\infty]\times\aleph_+)$ for some $c>0$.*

5. *There exists a probability measure $S(\cdot)$ on $\aleph_+$ and a sequence $b_n\to\infty$ such that for $(R,\boldsymbol{\Theta})=(\|\boldsymbol{Z}\|,\frac{\boldsymbol{Z}}{\|\boldsymbol{Z}\|})$, we have*

$$n\mathbb{P}\left[\left(\frac{R}{b_n},\boldsymbol{\Theta}\right)\in\cdot\right]\xrightarrow{v}c\nu_\alpha\times S \tag{6.9}$$

*in $M_+((0,\infty]\times\aleph_+)$ for some $c>0$.*

*Remark* 6.1. Normalization of all components by the same function means that marginal distributions are tail equivalent; that is [256, 260],

$$\lim_{x\to\infty} \frac{P[Z^{(i)} > x]}{P[Z^{(j)} > x]} =: r_{ij} \in [0, \infty],$$

for $1 \le i, j \le d$. For theoretical considerations, it is best to avoid cases where some marginal tails are heavier than others, corresponding to $r_{ij} = 0$ or $\infty$ for some $(i, j)$, and therefore it is frequently assumed that all components $\{Z^{(i)}, 1 \le i \le d\}$ are identically distributed. In practice, of course, the tails rarely look the same. More on this later in Section 6.5.6 (p. 203).

When $b(t) = t$ or $b_n = n$, we are in the *standard case* [95, 260] and all marginal distributions are tail equivalent to a standard Pareto distribution with $\alpha = 1$. In general, the possible choices of $b(t)$ include the following:

(i) $b(t) = (\frac{1}{1-F_{(1)}})^{\leftarrow}(t)$, where $F_{(1)}(x) = P[Z^{(1)} \le x]$ is the one-dimensional marginal distribution. This choice is sensible if the components of the vector are identically distributed.
(ii) $b(t) = (\frac{1}{1-F_R})^{\leftarrow}(t)$, where $F_R(x) = P[R \le x]$ is the distribution of $\|\boldsymbol{Z}\|$. Note that this choice of $b(\cdot)$ depends on the choice of norm $\|\cdot\|$.

Different choices of $b(\cdot)$ may introduce different constants $c$ in the limit statements.

The following lemma facilitates the proof of Theorem 6.1. Continue to assume that $\mathbb{E} = [\boldsymbol{0}, \boldsymbol{\infty}] \setminus \{\boldsymbol{0}\}$.

**Lemma 6.1.** *Suppose for $n \ge 0$ that $\mu_n \in M_+(\mathbb{E})$. Then*

$$\mu_n \stackrel{v}{\to} \mu_0 \quad \text{in } M_+(\mathbb{E}) \tag{6.10}$$

*iff*

$$\mu_n([\boldsymbol{0}, \boldsymbol{x}]^c) \to \mu_0([\boldsymbol{0}, \boldsymbol{x}]^c) \tag{6.11}$$

*for $\boldsymbol{x} \in [\boldsymbol{0}, \boldsymbol{\infty}) \setminus \{\boldsymbol{0}\}$, which are continuity points of the limit $\mu_0([\boldsymbol{0}, \cdot]^c)$.*

*Proof.* Theorem 3.2 (p. 52) shows that (6.10) implies (6.11), so assume (6.11) and we must prove (6.10). Let $f \in C_K^+(\mathbb{E})$. Then the support of $f$ is contained in $[\boldsymbol{0}, \boldsymbol{x}]^c$ for some continuity set $[\boldsymbol{0}, \boldsymbol{x}]^c$, and since we have convergence on this, set

$$\sup_n \mu_n(f) \le \sup_{\boldsymbol{x}\in\mathbb{E}} f(\boldsymbol{x}) \cdot \sup_n \mu_n([\boldsymbol{0}, \boldsymbol{x}]^c) < \infty.$$

This is true for any $f \in C_K^+(\mathbb{E})$, so $\{\mu_n\}$ is relatively compact from (3.16) (p. 51). If $\mu$ and $\mu'$ are two subsequential limits, then by (6.11) $\mu$ and $\mu'$ agree on the continuity sets $[\mathbf{0}, \boldsymbol{x}]^c$. Now argue that $\mu$ and $\mu'$ must agree on the $\pi$-system of rectangles whose vertices are continuity points of $\mu_0$ and which are bounded away from $\mathbf{0}$, and hence $\mu = \mu'$ on $\mathbb{E}$. □

*Proof of Theorem* 6.1.

1 → 2: Condition 1 says that $\bar{F} := 1 - F$ is a multivariate regularly varying function on the cone $[\mathbf{0}, \infty) \setminus \{\mathbf{0}\}$, and therefore $\bar{F}(t\mathbf{1})$ is a regularly varying function of $t$:

$$\bar{F}(t\mathbf{1}) \in RV_{-\alpha} \quad \text{for some } \alpha > 0.$$

Define $b(t)$ to satisfy

$$\bar{F}(b(t)\mathbf{1}) \sim t^{-1}, \quad t \to \infty.$$

Then replacing $t$ by $b(t)$ in 1 yields

$$t\mathbb{P}\left[\frac{\boldsymbol{Z}}{b(t)} \in [\mathbf{0}, \boldsymbol{x}]^c\right] \to \nu([\mathbf{0}, \boldsymbol{x}]^c).$$

Lemma 6.1 gives (6.6).

2 → 3: Obvious. Replace $t$ by $n$.

3 → 1: There exists a function $b(t) \in RV_{1/\alpha}$ such that $b(n) = b_n$. To see this, use marginal convergence: The assumed vague convergence in (6.7) allows us to insert into (6.7) relatively compact sets (the sets are bounded away from $\mathbf{0}$) of the form

$$[0, \infty] \times \cdots \times (x, \infty] \times [0, \infty] \times \cdots \times [0, \infty].$$

When we do this, we get marginal convergence:

$$n\mathbb{P}\left[\frac{Z^{(i)}}{b_n} > x\right] \to \nu([0, \infty] \times \cdots \times (x, \infty] \times [0, \infty] \times \cdots \times [0, \infty]).$$

Provided the limit is nonzero, which it must be for some $i$, we get that the marginal tail satisfies the sequential form of regular variation (see Proposition 2.3 (p. 21)). Hence the marginal tail is regularly varying and the quantile function $b(t)$, by inversion, is regularly varying. (See Proposition 2.6(v) (p. 32).) We can set $b(n) = b_n$.

Now for $\boldsymbol{x}$ a continuity point of $\nu([\mathbf{0}, \cdot]^c)$, we have by Theorem 3.2 (p. 52) or Lemma 6.1 that

$$n\mathbb{P}\left[\frac{\boldsymbol{Z}}{b_n} \in [\mathbf{0}, \boldsymbol{x}]^c\right] \to \nu([\mathbf{0}, \boldsymbol{x}]^c).$$

For any $t$, there exists an integer $n(t)$ such that

$$b_{n(t)} \leq t < b_{n(t)+1},$$

and so

$$
\begin{aligned}
b^{\leftarrow}(t)\mathbb{P}\left[\frac{\boldsymbol{Z}}{t} \in [\boldsymbol{0}, \boldsymbol{x}]^c\right] &\le b^{\leftarrow} \circ b(n(t)+1)\mathbb{P}\left[\frac{\boldsymbol{Z}}{b(n(t))} \in [\boldsymbol{0}, \boldsymbol{x}]^c\right] \\
&\sim n(t)\mathbb{P}\left[\frac{\boldsymbol{Z}}{b(n(t))} \in [\boldsymbol{0}, \boldsymbol{x}]^c\right] \\
&\to \nu([\boldsymbol{0}, \boldsymbol{x}]^c).
\end{aligned}
$$

A similar argument gives a lower bound, which gives 1 since $b^{\leftarrow}(\cdot)$ is regularly varying. Recall that the rephrasing of the definition of regular variation given in (6.3) allows normalization by any regularly varying function.

We summarize what we have proved so far: $1 \leftrightarrow 2 \leftrightarrow 3$. Assuming any one of these, the measure $\nu$ places no mass on the lines through $\infty$,

$$\nu(\mathbb{E} \setminus ([\boldsymbol{0}, \boldsymbol{\infty}) \setminus \{\boldsymbol{0}\})) = 0.$$

Reason: If there were mass on a line through $\infty$, one of the one-dimensional marginals would have mass at $\infty$. This, however, is impossible since

$$
\begin{aligned}
&\nu(\mathbb{E} \setminus ([\boldsymbol{0}, \boldsymbol{\infty}) \setminus \{\boldsymbol{0}\})) \\
&\le \lim_{x \to \infty} \sum_{i=1}^{d} \nu([0,\infty] \times \cdots \times (x,\infty] \times [0,\infty] \times \cdots \times [0,\infty]) \\
&= \lim_{x \to \infty} \sum_{i=1}^{d} c_i x^{-\alpha} = 0.
\end{aligned}
$$

The equivalence of 4 and 5 is similar to the equivalence of 2 and 3 and is omitted. It remains to show $3 \leftrightarrow 5$ and we content ourselves with showing $3 \to 5$ since the converse is very similar. We proceed in a series of steps.

STEP 1: *Restrict the space to the natural domain of the polar coordinate transformation.* We claim that (6.7) implies convergence on a smaller domain:

$$n\mathbb{P}\left[\frac{\boldsymbol{Z}}{b_n} \in \cdot\right] \xrightarrow{v} \nu \quad \text{in } M_+([\boldsymbol{0}, \boldsymbol{\infty}) \setminus \{\boldsymbol{0}\}). \tag{6.12}$$

To verify the claim (6.12), let $f \in C_K^+([\boldsymbol{0}, \boldsymbol{\infty}) \setminus \{\boldsymbol{0}\})$ and suppose $K \in \mathcal{K}([\boldsymbol{0}, \boldsymbol{\infty}) \setminus \{\boldsymbol{0}\})$ is the compact support. Then $K$ is also compact in $\mathbb{E}$ by Proposition 6.1. Extend $f$ to a function $\tilde{f}$ on $\mathbb{E}$ by

$$\tilde{f}(x) = \begin{cases} f(x) & \text{if } x \in K, \\ 0 & \text{if } x \in \mathbb{E} \setminus K. \end{cases}$$

Then $\tilde{f} \in C_K^+(\mathbb{E})$. Now 3 implies that

$$n\mathbf{E}\tilde{f}\left(\frac{\mathbf{Z}}{b_n}\right) \to \nu(\tilde{f}),$$

which is the same as

$$n\mathbf{E}f\left(\frac{\mathbf{Z}}{b_n}\right) \to \nu(f),$$

which implies (6.12).

See also Problem 6.3 (p. 206) for alternatives using the *restriction functional*.

STEP 2: *Apply the polar coordinate transform $T$; check the compactness criterion.* In the restricted space $[\mathbf{0}, \infty) \setminus \{\mathbf{0}\}$, we may now apply the polar coordinate transformation

$$T(x) = (r, \boldsymbol{a}) = \left(\|\boldsymbol{x}\|, \frac{\boldsymbol{x}}{\|\boldsymbol{x}\|}\right).$$

Note that

$$T : [\mathbf{0}, \infty) \setminus \{\mathbf{0}\} \mapsto (0, \infty) \times \aleph_+.$$

Let $K_2 \in \mathcal{K}((0, \infty) \times \aleph_+)$. Then $K_2$ is closed and contained in a set of the form

$$\{(r, \boldsymbol{a}) : \delta \le r \le M, \boldsymbol{a} \in \aleph_+\}$$

for small $\delta > 0$ and large $M$. Since $T$ is continuous, $T^{-1}(K_2)$ is closed and contained in

$$T^{-1}\{(r, \boldsymbol{a}) : \delta \le r \le M, \boldsymbol{a} \in \aleph_+\} = \{\boldsymbol{x} \in \mathbb{E} : \delta \le \|\boldsymbol{x}\| \le M\},$$

which is compact in $[\mathbf{0}, \infty) \setminus \{\mathbf{0}\}$. So $T^{-1}(K_2)$, being a closed subset of a compact set, is compact. Thus the compactness criterion of Proposition 5.5 (p. 141), is satisfied. Apply Proposition 5.5 to the convergence in (6.12), and we get

$$n\mathbb{P}\left[\left(\frac{R}{b_n}, \boldsymbol{\Theta}\right) \in \cdot\right] \stackrel{v}{\to} \nu \circ T^{-1}(\cdot) \quad \text{in } M_+((0, \infty) \times \aleph_+). \tag{6.13}$$

What is the form of $\nu \circ T^{-1}(\cdot)$? From (6.5) we have that $\nu([\mathbf{0}, \boldsymbol{x}]^c)$ is the limit function in the definition of multivariate regular variation, and hence $\nu([\mathbf{0}, \boldsymbol{x}]^c)$ has a scaling property

$$\nu([\mathbf{0}, s\boldsymbol{x}]^c) = s^{-\alpha}\nu([\mathbf{0}, \boldsymbol{x}]^c).$$

So for any rectangle $I$ bounded away from $\mathbf{0}$, we have

$$\nu(sI) = s^{-\alpha}\nu(I).$$

The class of rectangles is closed under finite intersections and generates the Borel $\sigma$-algebra $\mathcal{E}$, so by Dynkin's $\pi - \lambda$ theorem [264], for any $A \in \mathcal{E}$, we have

$$\nu(sA) = s^{-\alpha}\nu(A).$$

This means that for any $t > 0$ and measurable set $\Lambda \subset \aleph_+$, we have

$$\begin{aligned}
\nu\left\{\boldsymbol{x} \in [\mathbf{0}, \infty) \setminus \{\mathbf{0}\} : \|\boldsymbol{x}\| > t, \frac{\boldsymbol{x}}{\|\boldsymbol{x}\|} \in \Lambda\right\} &= \nu\left\{\boldsymbol{x} : \|t^{-1}\boldsymbol{x}\| > 1, \frac{t^{-1}\boldsymbol{x}}{\|t^{-1}\boldsymbol{x}\|} \in \Lambda\right\} \\
&= \nu\left\{t\boldsymbol{y} : \|\boldsymbol{y}\| > 1, \frac{\boldsymbol{y}}{\|\boldsymbol{y}\|} \in \Lambda\right\} \\
&= t^{-\alpha}\nu\left\{\boldsymbol{y} : \|\boldsymbol{y}\| > 1, \frac{\boldsymbol{y}}{\|\boldsymbol{y}\|} \in \Lambda\right\} \\
&=: t^{-\alpha}cS(\Lambda).
\end{aligned}$$

So homogenity implies the product form when the measure is applied to a pizza slice shaped polar set. Thus on $(0, \infty) \times \aleph_+$,

$$\nu \circ T^{-1} = c\nu_\alpha \times S. \tag{6.14}$$

STEP 3: *Extend to bigger space, where boundaries at infinity are included.* We now extend the convergence in (6.13) to $M_+((0, \infty] \times \aleph_+)$. To economize on notation for the proof, suppose the constant $c$ in (6.14) is 1. Let $f \in C_K^+((0, \infty] \times \aleph_+)$ and set $\|f\| = \sup_{(r,\boldsymbol{\theta}) \in (0,\infty] \times \aleph_+} f(r, \boldsymbol{\theta}) < \infty$. To relate this function $f$ to one defined on $C_K^+((0, \infty) \times \aleph_+)$, we perform a smooth truncation using the function

$$\phi_{+\delta,M}(t) = \begin{cases} 1 & \text{if } 0 < t \le M, \\ 0 & \text{if } t = M + \delta, \\ \text{linear interpolation} & \text{if } M < t \le M + \delta, \end{cases}$$

and we define

$$f_{M,\delta}(r, \theta) = f(r, \theta)\phi_{+\delta,M}(r) \in C_K^+((0, \infty) \times \aleph_+).$$

We have

$$\begin{aligned}
&\left| n\mathbf{E}f\left(\frac{R}{b_n}, \boldsymbol{\Theta}\right) - \nu_\alpha \times S(f)\right| \\
&\quad \le \left| n\mathbf{E}f\left(\frac{R}{b_n}, \boldsymbol{\Theta}\right) - n\mathbf{E}f_{M,\delta}\left(\frac{R}{b_n}, \boldsymbol{\Theta}\right)\right|
\end{aligned}$$

$$+ \left| n\mathbf{E} f_{M,\delta}\left(\frac{R}{b_n}, \mathbf{\Theta}\right) - \nu_\alpha \times S(f_{M,\delta}) \right| + |\nu_\alpha \times S(f_{M,\delta}) - \nu_\alpha \times S(f)|$$
$$= A + B + C.$$

Since $f_{M,\delta} \in C_K^+((0,\infty) \times \aleph_+)$, we have $\lim_{n\to\infty} B = 0$ from (6.13). The term $C$ is bounded by

$$\int |f(r,\theta)| 1 - \phi_{+,M}(r)|\nu_\alpha(dr)S(d\theta) \leq \|f\|\nu_\alpha(M,\infty] = \|f\|M^{-\alpha},$$

which can be made arbitrarily small by a suitable choice of $M$. The term $A$ is handled similarly after taking lim sup on $n$. This gives (6.9). □

*Remark* 6.2. We call the measure $\nu$ in, say, (6.6) the *limit measure*. The probability measure $S$ on $\aleph_+$ is called the *angular measure*. Theorem 6.1 shows that for a given $\alpha$, the class of limit measures is large since the class is in 1–1 correspondence with the set of probability measures on $\aleph_+$. If we take an independent pair $(R, \mathbf{\Theta})$ on $(0,\infty] \times \aleph_+$ with

$$\mathbb{P}[R > r] = r^{-\alpha}, \quad r > 1, \qquad \mathbb{P}[\mathbf{\Theta} \in \cdot] = S(\cdot),$$

then

$$t\mathbb{P}\left[\frac{R}{t^{1/\alpha}} > r, \mathbf{\Theta} \in \Lambda\right] = t(t^{1/\alpha}r)^{-\alpha}S(\Lambda) = \nu_\alpha \times S((r,\infty] \times \Lambda).$$

So any probability measure $S$ on $\aleph_+$ is a possible angular measure.

## 6.2 The Poisson transform

Multivariate regular variation of the probability distributions as given in the equivalences of Theorem 6.1 in either Cartesian or polar coordinates is equivalent to induced empirical measures weakly converging to Poisson random measure limits. We state the result next.

**Theorem 6.2.** *Suppose* $\{\mathbf{Z}, \mathbf{Z}_1, \mathbf{Z}_2, \dots\}$ *are iid; after transformation to polar coordinates, the sequence is* $\{(R, \mathbf{\Theta}), (R_1, \mathbf{\Theta}_1), (R_2, \mathbf{\Theta}_2), \dots\}$. *Any of the equivalences in Theorem* 6.1 (*p.* 173) *is also equivalent to the following:*

6. *There exists* $b_n \to \infty$ *such that*

$$\sum_{i=1}^{n} \epsilon_{\mathbf{Z}_i/b_n} \Rightarrow \text{PRM}(\nu) \tag{6.15}$$

*in* $M_p(\mathbb{E})$.

7. *There exists a sequence* $b_n \to \infty$ *such that*

$$\sum_{i=1}^{n} \epsilon_{(R_i/b_n, \Theta_i)} \Rightarrow \mathrm{PRM}(c\nu_\alpha \times S) \tag{6.16}$$

*in* $M_p((0,\infty] \times \aleph_+)$.

*These conditions imply that for any sequence* $k = k(n) \to \infty$ *such that* $n/k \to \infty$, *we have the following:*

8. *In* $M_+(\mathbb{E})$,

$$\frac{1}{k}\sum_{i=1}^{n} \epsilon_{\mathbf{Z}_i/b(\frac{n}{k})} \Rightarrow \nu \tag{6.17}$$

*and*

9. *In* $M_+((0,\infty] \times \aleph_+)$,

$$\frac{1}{k}\sum_{i=1}^{n} \epsilon_{(R_i/b(\frac{n}{k}), \mathbf{\Theta}_i)} \Rightarrow c\nu_\alpha \times S, \tag{6.18}$$

*and* 8 *or* 9 *is equivalent to any of* 1–7, *provided* $k(\cdot)$ *satisfies* $k(n) \sim k(n+1)$.

*Proof.* The bridge between the lists of equivalences in Theorems 6.2 and 6.1 is Theorem 5.3 (p. 138). □

Thus multivariate regular variation has an exact probabilistic equivalence in terms of convergence of empirical measures to a limiting Poisson random measure.

The following variant is needed for proving weak convergence of partial sum processes or maximal processes in the space $D[0,\infty)$.

**Theorem 6.3.** *Suppose* $\{\mathbf{Z}, \mathbf{Z}_1, \mathbf{Z}_2, \dots\}$ *are iid random elements of* $[\mathbf{0},\infty)$. *Then multivariate regular variation of the distribution of* $\mathbf{Z}$ *in* $\mathbb{E} = [\mathbf{0},\infty] \setminus \{\mathbf{0}\}$,

$$n\mathbb{P}\left[\frac{\mathbf{Z}}{b_n} \in \cdot\right] \xrightarrow{v} \nu,$$

*is also equivalent to*

$$\sum_{j} \epsilon_{(\frac{j}{n}, \mathbf{Z}_j/b_n)} \Rightarrow \mathrm{PRM}(\mathbb{LEB} \times \nu) \tag{6.19}$$

*in* $M_+([0,\infty) \times \mathbb{E})$.

*Proof.* We proceed in a series of steps to prove that regular variation implies (6.19). The converse is clear, for example, from Problem 3.7 (p. 65).

STEP 1. It suffices to prove (6.19) in $M_+([0, T] \times \mathbb{E})$ for any $T > 0$. To see this, observe that for $f \in C_K^+([0, \infty) \times \mathbb{E})$, with compact support in $[0, T] \times \mathbb{E}$, the Laplace functional of a random measure $M$ at $f$ is the same as the restriction of the random measure to $[0, T] \times \mathbb{E}$ evaluated on the restriction of $f$ to $[0, T] \times \mathbb{E}$.

For convenience, we restrict our attention to proving convergence in $M_+([0, 1] \times \mathbb{E})$.

STEP 2. Suppose $U_1, \ldots, U_n$ are iid $U(0, 1)$ random variables with order statistics

$$U_{1:n} \leq U_{2:n} \leq \cdots \leq U_{n:n},$$

which are independent of $\{\mathbf{Z}_j\}$. We claim that

$$\sum_{j=1}^{n} \epsilon_{(U_{j:n}, \mathbf{Z}_j/b_n)} \Rightarrow \mathrm{PRM}(\mathbb{LEB} \times \nu) \tag{6.20}$$

in $M_+([0, 1] \times \mathbb{E})$. A more general result is explored in Problem 6.7, but we can prove the simple result (6.20) as follows. First, we have, from the independence of $\{U_j\}$ and $\{\mathbf{Z}_n\}$, that

$$\sum_{j=1}^{n} \epsilon_{(U_{j:n}, \mathbf{Z}_j/b_n)} \stackrel{d}{=} \sum_{j=1}^{n} \epsilon_{(U_j, \mathbf{Z}_j/b_n)}$$

as random elements of $M_+([0, 1] \times \mathbb{E})$. Thus we need to prove

$$\sum_{j=1}^{n} \epsilon_{(U_j, \mathbf{Z}_j/b_n)} \Rightarrow \mathrm{PRM}(\mathbb{LEB} \times \nu) \quad (n \to \infty),$$

in $M_+([0, 1] \times \mathbb{E})$. However, because of independence,

$$n\mathbb{P}\left[\left(U_1, \frac{\mathbf{Z}_1}{b_n}\right) \in \cdot\right] = \mathbb{LEB} \times n\mathbb{P}\left[\frac{\mathbf{Z}_1}{b_n} \in \cdot\right] \Rightarrow \mathbb{LEB} \times \nu,$$

and therefore, from Theorem 5.3 (p. 138), the result follows.

STEP 3. Let $d(\cdot, \cdot)$ be the vague metric (cf. (3.14) (p. 51)) on $M_+([0, 1] \times \mathbb{E})$. From Slutsky's Theorem, Theorem 3.4 (p. 55), Step 2 implies the desired result if we prove that

$$d\left(\sum_{j=1}^{n} \epsilon_{(\frac{j}{n}, \mathbf{Z}_j/b_n)}, \sum_{j=1}^{n} \epsilon_{(U_{j:n}, \mathbf{Z}_j/b_n)}\right) \stackrel{P}{\to} 0 \tag{6.21}$$

as $n \to \infty$. From the definition of the vague metric in (3.14) (p. 51), it is enough to prove for $h \in C_K^+([0,1] \times \mathbb{E})$ that

$$\left| \sum_{j=1}^n h\left(\frac{j}{n}, \boldsymbol{Z}_j/b_n\right) - \sum_{j=1}^n h(U_{j:n}, \boldsymbol{Z}_j/b_n) \right| \stackrel{P}{\to} 0 \tag{6.22}$$

in $\mathbb{R}$. Suppose the compact support of $h$ is contained in $[0,1] \times \{\boldsymbol{x} : \|\boldsymbol{x}\| > \delta\}$ for some $\delta > 0$. Then the difference in (6.22) is bounded by

$$\sum_{j=1}^n \left| h\left(\frac{j}{n}, \boldsymbol{Z}_j/b_n\right) - h(U_{j:n}, \boldsymbol{Z}_j/b_n) \right| 1_{[\|\boldsymbol{Z}_j\|/b_n > \delta]}$$
$$\leq \omega_h\left(\sup_{j \leq n} |\frac{j}{n} - U_{j:n}|\right) \sum_{j=1}^n 1_{[\|\boldsymbol{Z}_j\|/b_n > \delta]},$$

where, as usual, $\omega_h(\eta)$ is the modulus of continuity of the uniformly continuous function $h$:

$$\omega_h(\eta) := \sup_{\|\boldsymbol{x}-\boldsymbol{y}\| \leq \eta} |h(\boldsymbol{x}) - h(\boldsymbol{y})|.$$

Now we know from Theorem 6.2 (p. 179) that

$$\sum_{j=1}^n 1_{[\|\boldsymbol{Z}_j\|/b_n > \delta]} = \sum_{j=1}^n \epsilon_{\boldsymbol{Z}_j/b_n}(\{\boldsymbol{x} \in \mathbb{E} : \|\boldsymbol{x}\| > \delta\})$$

converges and hence is stochastically bounded. So it is enough to prove that

$$\sup_{j \leq n} \left| \frac{j}{n} - U_{j:n} \right| \stackrel{P}{\to} 0 \quad (n \to \infty). \tag{6.23}$$

However, from the Glivenko–Cantelli theorem [264, Section 7.5], [24],

$$\sup_{0 \leq x \leq 1} \left| \frac{1}{n} \sum_{i=1}^n 1_{[U_j \leq x]} - x \right| \stackrel{\text{a.s.}}{\to} 0,$$

and hence, by inversion (see Proposition 3.2), inverses converge uniformly almost surely as well, which gives (6.23). □

A slightly more general formulation of Theorem 6.3 using the language of Theorem 5.3 is possible. The proof is the same.

**Corollary 6.1.** *Suppose for each* $n = 1, 2, \dots$ *that* $\{\boldsymbol{X}_{n,j}, j \geq 1\}$ *are iid random elements of a nice space* $(\mathbb{E}, \mathcal{E})$ *such that*

$$nP[\boldsymbol{X}_{n,1} \in \cdot] \stackrel{v}{\to} \nu.$$

*Then in* $M_p([0, \infty) \times \mathbb{E})$, *we have*

$$\sum_i \epsilon_{(i/n, \boldsymbol{X}_{n,i})} \Rightarrow \sum_i \epsilon_{(t_i, \boldsymbol{j}_i)} = \text{PRM}(\mathbb{LEB} \times \nu)$$

*as* $n \to \infty$.

## 6.3 Multivariate peaks over threshhold

The previous two sections described the connection between multivariate regular variation and Poisson processes and random measures. This connection leads to a dimensionless view of the peaks-over-threshold (POT) method in statistics. Assuming multivariate regular variation, the POT method assumes that the actual distribution of observations larger than a fixed threshold is the limit distribution if we send the threshold to infinity. Here are more details.

Suppose the multivariate regular variation condition of Theorem 6.1 holds. Set

$$\mathbb{E}^{>} := \{\boldsymbol{x} \in \mathbb{E} : \|\boldsymbol{x}\| \leq 1\}^c = \{\boldsymbol{x} \in \mathbb{E} : \|\boldsymbol{x}\| > 1\}.$$

Apply the restriction functional (see Problem 6.3 (p. 206)) to (6.15) in Theorem 6.2 (p. 179). So we restrict points to $\mathbb{E}^{>}$, which yields

$$\begin{aligned} \sum_{i=1}^n \epsilon_{\boldsymbol{Z}_i/b_n}(\mathbb{E}^{>} \cap \cdot) &= \sum_{i=1}^n 1_{[\|\boldsymbol{Z}_i\|/b_n > 1]} \epsilon_{\boldsymbol{Z}_i/b_n} \\ &\Rightarrow N_\infty(\mathbb{E}^{>} \cap \cdot) = \text{PRM}(\nu(\mathbb{E}^{>} \cap \cdot)), \end{aligned} \tag{6.24}$$

where $N_\infty(\cdot)$ is the limiting PRM in (6.15). The limiting point process in (6.24) is a Poisson random measure on $\mathbb{E}^{>}$ that has a finite total mean measure $\nu(\mathbb{E}^{>})$. Hence by the construction in Section 5.4.2 (p. 143), this limit can be constructed as follows. Let $\xi$ be a Poisson random variable with parameter $\nu(\mathbb{E}^{>})$ independent of the iid random vectors $\{\boldsymbol{X}_i, i \geq 1\}$, which have common distribution $\nu(\mathbb{E}^{>} \cap \cdot)/\nu(\mathbb{E}^{>})$. Then

$$N_\infty(\mathbb{E}^{>} \cap \cdot) \stackrel{d}{=} \sum_{i=1}^{\xi} \epsilon_{\boldsymbol{X}_i}.$$

**Peaks over threshold.** Consider the observations falling in $\mathbb{E}^>$ as the thresholded sample. The number of thresholded observations is approximately Poisson distributed with parameter $\nu(\mathbb{E}^>)$. These big observations relative to the threshold are approximately iid with distribution $\nu(\mathbb{E}^> \cap \cdot)/\nu(\mathbb{E}^>)$. The POT philosophy treats this approximate limit distribution as the actual distribution. In one dimension, this allows the use of the likelihood method.

Recall the procedure from Section 4.2 when $d = 1$: Assuming $b_n$ is the quantile function (which needs to be replaced by an order statistics estimator), the limit measure is $\nu_\alpha$, so $\nu_\alpha(1, \infty] = 1$, and the exceedances $\{X_i, i \geq 1\}$ are iid Pareto random variables on $[1, \infty)$ independent of the Poisson random variable $\xi$ with parameter $\nu_\alpha(1, \infty] = 1$. Thus, in one dimension, the POT method for estimating $\alpha$ would be as follows: Pick a threshold $T$ (which plays the role of $b_n$), look at the observations larger than $T$, and regard these observations normalized by $T$ as a random sample from the Pareto distribution with parameter $\alpha$. Use maximum likelihood to estimate $\alpha$.

For higher dimensions, this methodology can be mimicked if one assumes the limit measure $\nu$ is a member of a parametric family.

## 6.4 Why bootstrapping heavy-tailed phenomena is difficult

Sometimes, when estimating parameters in a complex model, one is confronted by the difficulty that the limit distribution of the centered estimator vector either is unknown, is too complicated to calculate explicitly, or, if known, still depends on the unknown parameters of the model. This prevents easy construction of confidence regions for the parameters. In classical contexts, the bootstrap [123–125, 166] was designed to overcome these difficulties. For heavy-tailed phenomena, the bootstrap has complexities preventing easy application. The root of the complexity is the distinction between (6.15) and (6.17) (p. 179). In (6.15), the limit is random, and in (6.17), the limit is deterministic. This subtlety distinguishes the bootstrap for heavy-tailed phenomena from classical problems.

### 6.4.1 An example to fix ideas

As an example which will fix ideas, suppose we have the stationary *autoregressive process of order* $p$, denoted by AR$(p)$, with nonnegative innovations $\{Z_t\}$ and with autoregressive coefficients $\phi_1, \dots, \phi_p$, $\phi_p \neq 0$. These processes are defined by the following relation:

$$X_t = \sum_{k=1}^{p} \phi_k X_{t-k} + Z_t; \quad t = 0, \pm 1, \pm 2, \dots, \tag{6.25}$$

where we assume that $\{Z_t\}$ is an independent and identically distributed (iid) sequence of nonnegative random variables with $\mathbb{P}[Z_1 > x] \in RV_{-\alpha}$, $\alpha > 0$. Assume that the order $p$ is known. Based on observation of $\{X_1, \ldots, X_n\}$, the task is to estimate the parameters.

The usual, classical method of estimating $\boldsymbol{\phi} = (\phi_1, \ldots, \phi_p)$, is the Yule–Walker method (see, for example, [31] for an excellent discussion of required background and [134] for the heavy-tailed case). However, one can sometimes do better by exploiting the special nature of the innovations. This was the motivation behind other methods discussed, for instance, in [2, 3, 67, 69, 131–133, 188, 221]. See also [129, Chapter 7], [192, 237].

Assuming the model is correct (a big assumption), the new estimators have excellent properties. Let $\hat{\boldsymbol{\phi}}(n)$ be an estimator of the vector of autoregressive coefficients; it is typical that $r(n)(\hat{\boldsymbol{\phi}}(n) - \boldsymbol{\phi})$ has a limit distribution for an appropriate choice of $\{r(n)\}$. However, this limit distribution may have the unfortunate characteristic that it depends on the unknown parameters $\boldsymbol{\phi}$ and $\alpha$, especially when $\alpha < 2$. For inference purposes, this is a serious difficulty, which we can try to overcome by using the bootstrap.

For the autoregressive model, a bootstrap procedure can be constructed as follows:

1. Assume we observe $X_1, \ldots, X_n$ from the autoregressive model (6.25).

2. Use your favorite method (Yule–Walker, linear programming, periodogram) to estimate the autoregressive coefficients and obtain a vector of estimates $\hat{\boldsymbol{\phi}}(n)$.

3. Use these estimates $\hat{\boldsymbol{\phi}}(n)$ to estimate the residuals
$$\hat{Z}_t(n) = X_t - \sum_{i=1}^{p} \hat{\phi}_i(n) X_{t-i}; \quad t = 1, \ldots, n.$$

4. Form the empirical measure generated by the estimated residuals
$$\hat{F}_n^{\text{resid}} = \frac{1}{n} \sum_{i=1}^{n} \epsilon_{\hat{Z}_i(n)}.$$

5. Resample: Draw an iid (bootstrap) sample $\{Z_t^*(n); t = 1, \ldots, m\}$ from the distribution $\hat{F}_n^{\text{resid}}$.

6. Construct a bootstrap time series $\{X_t^*(n); t = 1, \ldots, m\}$ by setting $X_0^*(n) = \cdots = X_{-p+1}^*(n) = 0$ and then using the recursion
$$X_t^*(n) = \sum_{i=1}^{p} \hat{\phi}_i(n) X_{t-i}^*(n) + Z_t^*(n).$$

7. Based on the bootstrapped time-series sample $\{X_t^*(n); t = 1, \ldots, m\}$, estimate autoregressive coefficients again; call these estimators $\hat{\boldsymbol{\phi}}^*(n)$.

8. The ideal bootstrap distribution is the sampling distribution (known in principle), conditional on $\hat{F}_n^{\text{resid}}$, of

$$r(m)(\hat{\boldsymbol{\phi}}^*(n) - \hat{\boldsymbol{\phi}}(n)).$$

It is difficult to compute the ideal bootstrap distribution, so, in practice, Monte Carlo is necessary. Also, a modification of this method is necessary to account for the fact that $r(m)$ is unknown. Assuming these hurdles are overcome, a confidence region for the parameters can be constructed.

However, the important issue is that in step 5, heavy-tail asymptotics require the bootstrap sample size to satisfy $m = m(n) \to \infty$, but $m/n \to 0$ as $n \to \infty$. Why? In connection with bootstrapping extremes and heavy-tailed phenomena, several authors have noticed that if the original sample is of size $n$, in order for the bootstrap asymptotics to work as desired, the bootstrap sample size should be of smaller order. See, for example, [9, 10, 79, 133, 147, 156, 185, 189, 205] and [21, Section 6].

### 6.4.2 Why the bootstrap sample size must be carefully chosen

We now now discuss understanding bootstrapping of heavy-tailed phenomena in the context of Theorems 6.1 and 6.2. We assume that $\{\mathbf{Z}, \mathbf{Z}_1, \mathbf{Z}_2, \ldots\}$ are iid random elements of $[\mathbf{0}, \infty)$ satisfying the regular variation condition (6.4) or one of its equivalent forms.

#### The bootstrap procedure

Assume we observe $\mathbf{Z}_1, \ldots, \mathbf{Z}_n$. The empirical distribution of the observed sample is

$$\hat{F}_n = \frac{1}{n} \sum_{i=1}^{n} \epsilon_{\mathbf{Z}_i}.$$

Sample $m$ times to get a *bootstrap* sample $\mathbf{Z}_1^*, \ldots, \mathbf{Z}_m^*$. For bootstrap asymptotics to work, we want the statistical characteristics of $\mathbf{Z}_1^*, \ldots, \mathbf{Z}_m^*$ to mimic those of the original sample. We take this to mean that the distribution of

$$\sum_{i=1}^{m} \epsilon_{\mathbf{Z}_i^*/q(m)}$$

for some scaling function $q(m) > 0$ should be close to the distribution of PRM($\nu$), according to (6.15) and Theorem 6.2. This will be the case only when $m = m(n) \to \infty$ and $m/n \to 0$.

So what makes the bootstrap procedure problematic in the heavy-tail case is the need to choose $m$. This can be as tricky as choosing a threshold or choosing $k$, the number of upper-order statistics used in, say, Hill estimation.

### What exactly is the bootstrap procedure?

Suppose we have the iid sequence $\{\boldsymbol{Z}_n, n \geq 1\}$ defined on some probability space $(\Omega, \mathcal{A}, \mathbb{P})$. Assume that the probability space is rich enough to support an array $\{(I_1^{(n)}, \dots, I_m^{(n)}), n \geq 1\}$ that is independent of $\{\boldsymbol{Z}_n, n \geq 1\}$ and has the property that for each $n = 1, 2, \dots$, $I_1^{(n)}, \dots, I_m^{(n)}$ are iid and uniformly distributed on $\{1, \dots, n\}$. We imagine repeated multinomial trials, and

$$P_j(n) := \sum_{i=1}^{m} 1_{[I_i^{(n)}=j]}, \quad j = 1, \dots, n,$$

is a multinomial random vector corresponding to $m$ trials and possible outcomes $1, \dots, n$.

A bootstrap sample $\boldsymbol{Z}_1^*, \dots, \boldsymbol{Z}_m^*$ of size $m$ is obtained by sampling with replacement $m$ times from the population $1, \dots, n$. If the $i$th sample yields $j$, then $\boldsymbol{Z}_i^* = \boldsymbol{Z}_j$. Another way to think about this is

$$\boldsymbol{Z}_i^* = \boldsymbol{Z}_{I_i^{(n)}}, \quad i = 1, \dots, m,$$

and then for a scaling function $q(m)$,

$$\sum_{i=1}^{m} \epsilon_{\boldsymbol{Z}_i^*/q(m)} = \sum_{j=1}^{n} \left( \sum_{i=1}^{m} 1_{[I_i^{(n)}=j]} \right) \epsilon_{\boldsymbol{Z}_j/q(m)} = \sum_{j=1}^{n} P_j(n) \epsilon_{\boldsymbol{Z}_j/q(m)}. \tag{6.26}$$

Let PR($\mathbb{S}$) be the set of all probability measures on the Borel $\sigma$-algebra $\mathcal{S}$ of subsets of the complete, separable metric space $\mathbb{S}$. There is a notion of convergence, namely weak convergence, in PR($\mathbb{S}$), and this convergence concept is compatible with a metric that turns PR($\mathbb{S}$) into a complete, separable metric space [25, p. 72]. In particular, we need PR($M_+(\mathbb{E})$) and PR($M_p(\mathbb{E})$) because bootstrap asymptotics are usually considered conditional on the sample. So we will consider

$$\mathbb{P}\left[ \sum_{i=1}^{m} \epsilon_{\boldsymbol{Z}_i^*/b_m} \in \cdot \mid \boldsymbol{Z}_1, \dots, \boldsymbol{Z}_n \right],$$

which is a random element of PR($M_p(\mathbb{E})$).

**When bootstrap asymptotics work**

If $m = m(n) \to \infty$ and $m/n \to 0$, then bootstrap asymptotics work in the sense given in the next propositon.

**Proposition 6.2.** *Suppose $m(n) \to \infty$, $m/n \to 0$ as $n \to \infty$. Then as $n \to \infty$,*

$$\mathbb{P}\left[\sum_{i=1}^{m} \epsilon_{\mathbf{Z}_i^*/b_m} \in \cdot \mid \mathbf{Z}_1, \ldots, \mathbf{Z}_n\right] \stackrel{P}{\to} \mathbb{P}[\mathrm{PRM}(\nu) \in \cdot]$$

*in* $\mathrm{PR}(M_+(E))$. *Hence, taking expectations,*

$$\mathbb{P}\left[\sum_{i=1}^{m} \epsilon_{\mathbf{Z}_i^*/b_m} \in \cdot\right] \Rightarrow \mathbb{P}[\mathrm{PRM}(\nu) \in \cdot]$$

*in* $\mathrm{PR}(M_+(\mathbb{E}))$, *that is, the distribution of* $\sum_{i=1}^{m} \epsilon_{\mathbf{Z}_i^*/b_m}$ *converges weakly to* $\mathrm{PRM}(\nu)$.

Proposition 6.2 provides us with the motivation to subsample—only then will the bootstrap distribution of the point process approximate the true asymptotic distribution of the original point process.

*Proof.* First, observe using (6.26) that

$$\begin{aligned}\mathbf{E}\left(\sum_{i=1}^{m} \epsilon_{\mathbf{Z}_i^*/b_m} | \mathbf{Z}_1, \ldots, \mathbf{Z}_n\right) &= \sum_{j=1}^{n} \mathbf{E}\left(\sum_{i=1}^{m} 1_{[I_i^{(n)}=j]}\right) \epsilon_{\mathbf{Z}_j/b(m)} \\ &= \frac{m}{n} \sum_{j=1}^{n} \epsilon_{\mathbf{Z}_j/b(m)} \stackrel{P}{\to} \nu\end{aligned}$$

from (5.16) of Theorem 5.3 (p. 139) or (6.17) of Theorem 6.2 (p. 179), with $k$ playing the role of $n/m$.

For any subsequence $\{n''\}$ of $\{n\}$, choose a further subsequence $\{n'\}$ along which

$$\mathbb{E}\left(\sum_{i=1}^{m(n')} \epsilon_{\mathbf{Z}_i^*/b_{m(n')}} \mid \mathbf{Z}_1, \ldots, \mathbf{Z}_{n'}\right) \stackrel{\text{a.s.}}{\to} \nu.$$

Therefore, for almost all $\omega$, mean measures converge to $\nu$ in $M_+(\mathbb{E})$ and by Theorem 5.3(i) (p. 138), we conclude that for such $\omega$,

$$\mathbb{P}\left[\sum_{i=1}^{m(n')} \epsilon_{\mathbf{Z}_i^*/b_{m(n')}} \in \cdot \mid \mathbf{Z}_1, \ldots, \mathbf{Z}_{n'}\right] \Rightarrow \mathbb{P}[\mathrm{PRM}(\nu) \in \cdot]$$

weakly. By the usual subsequence argument for convergence in probability, we conclude that

$$\mathbb{P}\left[\sum_{i=1}^{m} \epsilon_{\mathbf{Z}_i^*/b_m} \in \cdot \mid \mathbf{Z}_1, \ldots, \mathbf{Z}_n\right] \xrightarrow{P} \mathbb{P}[\mathrm{PRM}(\nu) \in \cdot] \quad \text{in } \mathrm{PR}(M_p(\mathbb{E})),$$

as required. □

### When bootstrap asymptotics do not work

What happens in the limit to the full-sample bootstrap random point process $\sum_{i=1}^{n} \epsilon_{\mathbf{Z}_i^*/b_n}$? We give the answer in the next result, which shows that the empirical measure of the scaled bootstrap sample is not approximated by the limiting PRM$(\nu)$; the limit is a random measure.

**Proposition 6.3.** *Assume the regular variation condition, say,* (6.4), *holds with limit measure* $\nu$. *Suppose we represent* PRM$(\nu)$ *as* $N_\infty := \sum_i \epsilon_{\mathbf{J}_i}$. *(For instance, when* $d = 1$ *and* $\nu = \nu_\alpha$, *we could represent* PRM$(\nu_\alpha)$ *as* $\sum_i \epsilon_{\Gamma_i^{-1/\alpha}}$, *where* $\{\Gamma_i\}$ *are homogeneous, unit rate, Poisson points.) Let* $\{\xi_i, i \geq 1\}$ *be iid, unit mean, Poisson random variables that are independent of* $\{\mathbf{J}_i\}$. *Then in* PR$(M_p(\mathbb{E}))$,

$$\mathbb{P}\left[\sum_{i=1}^{n} \epsilon_{\mathbf{Z}_i^*/b_n} \in \cdot \mid \mathbf{Z}_1, \ldots, \mathbf{Z}_n\right] \Rightarrow \mathbb{P}\left[\sum_{i} \xi_i \epsilon_{\mathbf{J}_i} \in \cdot \mid \mathbf{J}_i, i \geq 1\right],$$

*and taking expectations,*

$$\mathbb{P}\left[\sum_{i=1}^{n} \epsilon_{\mathbf{Z}_i^*/b_n} \in \cdot\right] \Rightarrow \mathbb{P}\left[\sum_{i} \xi_i \epsilon_{\mathbf{J}_i} \in \cdot\right].$$

*Remark* 6.3. We emphasize that the conditional probability is a random element of PR$(M_p(\mathbb{E}))$. Also, the limit is not PRM$(\nu)$, even unconditionally, since it represents a cluster process. We confirm this by computing the Laplace functional of the limit. For $f \in C_K^+(\mathbb{E})$,

$$\begin{aligned}
\mathbf{E}\left(\exp\left\{-\sum_i \xi_i f(\mathbf{J}_i)\right\}\right) &= \mathbf{E}\left(\mathbf{E}\left(\exp\left\{-\sum_i \xi_i f(\mathbf{J}_i)\right\} \mid \mathbf{J}_i, i \geq 1\right)\right) \\
&= \mathbf{E}\prod_i \mathbf{E}(\exp\{-f(\mathbf{J}_i)\xi_1\} \mid \mathbf{J}_i, i \geq 1) \\
&= \mathbf{E}\prod_i \exp\{e^{-f(\mathbf{J}_i)} - 1\}
\end{aligned}$$

$$
\begin{aligned}
&= \mathbf{E}\exp\left\{-\int_{\mathbb{E}}(1-e^{-f(x)})N_\infty(dx)\right\} && (6.27)\\
&= \mathbf{E}\exp\{-N_\infty(1-e^{-f})\}\\
&= \exp\left\{-\int_{\mathbb{E}}(1-e^{-(1-e^{-f(x)})})\nu(dx)\right\},
\end{aligned}
$$

where in the last step, we used the fact that the Laplace functional of $N_\infty$ at the function $1-e^{-f}$ has a known form given by (5.12) (p. 135).

*Proof.* For $1 \le n \le \infty$, let $M_p^{(n)}(\mathbb{E}) \subset M_p(\mathbb{E})$ be point measures on $\mathbb{E}$ having $n$ points:

$$M_p^{(n)}(\mathbb{E}) = \{m \in M_p(\mathbb{E}) : m(\mathbb{E}) = n\}.$$

With $m = n$ in (6.26), recall that $P_j(n) := \sum_{i=1}^n 1_{[I_i^{(n)}=j]}$ is the multinomial number of $j$s sampled in $n$ trials. Define $h_n : M_p^{(n)}(\mathbb{E}) \mapsto \mathrm{PR}(M_p(\mathbb{E}))$ by

$$h_n\left(\sum_{i=1}^n \epsilon_{y_i^{(n)}}\right) = \mathbb{P}\left[\sum_{i=1}^n P_i(n)\epsilon_{y_i^{(n)}} \in \cdot\right].$$

The map $h_n$ is well defined. Similarly, define $h : M_p^{(\infty)}(\mathbb{E}) \mapsto \mathrm{PR}(M_p(\mathbb{E}))$ by

$$h\left(\sum_i \epsilon_{y_i^{(\infty)}}\right) = \mathbb{P}\left[\sum_i \xi_i \epsilon_{y_i^{(\infty)}} \in \cdot\right],$$

where $\{\xi_i, i \ge 1\}$ are iid Poisson random variables with parameter 1.

For $m_n = \sum_{i=1}^n \epsilon_{y_i^{(n)}} \in M_p^{(n)}(\mathbb{E})$, $1 \le n \le \infty$, we claim that if $m_n \xrightarrow{v} m_\infty$, then

$$h_n(m_n) \Rightarrow h(m_\infty) \tag{6.28}$$

in $\mathrm{PR}(M_p(\mathbb{E}))$; that is, weak convergence of the probability measures takes place. To prove this, we may show that the Laplace functional of the probability measure on the left in (6.28) converges to the Laplace functional of the probability measure on the right. Thus we write ($f \in C_K^+(\mathbb{E})$)

$$
\begin{aligned}
\int_{M_p(\mathbb{E})} & e^{-m(f)}\mathbb{P}\left[\sum_{i=1}^n P_i(n)\epsilon_{y_i^{(n)}} \in dm\right]\\
&= \mathbf{E}\exp\left\{-\sum_{i=1}^n P_i(n) f(y_i^{(n)})\right\}
\end{aligned}
$$

$$= \mathbf{E}\exp\left\{-\sum_{j=1}^{n}\left(\sum_{i=1}^{n}1_{[I_j^{(n)}=i]}f(y_i^{(n)})\right)\right\}$$

$$= \left(\mathbf{E}\exp\left\{-\left(\sum_{i=1}^{n}1_{[I_1^{(n)}=i]}f(y_i^{(n)})\right)\right\}\right)^n = \left(\mathbf{E}\exp\left\{-f(y_{I_1^{(n)}}^{(n)})\right\}\right)^n,$$

and remembering that $I_1^{(n)}$ is uniform on $1, \dots, n$, this is

$$= \left(1 - \frac{\sum_{i=1}^{n}[1 - e^{-f(y_i^{(n)})}]}{n}\right)^n = \left(1 - \frac{\int_{\mathbb{E}}[1 - e^{-f}]dm_n}{n}\right)^n$$

$$\to \exp\left\{-\int_{\mathbb{E}}[1 - e^{-f}]dm_\infty\right\}$$

$$= \int_{M_p(\mathbb{E})} e^{-m(f)}\mathbb{P}\left[\sum_i \xi_i \epsilon_{y_i^{(\infty)}} \in dm\right],$$

which is the Laplace functional of $h(m_\infty)$. (See the calculation leading to (6.27) (p. 190).)

Now we know from (6.15) (p. 179) that

$$M_n := \sum_{i=1}^{n}\epsilon_{\mathbf{Z}_i/b_n} \Rightarrow M_\infty := \sum_i \epsilon_{\boldsymbol{J}_i}$$

in $M_p(\mathbb{E})$. Therefore, by the second continuous mapping theorem (see Problem 3.19 (p. 69)), replacing $m_n$ by $M_n$ in (6.28) yields

$$h_n(M_n) = \mathbb{P}\left[\sum_{i=1}^{n}\epsilon_{\mathbf{Z}_i^*/b_n} \in \cdot \mid \mathbf{Z}_1, \dots, \mathbf{Z}_n\right]$$

$$\Rightarrow h(M_\infty) = \mathbb{P}\left[\sum_i \xi_i \epsilon_{\boldsymbol{J}_i} \in \cdot \mid \boldsymbol{J}_i, i \geq 1\right]$$

in $\mathrm{PR}(M_p(\mathbb{E}))$, as asserted. □

## 6.5 Multivariate regular variation: Examples, comments, amplification

Here we give some examples and further information about the concept of multivariate regular variation of distribution tails.

### 6.5.1 Two examples

Consider the following two cases, which represent opposite ends of the dependence spectrum.

#### Independence and asymptotic independence

Suppose $\{\boldsymbol{Z}_n, n \geq 1\}$ are iid random vectors in $\mathbb{R}_+^d$ such that for each $n$, the vector $\boldsymbol{Z}_n = (Z_n^{(1)}, \dots, Z_n^{(d)})$ consists of iid nonnegative components with

$$\bar{F}_{(1)}(x) := \mathbb{P}[Z_1^{(1)} > x] \sim x^{-\alpha} L(x), \quad x \to \infty, \quad \alpha > 0.$$

Then the vector $\boldsymbol{Z}_1$ has multivariate regularly varying tail probabilities. Define

$$b_n = \left(\frac{1}{1 - F_{(1)}}\right)^{\leftarrow}(n).$$

Then we have

$$n\mathbb{P}\left[\frac{\boldsymbol{Z}_1}{b_n} \in \cdot\right] \xrightarrow{v} \nu, \tag{6.29}$$

and $\nu(\cdot)$ is given by

$$\nu(dx^{(1)}, \dots, dx^{(d)}) = \sum_{j=1}^{d} \epsilon_0(dx^{(1)}) \times \cdots \times \epsilon_0(dx^{(j-1)}) \times \nu_\alpha(dx^{(j)}) \times \cdots \times \epsilon_0(dx^{(d)}). \tag{6.30}$$

That is, for any $\delta > 0$,

$$\nu\left(\bigcup_{i \neq j} \{\boldsymbol{x} \in \mathbb{E} : x^{(i)} \wedge x^{(j)} > \delta\}\right) = 0.$$

The measure $\nu$ spreads mass onto each axis according to the one-dimensional measure $\nu_\alpha$ but assigns no mass off the axes.

To see this, it suffices by Lemma 6.1 to show that for $\boldsymbol{x} > \boldsymbol{0}$ (if one or more components of $\boldsymbol{x}$ are 0, a slightly different argument is needed),

$$n\mathbb{P}\left[\frac{\boldsymbol{Z}_1}{b_n} \in [\boldsymbol{0}, \boldsymbol{x}]^c\right] \to \nu([\boldsymbol{0}, \boldsymbol{x}]^c) = \sum_{i=1}^{d} (x^{(i)})^{-\alpha} \tag{6.31}$$

for $\boldsymbol{x} \in [\boldsymbol{0}, \boldsymbol{\infty}) \setminus \{\boldsymbol{0}\}$. We do this readily using inclusion/exclusion. Write

$$
\begin{aligned}
n\mathbb{P}\left[\frac{\boldsymbol{Z}_1}{b_n} \in [\boldsymbol{0}, \boldsymbol{x}]^c\right] &= n\mathbb{P}\left\{\bigcup_{j=1}^{d}\left[\frac{Z_1^{(j)}}{b_n} > x^{(j)}\right]\right\} \\
&= \sum_{j=1}^{d} n\mathbb{P}\left[\frac{Z_1^{(j)}}{b_n} > x^{(j)}\right] - \sum_{1\le i<j\le d} n\mathbb{P}\left[\frac{Z_1^{(i)}}{b_n} > x^{(i)}, \frac{Z_1^{(j)}}{b_n} > x^{(j)}\right] \\
&\quad + \sum_{1\le i<j<k\le d} n\mathbb{P}\left[\frac{Z_1^{(i)}}{b_n} > x^{(i)}, \frac{Z_1^{(j)}}{b_n} > x^{(j)}, \frac{Z_1^{(k)}}{b_n} > x^{(k)}\right] \\
&\quad - \cdots (-1)^{d+1} n\mathbb{P}\left\{\bigcap_{i=1}^{d}\left[\frac{Z_1^{(i)}}{b_n} > x^{(i)}\right]\right\}.
\end{aligned}
$$

All terms but the first go to zero. For the first,

$$
n\mathbb{P}\left[\frac{Z_1^{(j)}}{b_n} > x^{(j)}\right] \to (x^{(j)})^{-\alpha},
$$

and other terms are bounded by an expression of the form $(i \neq j)$

$$
\begin{aligned}
n\mathbb{P}\left[\frac{Z_1^{(i)}}{b_n} > x^{(i)}, \frac{Z_1^{(j)}}{b_n} > x^{(j)}\right] &= n\mathbb{P}\left[\frac{Z_1^{(i)}}{b_n} > x^{(i)}\right]\mathbb{P}\left[\frac{Z_1^{(j)}}{b_n} > x^{(j)}\right] \\
&\sim (x^{(i)})^{-\alpha}\mathbb{P}\left[\frac{Z_1^{(j)}}{b_n} > x^{(j)}\right] \to 0.
\end{aligned}
$$

What is the form of the angular measure $S$? Let

$$
\boldsymbol{e}_i = (0, \dots, 1, \dots, 0), \quad i = 1, \dots, d,
$$

be the basis vectors and suppose the norm is defined so that

$$
\|\boldsymbol{e}_i\| = 1, \quad i = 1, \dots, d.
$$

This amounts to a normalization. Then $\nu$ concentrates on the lines

$$
\bigcup_{i=1}^{d}\{t\boldsymbol{e}_i, t > 0\}
$$

and

$$
\nu\left(\mathbb{E} \setminus \bigcup_{i=1}^{d}\{t\boldsymbol{e}_i, t > 0\}\right) = 0.
$$

We know that $\nu \circ T^{-1} = c\nu_\alpha \times S$, where $T$ is the polar coordinate transformation, and

$$S = \frac{\nu \circ T^{-1}((1,\infty] \times \cdot)}{\nu \circ T^{-1}((1,\infty] \times \aleph_+)}.$$

Call the denominator $1/c$. For a measurable set $\Lambda \subset \aleph_+$, we have

$$\begin{aligned} S(\Lambda) &= c\nu\left\{\boldsymbol{x} : \|\boldsymbol{x}\| > 1, \frac{\boldsymbol{x}}{\|\boldsymbol{x}\|} \in \Lambda\right\} \\ &= c\sum_{i=1}^{d} \nu\left(\left\{\boldsymbol{x} : \|\boldsymbol{x}\| > 1, \frac{\boldsymbol{x}}{\|\boldsymbol{x}\|} \in \Lambda\right\} \cap \{t\boldsymbol{e}_i : t > 0\}\right) \\ &= c\sum_{i:\boldsymbol{e}_i \in \Lambda} \nu(\{t\boldsymbol{e}_i : t > 1\}). \end{aligned}$$

So $S$ concentrates on $\{\boldsymbol{e}_i, i = 1, \dots, d\}$, and $S$ is of the form

$$S = \sum_{i=1}^{d} p_i \epsilon_{\boldsymbol{e}_i},$$

where $(p_1, \dots, p_d)$ is a probability vector whose components sum to 1.

The equivalence for (6.29) in terms of convergence to a Poisson process is as follows: We have

$$\sum_{i=1}^{n} \epsilon_{\boldsymbol{Z}_i/b_n} \Rightarrow N = \text{PRM}(\nu).$$

The limiting PRM has all its points on the axes and can be represented as a superposition. Let

$$N_i = \sum_k \epsilon_{j_k(i)}, \quad i = 1, \dots, d,$$

be $d$ iid PRM$(\nu_\alpha)$. Then we have

$$N \stackrel{d}{=} \sum_{i=1}^{d} \sum_k \epsilon_{j_k(i)\boldsymbol{e}_i}.$$

All the points of $N$ lie on the axes, and the way we construct $N$ is to go to the first axis and drop down Poisson points, then go to the second axis and drop an independent collection of Poisson points, and so on.

**Asymptotic independence.** Suppose $\mathbf{Z}$ is a random vector in $\mathbb{R}_+^d$ with distribution tail that is regularly varying and that (6.29) and (6.30) (p. 192) hold even though we do not assume that $\mathbf{Z}$ has independent components. Then we say $\mathbf{Z}$ possesses *asymptotic independence*.

One reason for the name is that for $d = 2$, observe that

$$\begin{aligned}
\lim_{t\to\infty} \mathbb{P}[Z^{(2)} > t | Z^{(1)} > t] &= \lim_{t\to\infty} \frac{\mathbb{P}[Z^{(1)} > t, Z^{(2)} > t]}{\mathbb{P}[Z^{(1)} > t]} \\
&= \lim_{n\to\infty} \frac{\mathbb{P}[Z^{(1)} > b_n, Z^{(2)} > b_n]}{\mathbb{P}[Z^{(1)} > b_n]} \\
&= \lim_{n\to\infty} cn\mathbb{P}[Z^{(1)} > b_n, Z^{(2)} > b_n] \\
&= c\nu(\mathbf{1}, \infty] = 0
\end{aligned}$$

for some positive constant $c$. This follows because the $\nu$ in (6.30) puts zero mass in the interior of the positive quadrant. Asymptotic independence means that if one component is large, there is negligible probability of the other component also being large.

Another motivation for the definition comes from classical extreme-value theory. See [260, p. 296] and [90].

### Repeated components and asymptotic full dependence

Now suppose that $\{\mathbf{Z}_n, n \geq 1\}$ are iid random vectors in $\mathbb{R}_+^d$ such that for each $n$, we have that $\mathbf{Z}_n = (Z_n^{(1)}, \dots, Z_n^{(1)})$ is a vector with each component the same random variable. Assume

$$\bar{F}_{(1)}(x) = \mathbb{P}[Z_1^{(1)} > x] \sim x^{-\alpha} L(x), \quad x \to \infty, \quad \alpha > 0.$$

Then $\mathbf{Z}_1$ has multivariate regularly varying tail probabilities. Define

$$b_n = \left(\frac{1}{1 - F_{(1)}}\right)^{\leftarrow} (n).$$

Then with $f \in C_K^+(\mathbb{E})$ and the support of $f$ in $[\mathbf{0}, \delta\mathbf{1}]^c$, for some $\delta > 0$, we have

$$n\mathbf{E}f\left(\frac{\mathbf{Z}_1}{b_n}\right) = \int f(x, \dots, x)\mathbb{P}\left[\frac{Z_1^{(1)}}{b_n} \in dx\right] = \int f(x\mathbf{1})n\mathbb{P}\left[\frac{Z_1^{(1)}}{b_n} \in dx\right],$$

and since $f(x\mathbf{1}) \in C_K^+(0, \infty]$, this converges to

$$\to \int f(x\mathbf{1})\nu_\alpha(x) = \int_{\mathbb{E}} f(z)\nu(dz),$$

where $\nu$ concentrates on $\{t\mathbf{1}, t > 0\}$. This means that

$$\begin{aligned}
\nu([\mathbf{0}, \boldsymbol{x}]^c) &= \nu\left(\bigcup_{i=1}^{d}\{\boldsymbol{y} : y^{(i)} > x^{(i)}\}\right) \\
&= \nu\left(\bigcup_{i=1}^{d}\{\boldsymbol{y} : y^{(i)} > x^{(i)}\} \cap \{t\mathbf{1} : t > 0\}\right) \\
&= \nu\left(\bigcup_{i=1}^{d}\{t\mathbf{1} : t > x^{(i)}\}\right) \\
&= \nu\left\{t\mathbf{1} : t > \bigwedge_{i=1}^{d} x^{(i)}\right\} \\
&= \nu_\alpha\left(\bigwedge_{i=1}^{d} x^{(i)}, \infty\right] = \left(\bigwedge_{i=1}^{d} x^{(i)}\right)^{-\alpha}. \qquad (6.32)
\end{aligned}$$

What will be the angular measure $S$ in this case? It is the measure concentrating all mass on $\mathbf{1}/\|\mathbf{1}\|$.

The limiting Poisson process for the empirical measure has all its points on the diagonal and has the following structure. Let

$$\sum_k \epsilon_{j_k}$$

be PRM$(\nu_\alpha)$ on $(0, \infty]$. Then

$$\sum_{i=1}^{n} \epsilon_{\mathbf{Z}_i/b_n} \Rightarrow \sum_k \epsilon_{j_k \mathbf{1}}.$$

**Asymptotic full dependence.** Suppose $\mathbf{Z}$ is a random vector whose distribution concentrates on the positive quadrant, and suppose $\mathbf{Z}$ does not consist of only repeated components. If the distribution tail is regularly varying with limit measure $\nu$ of the form (6.32) then the distribution possesses *asymptotic full dependence*. This concept is appropriate, for instance, for modeling insurance claims for house structural damage and personal property damage per fire incident. One expects the components to be highly dependent.

### 6.5.2 A general representation for the limiting measure $\nu$

The limit measure $\nu$ in the definition of regular variation has the following representation in terms of the angular measure $S$ [95].

**Proposition 6.4.** *Suppose that* $\mathbb{E} = [\mathbf{0}, \boldsymbol{\infty}] \setminus \{\mathbf{0}\}$. *As before, suppose that* $T$ *is the polar coordinate transformation and*

$$\nu \circ T^{-1} = c\nu_\alpha \times S,$$

*where* $\nu_\alpha(x, \infty] = x^{-\alpha}$, $x > 0$, *and* $S$ *is a probability measure on* $\aleph_+$. *Then for* $\boldsymbol{x} \in [\mathbf{0}, \boldsymbol{\infty}) \setminus \{\mathbf{0}\}$,

$$\nu([\mathbf{0}, \boldsymbol{x}]^c) = c \int_{\aleph_+} \bigvee_{i=1}^{d} \left( \frac{x^{(i)}}{a^{(i)}} \right)^{-\alpha} S(d\boldsymbol{a}). \tag{6.33}$$

*Proof.* We have

$$\nu([\mathbf{0}, \boldsymbol{x}]^c) = \nu \circ T^{-1}(T([\mathbf{0}, \boldsymbol{x}]^c)).$$

Now

$$\begin{aligned} T([\mathbf{0}, \boldsymbol{x}]^c) &= T\{\boldsymbol{y} \in [\mathbf{0}, \boldsymbol{\infty}) \setminus \{\mathbf{0}\} : y^{(i)} > x^{(i)} \text{ for some } i\} \\ &= \{(r, \boldsymbol{a}) \in (0, \infty) \times \aleph_+ : (r\boldsymbol{a})^{(i)} > x^{(i)} \text{ for some } i\} \\ &= \left\{ (r, \boldsymbol{a}) : r > \frac{x^{(i)}}{a^{(i)}} \text{ for some } i \right\} \\ &= \left\{ (r, \boldsymbol{a}) : r > \bigwedge_{i=1}^{d} \frac{x^{(i)}}{a^{(i)}} \right\}. \end{aligned}$$

Thus

$$\begin{aligned} \nu([\mathbf{0}, \boldsymbol{x}]^c) &= \nu \circ T^{-1} \left\{ (r, \boldsymbol{a}) : r > \bigwedge_{i=1}^{d} \frac{x^{(i)}}{a^{(i)}} \right\} \\ &= c\nu_\alpha \times S \left\{ (r, \boldsymbol{a}) : r > \bigwedge_{i=1}^{d} \frac{x^{(i)}}{a^{(i)}} \right\} \\ &= \int_{\boldsymbol{a} \in \aleph_+} \left[ \int_{[r > \wedge_{i=1}^{d} \frac{x^{(i)}}{a^{(i)}}]} \nu_\alpha(dr) \right] S(d\boldsymbol{a}) \\ &= \int_{\boldsymbol{a} \in \aleph_+} \left( \bigwedge_{i=1}^{d} \frac{x^{(i)}}{a^{(i)}} \right)^{-\alpha} S(d\boldsymbol{a}) = \int_{\boldsymbol{a} \in \aleph_+} \bigvee_{i=1}^{d} \left( \frac{x^{(i)}}{a^{(i)}} \right)^{-\alpha} S(d\boldsymbol{a}). \end{aligned}$$

□

### 6.5.3 A general construction of a multivariate regularly varying distribution

Suppose $R > 0$ is a nonnegative random variable with a regularly varying tail

$$n\mathbb{P}[R > b_n x] \to x^{-\alpha}, \quad x > 0, \alpha > 0.$$

Suppose further that $\boldsymbol{\Theta}$ is a random element of $\aleph_+$ with distribution $S$ and independent of $R$. Then $\boldsymbol{X} = R\boldsymbol{\Theta}$ is multivariate regularly varying with limit measure $\nu$ given by (6.33) since

$$n\mathbb{P}\left[\left(\frac{R}{b_n}, \boldsymbol{\Theta}\right) \in \cdot\right] = n\mathbb{P}\left[\frac{R}{b_n} \in \cdot\right]\mathbb{P}[\boldsymbol{\Theta} \in \cdot] \stackrel{v}{\to} \nu_\alpha \times S.$$

This follows from the equivalence of (6.7) and (6.9) in Theorem 6.1 (p. 173).

*Example* 6.1 (*bivariate Cauchy density*). Consider the bivariate Cauchy density

$$F'(x, y) = \frac{1}{2\pi}(1 + x^2 + y^2)^{-3/2}, \quad (x, y) \in \mathbb{R}^2,$$

of a random vector $\boldsymbol{X}$. Transforming to the usual polar coordinates,

$$r^2 = x^2 + y^2, \qquad \theta = \arctan(y/x),$$

we get

$$P[\|\boldsymbol{X}\| \in dr, \Theta(\boldsymbol{X}) \in d\theta] = F'(r\cos\theta, r\sin\theta) = r(1 + r^2)^{-3/2}dr\frac{1}{2\pi}d\theta.$$

Therefore, $\Theta(\boldsymbol{X})$ is uniform on $[0, 2\pi)$, and

$$\mathbb{P}[\|\boldsymbol{X}\| > r] = (1 + r^2)^{-1/2} \sim r^{-1}, \quad r \to \infty,$$

so $\mathbb{P}[\boldsymbol{X}\| > r]$ is regularly varying with index $-1$. Furthermore, $R = \|\boldsymbol{X}\|$ and $\Theta(\boldsymbol{X})$ are independent.

Note this gives multivariate regular variation on the cone $\mathbb{R}^2$, and

$$n\mathbb{P}\left[\left(\frac{\|\boldsymbol{X}\|}{n}, \Theta(\boldsymbol{X})\right) \in \cdot\right] \stackrel{v}{\to} \nu_1 \times U,$$

where $U$ is uniform on $[0, 2\pi)$ and $\nu_1(x, \infty] = x^{-1}$, $x > 0$. This limit has a density

$$r^{-2}dr\frac{d\theta}{2\pi} = r^{-3}rdr\frac{d\theta}{2\pi}, \quad r > 0, \quad \theta \in [0, 2\pi),$$

and so the limit measure $\nu$ has density

$$(x^2 + y^2)^{-3/2}dxdy, \quad (x, y) \in \mathbb{R}^2.$$

### 6.5.4 Regularly varying densities

As suggested by Example 6.1, most multivariate distributions are specified by densities, not distribution functions. It would be convenient to have workable criteria guaranteeing that regular variation of a multivariate density implies regular variation of the probability distribution tail.

When $d = 1$, regular variation of the density implies the distribution tail is a regularly varying function because of Karamata's theorem. This is not always true in higher dimensions and some regularity is needed. Intricacies are discussed in [91, 92, 97, 98]. Roughly speaking, multivariate regular variation knits together one-dimensional regular variation along rays but does not control what happens as we hop from ray to ray. Imposing a uniformity condition as we move across rays overcomes this difficulty.

It is convenient to return to the assumption that $\mathbb{E} = [\mathbf{0}, \infty] \setminus \{\mathbf{0}\}$.

**Theorem 6.4 ([98]).** *Suppose $F$ is a probability distribution on $\mathbb{E}$ with density $F'$ that is regularly varying with the limit function $\lambda(\cdot)$ on $[\mathbf{0}, \infty) \setminus \{\mathbf{0}\}$. That is, we suppose for some regularly varying function $V(t) \in \mathrm{RV}_\rho$, $\rho < 0$, we have for $\boldsymbol{x} \in [\mathbf{0}, \infty) \setminus \{\mathbf{0}\}$,*

$$\lim_{t\to\infty} \frac{F'(t\boldsymbol{x})}{t^{-d}V(t)} = \lambda(\boldsymbol{x}) > 0. \tag{6.34}$$

*Necessarily $\lambda$ satisfies $\lambda(t\boldsymbol{x}) = t^{\rho-d}\lambda(\boldsymbol{x})$ for $\boldsymbol{x} \in [\mathbf{0}, \infty) \setminus \{\mathbf{0}\}$. Further, suppose that $\lambda$ is bounded on $\aleph_+$ and that the following uniformity condition holds:*

$$\lim_{t\to\infty} \sup_{\boldsymbol{x}\in\aleph_+} \left| \frac{F'(t\boldsymbol{x})}{t^{-d}V(t)} - \lambda(\boldsymbol{x}) \right| = 0. \tag{6.35}$$

*It then follows that for any $\delta > 0$,*

$$\lim_{t\to\infty} \sup_{\|\boldsymbol{x}\|>\delta} \left| \frac{F'(t\boldsymbol{x})}{t^{-d}V(t)} - \lambda(\boldsymbol{x}) \right| = 0. \tag{6.36}$$

*Furthermore, $\lambda(\cdot)$ is integrable on $[\mathbf{0}, \boldsymbol{x}]^c$ for $\boldsymbol{x} > \mathbf{0}$ and $1 - F$ is a regularly varying function on $(\mathbf{0}, \infty)$, which takes the form*

$$\lim_{t\to\infty} \frac{1 - F(t\boldsymbol{x})}{V(t)} = \int_{[\mathbf{0},\boldsymbol{x}]^c} \lambda(\boldsymbol{y})d\boldsymbol{y}. \tag{6.37}$$

*Example* 6.2. Consider the following examples:

1. *Two-dimensional Cauchy density*: Return to Example 6.1,

$$F'(x, y) = \frac{1}{2\pi}(1 + x^2 + y^2)^{-3/2}, \quad (x, y) \in \mathbb{R}^2.$$

Let the norm be the usual Euclidean norm

$$\|(x, y)\| = \sqrt{x^2 + y^2}.$$

Then the density is, for $\boldsymbol{x} \in \mathbb{E}$,

$$F'(\boldsymbol{x}) = \frac{1}{2\pi}(1 + \|\boldsymbol{x}\|^2)^{-3/2}, \quad \boldsymbol{x} \in \mathbb{R}^2.$$

Therefore, as $t \to \infty$,

$$\begin{aligned} \frac{F'(t\boldsymbol{x})}{F'(t\boldsymbol{1})} &= \frac{(1 + t^2\|\boldsymbol{x}\|^2)^{-3/2}}{(1 + 2t^2)^{-3/2}} \\ &\sim \frac{t^{-3}(\|\boldsymbol{x}\|)^{-3}}{2^{-3/2}t^{-3}} \\ &\to 2^{3/2}\|\boldsymbol{x}\|^{-3} =: \lambda(\boldsymbol{x}). \end{aligned}$$

The uniformity condition (6.35) is easy to check since

$$\sup_{\boldsymbol{x}:\|\boldsymbol{x}\|=1} \left| \frac{F'(t\boldsymbol{x})}{F'(t\boldsymbol{1})} - \lambda(\boldsymbol{x}) \right| = \left| \left( \frac{1 + t^2}{1 + 2t^2} \right)^{-3/2} - 2^{3/2} \right| \to 0$$

as $t \to \infty$.

2. The bivariate $t$-density: On $\mathbb{E}$ define

$$F'(x, y) = c(1 + x^2 + 2\rho xy + y^2)^{-2}, \quad (x, y) \in \mathbb{E}, \quad -1 < \rho < 1.$$

Define the norm (cleverly)

$$\|(x, y)\|^2 = x^2 + 2\rho xy + y^2 = (x + \rho y)^2 + (1 - \rho^2)y^2.$$

Then the density is of the form

$$F'(\boldsymbol{x}) = c(1 + \|\boldsymbol{x}\|^2)^{-2}, \quad \boldsymbol{x} \in \mathbb{E},$$

and we may proceed to check the conditions of Theorem 6.4 as in the bivariate Cauchy case.

Note this method works whenever the density is of the form

$$F'(\boldsymbol{x}) = c(1 + \|\boldsymbol{x}\|^{\gamma})^{-\beta}, \quad \boldsymbol{x} \in \mathbb{E}.$$

What suggested the form of the norm for the bivariate $t$-density? If $F'$ is regularly varying with limit function $\lambda$, then

$$\lambda(t\boldsymbol{x}) = t^{-\alpha}\lambda(\boldsymbol{x}), \quad \boldsymbol{x} > \boldsymbol{0},$$

so that

$$\lambda^{-1/\alpha}(t\boldsymbol{x}) = t\lambda^{-1/\alpha}(\boldsymbol{x}),$$

which is the scaling a norm should have. So we could try and set

$$\|\boldsymbol{x}\| = \lambda^{-1/\alpha}(\boldsymbol{x})$$

and hope this defines a norm.

For the proof of Theorem 6.4, see [98] or [260, p. 284].

### 6.5.5 Beyond the nonnegative orthant

Up to now, we have typically assumed the state space was the nonnegative orthant. However, for certain problems, the natural state space is not the nonnegative orthant. This is true for weak convergence problems for partial sums and applications in finance, where negative values of returns are important since they indicate losses. In extreme-value theory, the natural state space is typically a rectangle of the form $[\boldsymbol{x}_l, \boldsymbol{x}_r] \setminus \{\boldsymbol{x}_l\}$, where $-\infty \le \boldsymbol{x}_l < \boldsymbol{x}_r \le \infty$.

If $\mathbb{E}$ is a closed cone in $[-\infty, \infty]\setminus\{\boldsymbol{0}\}$, the most useful examples are $\mathbb{E} = [\boldsymbol{0}, \infty]\setminus\{\boldsymbol{0}\}$, as we have already considered, and $\mathbb{E} = [-\infty, \infty] \setminus \{\boldsymbol{0}\}$. Item 1 of Theorem 6.1 (p. 173), in terms of multivariate regular variation of functions, no longer has an easy analogue, except in $d = 1$. One could express the correct analogue of item 1 in terms of the multivariate distribution function tail of the positive and negative parts of the components of $\boldsymbol{Z}$ by considering the regular variation of the $2d$-dimensional vector

$$((Z_1^{(1)})^+, (Z_1^{(1)})^-, (Z_1^{(2)})^+, (Z_1^{(2)})^-, \dots, (Z_1^{(d)})^+, (Z_1^{(d)})^-),$$

but this would be awkward and not very elegant.

There is no trouble extending items 2–7 of Theorem 6.1 to the more general cone, provided one works with measures and $\aleph_+$ is replaced by $\aleph \cap \mathbb{E}$. If $\mathbb{E} = [-\infty, \infty] \setminus \{\boldsymbol{0}\}$, then

$$\aleph \cap \mathbb{E} = \{\boldsymbol{x} \in \mathbb{R}^d : \|\boldsymbol{x}\| = 1\}.$$

If $d = 1$, then

$$n\mathbb{P}\left[\frac{Z_1}{b_n} \in \cdot\right] \xrightarrow{v} \nu \quad \text{in } M_+([-\infty, \infty] \setminus \{0\}) \tag{6.38}$$

is the basic condition. This means that for $x > 0$,

$$n\mathbb{P}\left[\frac{Z_1}{b_n} > x\right] \to \nu(x, \infty] \quad (n \to \infty), \tag{6.39}$$

which is sequential regular variation, so

$$\nu(x, \infty] = c_+ x^{-\alpha}, \quad c_+ \geq 0.$$

Also, for $x > 0$,

$$n\mathbb{P}\left[\frac{Z_1}{b_n} < -x\right] = n\mathbb{P}\left[\frac{-Z_1}{b_n} > x\right] \to \nu[-\infty, -x) \quad (n \to \infty), \tag{6.40}$$

and again by sequential regular variation of functions, we get

$$\nu[-\infty, -x) = c_- x^{-\alpha}, \quad c_- \geq 0.$$

The $\alpha$ for the right tail must be the same as for the left tail since the same $b_n$ successfully scales both tails and $b_n$ relates to $\alpha$ through the fact that $b_n = b(n)$, where the function $b(\cdot) \in \mathrm{RV}_{1/\alpha}$. (See, for example, Proposition 2.6 (p. 32) or (2.12) (p. 23).) We therefore conclude that

$$\nu(dx) = c_+ \alpha x^{-\alpha-1} dx 1_{(0,\infty]}(x) + c_- \alpha |x|^{-\alpha-1} dx 1_{[-\infty,0)}(x).$$

Sometimes (6.39) and (6.40) are written together as

$$P[|Z_1| > x] \in \mathrm{RV}_{-\alpha},$$

along with the tail-balancing condition ($0 \leq p \leq 1$)

$$\frac{\mathbb{P}[Z_1 > x]}{\mathbb{P}[|Z_1| > x]} \to p, \qquad \frac{\mathbb{P}[Z_1 < -x]}{\mathbb{P}[|Z_1| > x]} \to 1 - p =: q$$

as $x \to \infty$.

For this $d = 1$ case,

$$\aleph = \{-1, 1\}$$

and

$$S(\{1\}) = \frac{c_+}{c_+ + c_-}, \qquad S(\{-1\}) = \frac{c_-}{c_+ + c_-}.$$

### 6.5.6 Standard vs. nonstandard regular variation

The phrasing of the regular variation condition in Theorem 6.1 assumes tail equivalence for the distribution tails of the components. The requirement that

$$n\mathbb{P}\left[\frac{\mathbf{Z}_1}{b_n} \in \cdot\right] \xrightarrow{v} \nu \quad \text{in } M_+(\mathbb{E})$$

implies that

$$n\mathbb{P}\left[\frac{Z_1^{(i)}}{b_n} \in \cdot\right] \xrightarrow{v} c_i \nu_\alpha \quad \text{in } M_+((0,\infty])$$

for $c_i \geq 0$ and $i = 1, \dots, d$. We have not ruled out the possibility that for some (but not all) $i$, $c_i$ could be zero. For those components with $c_i > 0$, the $\alpha$s are the same because $b_n = b(n)$ and $b(\cdot) \in \mathrm{RV}_{1/\alpha}$. The marginal convergences with the same scaling function $b_n$, in turn, imply that for $1 \leq i < j \leq d$,

$$\lim_{x\to\infty} \frac{\mathbb{P}[Z_1^{(i)} > x]}{\mathbb{P}[Z_1^{(j)} > x]} = \frac{c_i}{c_j}.$$

However, in practice, one rarely observes components having the same tail indices and one needs a broader understanding of multivariate heavy tails.

*Example* 6.3. Let $P$ be a nonnegative random variable with unit Pareto distribution and consider $\mathbf{Z} = (P, P^2)$. If in the definition of regular variation we insist on normalizing both components with the same scaling, then

$$n\mathbb{P}\left[\left(\frac{P}{n^2}, \frac{P^2}{n^2}\right) \in \cdot\right] \xrightarrow{v} \epsilon_0 \times \nu_\alpha.$$

A more subtle normalization would reveal more structure; in particular,

$$n\mathbb{P}\left[\left(\frac{P}{n}, \frac{P^2}{n^2}\right) \in \cdot\right] \xrightarrow{v} \nu_\alpha \circ T_{1,1^2}^{-1},$$

where $T_{1,1^2} : (0,\infty] \mapsto (0,\infty] \times (0,\infty]$ and is defined by $T_{1,1^2}(x) = (x, x^2)$.

In heavy-tail analysis, one wishes to rule out degeneracies coming from a one-dimensional marginal distribution that is not heavy tailed. However, the degeneracy of Example 6.3 is fairly natural. When estimating the $\alpha$s of heavy-tailed multivariate data, one never gets equal $\alpha$s for all the components. Examples include the following:

- exchange rate returns of Germany, France, and Japan against the US dollar;

- heavy-tailed Internet data of the form $\{(L_j, F_j, R_j), 1 \le j \le n\}$, where

$$\begin{aligned} L_j &= \text{download time of } j\text{th file},\\ F_j &= \text{size of } j\text{th file},\\ R_j &= \text{transferred rate of } j\text{th file}. \end{aligned}$$

(See p. 238.) So, while the theory is most elegantly developed using the single normalization of Theorem 6.1, in practice this is not adequate and sensitivity to tails with different weights is needed. The next result shows that a broader definition is possible but that a monotone transformation brings the broader definition back to what we call the *standard case* [95, 260], which is the case of Theorem 6.1 with $b_n = n$. Eventually, we will address how to deal with this in a statistical context.

**Theorem 6.5.** *As usual, assume* $\mathbf{Z} = (Z^{(1)}, \ldots, Z^{(d)})$ *is a vector with nonnegative components and* $\mathbb{E} = [\mathbf{0}, \infty] \setminus \{\mathbf{0}\}$. *Suppose for* $1 \le i \le d$ *that there exist sequences* $\{b_n^{(i)}, n \ge 1\}$, *with* $\lim_{n\to\infty} b_n^{(i)} = \infty$, $i = 1, \ldots, d$, *such that we have the following:*

(i) Marginal regular variation: *For each* $i = 1, \ldots, d$,

$$nP\left[\frac{Z^{(i)}}{b_n^{(i)}} \in \cdot\right] \stackrel{v}{\to} \nu_{\alpha_i}, \quad \alpha_i > 0, \tag{6.41}$$

*in* $M_+(0, \infty]$

*and*

(ii) Nonstandard global regular variation: *There exists a measure* $\nu$ *on Borel subsets of* $\mathbb{E}$ *such that*

$$nP\left[\left(\frac{Z^{(i)}}{b_n^{(i)}}, i = 1, \ldots, d\right) \in \cdot\right] \stackrel{v}{\to} \nu \tag{6.42}$$

*in* $M_+(\mathbb{E})$.

*Let* $\bar{F}_{(i)}(x) = \mathbb{P}[Z^{(i)} > x]$ *be the* $i$*th marginal distribution tail, and from* (6.41), *we can define*

$$b^{(i)}(x) = \left(\frac{1}{1 - F_{(i)}(\cdot)}\right)^{\leftarrow}(x), \quad x > 1,$$

*and set* $b_n^{(i)} = b^{(i)}(n)$. *Then we have the following:*

(i) Standard global regular variation:

$$nF_*(n\cdot) := n\mathbb{P}\left[\left(\frac{(b^{(i)})^{\leftarrow}(Z^{(i)})}{n}, i = 1, \ldots, n\right) \in \cdot\right]$$

$$\stackrel{v}{\to} \nu_*(\cdot) \quad in\ M_+(\mathbb{E}), \tag{6.43}$$

*where*

$$\nu_*(t\cdot) = t^{-1}\nu_*(\cdot) \tag{6.44}$$

*on Borel subsets of* $\mathbb{E}$,

*and*

(ii) Standard marginal convergence: *For* $i = 1, \dots, d$,

$$n\mathbb{P}\left[\frac{(b^{(i)})^{\leftarrow}(Z^{(i)})}{n} > x\right] \to x^{-1}, \quad x > 0. \tag{6.45}$$

The marginal condition (6.41) rules out tails that are not heavy. The global condition (6.42) describes dependence among the components. The standard case is where we have tail-equivalent marginal tails, each of which is regularly varying with index $-1$ and normalization by the same constant $b(n) = n$ is adequate.

*Proof.* Observe that

$$\begin{aligned} n\mathbb{P}\left\{\left[\frac{(b^{(i)})^{\leftarrow}(Z^{(i)})}{n} \le \boldsymbol{x}\right]^c\right\} &= n\mathbb{P}\left\{\left[\boldsymbol{Z} \le (b^{(i)}(nx^{(i)}), i = 1, \dots, d)\right]^c\right\} \\ &= n\mathbb{P}\left\{\left[\frac{Z^{(i)}}{b_n^{(i)}} \le \frac{b^{(i)}(nx^{(i)})}{b^{(i)}(n)}, i = 1, \dots, d\right]^c\right\}. \end{aligned}$$

Since

$$\frac{b^{(i)}(nx^{(i)})}{b^{(i)}(n)} \to (x^{(i)})^{1/\alpha_i},$$

we get, as $n \to \infty$,

$$\begin{aligned} n\mathbb{P}\left\{\left[\frac{(b^{(i)})^{\leftarrow}(Z^{(i)})}{n} \le \boldsymbol{x}\right]^c\right\} &\to \nu\left(\left[\boldsymbol{0}, \left(\lim_{n\to\infty} \frac{b^{(i)}(nx^{(i)})}{b^{(i)}(n)}, i = 1, \dots, d\right)\right]^c\right) \\ &= \nu([\boldsymbol{0}, \boldsymbol{x}^{1/\boldsymbol{\alpha}}]^c) =: \nu_*([\boldsymbol{0}, \boldsymbol{x}]^c). \end{aligned} \tag{6.46}$$

This completes the proof and defines $\nu_*$ appearing in (6.43). □

Note the relation between $\nu$ and the standardized $\nu_*$ given in (6.46). How to transform to the standard case in a statistical context is a significant problem that we will discuss in Chapter 9.

## 6.6 Problems

**6.1 (Deleting axes).** As another use of Proposition 6.1, suppose $\mathbb{X} = [\mathbf{0}, \boldsymbol{\infty}] \setminus \{\mathbf{0}\}$, and define the cone $\mathbb{X}^0$ by

$$\mathbb{X}^0 := \{s \in \mathbb{X} : \text{for some } 1 \le i < j \le d,\, s^{(i)} \wedge s^{(j)} > 0\},$$

where we write the vector $\boldsymbol{s} = (s^{(1)}, \dots, s^{(d)})$. An alternative description: For $i = 1, \dots, d$, define the basis vectors

$$\boldsymbol{e}_i = (0, \dots, 0, 1, 0, \dots, 0),$$

so that the axes originating at $\mathbf{0}$ are $\mathbb{L}_i := \{t\boldsymbol{e}_i, t > 0\}, i = 1, \dots, d$. Then we also have

$$\mathbb{X}^0 = \mathbb{X} \setminus \bigcup_{i=1}^{d} L_i .$$

If $d = 2$, we have $\mathbb{X}^0 = (0, \infty]^2$. What are the relatively compact subsets of $\mathbb{X}^0$? (Such a space is useful in consideration of asymptotic independence. See [217].)

**6.2 ([77]).** Suppose $Y_1, \dots, Y_k$ are nonnegative random variables (but not necessarily independent or identically distributed). If $Y_1$ has distribution $F$ satisfying $\bar{F} \in RV_{-\alpha}$ and if as $x \to \infty$,

$$\frac{P[Y_i > x]}{1 - F(x)} \to c_i, \quad i = 1, \dots, k,$$

and

$$\frac{P[Y_i > x, Y_j > x]}{1 - F(x)} \to 0, \quad i \neq j,$$

then

$$\frac{P\left[\sum_{i=1}^{k} Y_i > x\right]}{1 - F(x)} \to c_1 + \dots + c_k .$$

**6.3 (The restriction functional [130]).** Suppose $\mathbb{E}'$ is a measurable subset of $\mathbb{E}$, and give $\mathbb{E}'$ the relative topology inherited from $\mathbb{E}$. For a set $B \subset \mathbb{E}'$, denote by $\partial_{\mathbb{E}'} B$ the boundary of $B$ in $\mathbb{E}'$, and denote by $\partial_{\mathbb{E}} B$ the boundary of $B$ in $\mathbb{E}$.

(a) Define

$$\hat{T} : M_+(\mathbb{E}) \mapsto M_+(\mathbb{E}')$$

by

$$\hat{T}\mu = \mu(\cdot \cap \mathbb{E}').$$

If $\mu \in M_+(\mathbb{E})$ and $\mu(\partial_{\mathbb{E}}\mathbb{E}') = 0$, then $\hat{T}$ is continuous at $\mu$, so that if $\mu_n \xrightarrow{v} \mu$ in $M_+(\mathbb{E})$, then $\hat{T}\mu_n \xrightarrow{v} \hat{T}\mu$ in $M_+(\mathbb{E}')$.

(b) The same conclusion holds if we define $\hat{T}$ the same way but consider it as a mapping

$$\hat{T} : M_+(\mathbb{E}) \mapsto M_+(\mathbb{E}).$$

(c) Conversely, suppose that $\mu_n \in M_+(\mathbb{E})$ for $n \geq 0$ and that $\mu_n \stackrel{v}{\to} \mu_0$ in $M_+(\mathbb{E}')$. If

$$\mu_n((\mathbb{E}')^c) = 0, \quad n \geq 0,$$

and $\mu_0(\partial_{\mathbb{E}}\mathbb{E}') = 0$, then $\mu_n \stackrel{v}{\to} \mu_0$ in $M_+(\mathbb{E})$ as well.

**6.4 (Stochastic analogue of Karamata's theorem [133]).** Suppose $\{Z_n, n \geq 1\}$ are iid nonnegative random variables with common distribution $F$ satisfying $\bar{F} \in RV_{-\alpha}$. Then in $C[0, \infty)$ we have for any $\beta > \alpha$, as $m = m(n) \to \infty$, $m/n \to 0$,

$$\int_0^x \frac{m}{n} \sum_{t=1}^n \epsilon_{Z_t/b(m)}(u, \infty] u^{\beta-1} du \stackrel{P}{\to} \int_0^x \nu_\alpha(u, \infty] u^{\beta-1} du = \frac{x^{\beta-\alpha}}{\beta - \alpha}.$$

**6.5 (Regular variation at 0 [133]).** Suppose $\{\mathbf{Z}_n, n \geq 1\}$ are iid random vectors in $[0, \infty)^d$ and the distribution function of $\mathbf{Z}_1$ is regularly varying at $\mathbf{0}$. Formulate this as vague convergence of measures and verify the analogue of Theorem 6.2.

Suppose $d = 1$ and $P[Z_1 \leq x] \in RV_\alpha$ at 0. Then in $C[0, \infty)$, we have for appropriately chosen scaling constants $a(m)$ and $\beta > \alpha$,

$$\int_0^x \frac{m}{n} \sum_{t=1}^n \epsilon_{Z_t/a(m)}[0, u^{-1}) u^{\beta-1} du \stackrel{P}{\to} \int_0^x \nu[0, u^{-1}) u^{\beta-1} du = \frac{x^{\beta-\alpha}}{\beta - \alpha}.$$

**6.6 (Π-variation [104, 259]).** A measurable function $U : (0, \infty) \mapsto (0, \infty)$ is called Π-varying (written $U \in \Pi$) [26, 90, 102, 144, 260] if there exists $g \in RV_0$ such that for all $x > 0$,

$$\lim_{t \to \infty} \frac{U(tx) - U(t)}{g(t)} = \log x. \tag{6.47}$$

1. Suppose

   $$U(x) = \int_0^x u(s) ds, \quad x > 0, \quad u(\cdot) \in RV_{-1}.$$

   Show that $U \in \Pi$.

2. Suppose $U$ is nondecreasing. Show $U \in \Pi$ iff there exists $a(n) \to \infty$ and

   $$\frac{n}{a(n)} U(a(n)\cdot) \stackrel{v}{\to} L(\cdot),$$

   where $L$ is the measure satisfying $L(a, b] = \log b/a$, $0 < a < b < \infty$.

3. Suppose $U$ is nondecreasing and

$$\sum_k \epsilon_{(t_k,u_k)}$$

is PRM($\mathbb{LEB} \times U$). $U \in \Pi$ iff there exists $a(\cdot) \in \mathrm{RV}_1$ such that

$$\frac{1}{a(n)} \sum_k \epsilon_{(nt_k,u_k/a(n))} \Rightarrow \mathbb{LEB} \times L.$$

4. Suppose $U$ is nondecreasing and

$$\sum_k \epsilon_{(t_k,u_k)}$$

is PRM($\mathbb{LEB} \times U$). $U \in \Pi$ iff there exists $a(\cdot) \in \mathrm{RV}_1$ such that

$$\sum_k \epsilon_{(n^{-1}a(n)t_k,u_k/a(n))} \Rightarrow \mathrm{PRM}(\mathbb{LEB} \times L).$$

**6.7 ([259]).** Suppose $\{Y_{n,k}\}$ are random elements of a nice space $\mathbb{E}$ and

$$\sum_{k=1}^{n} \epsilon_{Y_{n,k}} \Rightarrow \mathrm{PRM}(\mu)$$

in $M_p(\mathbb{E})$, where $\mu \in M_+(\mathbb{E})$. If $\{X_k\}$ is an iid sequence of random elements of a nice space $\mathbb{E}'$, and for each $n$ the families $\{X_k\}$ and $\{Y_{n,k}, k \geq 1\}$ are independent, then

$$\sum_{k=1}^{n} \epsilon_{(Y_{n,k},X_k)} \Rightarrow \mathrm{PRM}(\mu \times \mathbb{P}[X_1 \in \cdot])$$

in $M_p(\mathbb{E} \times \mathbb{E}')$.

**6.8 ([13]).** Suppose $\mathbf{Z} \in \mathbb{R}_+^d$ is a random vector. Show that it has a distribution with a regularly varying tail iff for some $\alpha > 0$ and some random vector $\mathbf{\Theta} \in \aleph_+$, we have for all $x > 0$,

$$\frac{\mathbb{P}\left[\|\mathbf{Z}\| > tx, \frac{\mathbf{Z}}{\|\mathbf{Z}\|} \in \cdot\right]}{\mathbb{P}[\|\mathbf{Z}\| > t]} \xrightarrow{v} x^{-\alpha}\mathbb{P}[\mathbf{\Theta} \in \cdot]$$

as $t \to \infty$, where vague convergence is in $M_+(\aleph)$.

**6.9.** Suppose $\{\boldsymbol{Z}_n, n \geq 1\}$ are iid random vectors in $[0, \infty)^d$ with a distribution satisfying the standard form of regular variation (see (6.43) (p. 205)). Then

$$\lim_{n\to\infty} \mathbb{P}\left[\bigvee_{i=1}^{n} \frac{\boldsymbol{Z}_i}{n} \leq \boldsymbol{x}\right] = G(\boldsymbol{x}), \quad \boldsymbol{x} \geq \boldsymbol{0},$$

weakly, where $G$ is a product probability distribution iff the distribution of $\boldsymbol{Z}_1$ exhibits asymptotic independence. (See, for example, [260, p. 296] and [90].)

**6.10 ([89]).** Let $U$ be uniform on $(0, 1)$. Prove that

$$\boldsymbol{Z} = \left(\frac{1}{U}, \frac{1}{1-U}\right)$$

possesses asymptotic independence.

**6.11 (Normal dependence model [260, p. 297], [90, 279]).** Suppose $(N_1, N_2)$ is a normal random vector with zero means, unit variances and correlation $\rho < 1$. Define

$$\boldsymbol{Z} = \left(\frac{1}{\Phi(N_1)}, \frac{1}{\Phi(N_2)}\right),$$

and show that $\boldsymbol{Z}$ possesses asymptotic independence.

**6.12 (Pairwise asymptotic independence [260, p. 296], [95, 140, 142, 143, 210]).** Suppose $\boldsymbol{Z}$ is a $\mathbb{R}_+^d$-valued vector with a regularly varying distribution tail. Prove that $\boldsymbol{Z}$ possesses asymptotic independence iff for any $i \neq j$, the pair $(Z^{(i)}, Z^{(j)})$ possesses asymptotic independence.

**6.13 (Sample range [103]).** Suppose $\{Z_i, i \geq 1\}$ are iid with common distribution $F$ satisfying regular variation on $[-\infty, \infty] \setminus \{0\}$; that is,

$$n\mathbb{P}\left[\frac{Z_1}{b_n} \in \cdot\right] \xrightarrow{v} \nu(\cdot)$$

in $M_+([-\infty, \infty] \setminus \{0\})$ as in (6.38), (6.39), (6.40). Prove that

$$\left(b_n^{-1} \bigvee_{i=1}^{n} Z_i, b_n^{-1} \bigwedge_{i=1}^{n} Z_i\right)$$

has a limit distribution, and hence so does the sample range

$$R_n := b_n^{-1}\left(\bigvee_{i=1}^{n} Z_i - \bigwedge_{i=1}^{n} Z_i\right).$$

Are $Z_1$ and $-Z_1$ asymptotically independent using some sensible definition?

**6.14 (Binding vectors).** Suppose $\boldsymbol{X}$ and $\boldsymbol{Y}$ are independent random vectors with non-negative components defined on the same probability space. If $\boldsymbol{X}$ is regularly varying in $\mathbb{E}_{d_1}$ with limit measure $\nu_1$, and $\boldsymbol{Y}$ is regularly varying in $\mathbb{E}_{d_2}$ with limit measure $\nu_2$, show that $(\boldsymbol{X}, \boldsymbol{Y})$ is regularly varying in $\mathbb{E}_{d_1+d_2}$. What is the limit measure? (If you need help, peek ahead to Lemma 7.2.)

**6.15 (Continuity of the limit function [97, 284]).** Suppose $f : [\mathbf{0}, \infty) \mapsto (0, \infty)$ is nondecreasing and regularly varying on the cone $(\mathbf{0}, \infty)$ with limit function $\lambda$. Show that $\lambda(\cdot)$ is continuous on $(\mathbf{0}, \infty)$.

**6.16 ($m$-dependence and the Poisson transform).** The Poisson transform given in (6.15) also applies to $m$-dependent stationary sequences whose one-dimensional marginal distributions are regularly varying. Suppose that $\{X_n\}$ are $m$-dependent random elements of a nice space $\mathbb{E}$, in the sense that random variables more than $m$ apart in the sequence are independent. More precisely, let

$$\mathcal{B}_j^k = \sigma(X_j, \dots, X_k)$$

be the $\sigma$-algebra generated by $X_j, \dots, X_k$ for $k \leq j$. Assume $\mathcal{B}_{j_1}^{k_1}, \dots, \mathcal{B}_{j_l}^{k_l}$ are independent if $k_{i-1} + m < j_i$ for $i = 2, \dots, l$. (Independent random variables are 0-*dependent*.)

Now assume that for each $n = 1, 2, \dots,$ the array $\{X_{n,i}, i \geq 1\}$ are stationary and $m$-dependent and that

$$n\mathbb{P}[X_{n,1} \in \cdot] \stackrel{v}{\to} \nu \tag{6.48}$$

for a Radon limit measure $\nu$ and

$$\lim_{k\to\infty} \limsup_{n\to\infty} n \sum_{i=2}^{[n/k]} \mathbf{E}(g(X_{n,1})g(X_{n,i})) = 0 \tag{6.49}$$

for any $g \in C_K^+(\mathbb{E})$, $g \leq 1$.

Prove that [76]

$$\sum_{i=1}^{n} \epsilon_{X_{n,i}} \Rightarrow \mathrm{PRM}(\nu).$$

*Hint*: Use the big block–little block method and Laplace functionals. Examples of the big block–little block method are in [197, 264].

**6.17 (Basic convergence with a time coordinate).** Theorem 6.3 (p. 180) extends (6.15) (p. 179) of Theorem 6.2. Provide a similar extension of (6.17) (p. 180).

# 7

# Weak Convergence and the Poisson Process

This chapter exploits connections between regular variation and the Poisson process given in Theorems 6.2 (p. 179) and 6.3 (p. 180) to understand several limit theorems and also to understand how regular variation of distributions of random vectors is transmitted by various transformations on the vectors. The fundamental philosophy is that we should capitalize on the equivalence between the analytical concept of regular variation and the probabilistic notion of convergence of empirical measures to limiting Poisson random measures.

## 7.1 Extremes

Regular variation is equivalent to scaled extremes converging weakly. If necessary, vector notation may be reviewed in Appendix 10 (p. 359). Remember that maxima of collections of vectors are taken componentwise. Suppose $\{\mathbf{Z}_n, n \geq 1\}$ are iid random vectors in $[0, \infty)^d$ with common distribution $F$. We take $\mathbb{E} = [\mathbf{0}, \boldsymbol{\infty}] \setminus \{\mathbf{0}\}$. Theorems 6.1 (p. 173) and 6.2 (p. 179) state the regular variation condition in its equivalent forms. The representation of the limit measure in the definition of regular variation in terms of the angular measure is given in Proposition 6.4 (p. 197).

### 7.1.1 Weak convergence of multivariate extremes: The timeless result

Here is the equivalence between regular variation and convergence of extremes.

**Proposition 7.1.** *Suppose $\{\mathbf{Z}_n, n \geq 1\}$ are iid random vectors in $\mathbb{E}$. Then regular variation of the distribution of $\mathbf{Z}_1$ with limit measure $\nu$,*

$$n\mathbb{P}\left[\frac{\mathbf{Z}_1}{b_n} \in \cdot\right] \xrightarrow{v} \nu$$

*in* $M_+(\mathbb{E})$, *is equivalent to*

$$\bigvee_{i=1}^{n} \frac{\boldsymbol{Z}_i}{b_n} \Rightarrow \boldsymbol{Y}_0, \tag{7.1}$$

*where* $\boldsymbol{Y}_0$ *is a random vector with distribution* $F_0$ *given by*

$$F_0(\boldsymbol{x}) := e^{-\nu([\boldsymbol{0},\boldsymbol{x}]^c)}, \quad \boldsymbol{x} \geq \boldsymbol{0}.$$

*Proof.* Given the multivariate regular variation condition, write (6.15) as

$$N_n = \sum_{i=1}^{n} \epsilon_{\boldsymbol{Z}_i/b_n} \Rightarrow N_0 = \sum_k \epsilon_{\boldsymbol{j}_k} = \text{PRM}(\nu) \tag{7.2}$$

in $M_p(\mathbb{E})$. Write

$$\mathbb{P}\left[\bigvee_{i=1}^{n} \frac{\boldsymbol{Z}_i}{b_n} \leq \boldsymbol{x}\right] = \mathbb{P}[N_n([\boldsymbol{0},\boldsymbol{x}]^c) = 0] \to \mathbb{P}[N_0([\boldsymbol{0},\boldsymbol{x}]^c) = 0] = e^{-\nu([\boldsymbol{0},\boldsymbol{x}]^c)},$$

since $N_0([\boldsymbol{0},\boldsymbol{x}]^c)$ is a Poisson random variable.

Conversely, if (7.1) holds, we have

$$F^n(b_n\boldsymbol{x}) \to e^{-\nu([\boldsymbol{0},\boldsymbol{x}]^c)},$$

and using the same argument that led to (2.10) (p. 23), we get

$$n\bar{F}(b_n\boldsymbol{x}) = n\mathbb{P}\left[\frac{\boldsymbol{Z}_1}{b_n} \in [\boldsymbol{0},\boldsymbol{x}]^c\right] \to \nu([\boldsymbol{0},\boldsymbol{x}]^c).$$

Finish using Lemma 6.1 (p. 174). □

Variants of this result can be constructed for weak convergence of the (largest, second largest, ..., $k$th largest) of the sample.

### 7.1.2 Weak convergence of multivariate extremes: Functional convergence to extremal processes

Suppose that

$$N_0^{\#} = \sum_k \epsilon_{(t_k,\boldsymbol{j}_k)} = \text{PRM}(\mathbb{LEB} \times \nu)$$

is a Poisson random measure on $[0,\infty) \times \mathbb{E}$ with mean measure $\mathbb{LEB} \times \nu$. As in Section 5.6.1 (p. 161), we define the extremal process

$$\boldsymbol{Y}_0(t) = \bigvee_{t_k \le t} \boldsymbol{j}_k, \quad t \ge 0.$$

So $\boldsymbol{Y}_0(t)$ looks at points whose time coordinate is prior to $t$ and then takes the biggest $\boldsymbol{j}$ satisfying the time constraint. Its marginal distribution is given by (5.52) (p. 161).

Recall that Theorem 6.3 (p. 180) gave a point process equivalence to regular variation that added a time coordinate: As $n \to \infty$, the regular variation condition

$$n\mathbb{P}\left[\frac{\boldsymbol{Z}_1}{b_n} \in \cdot\right] \xrightarrow{v} \nu$$

is equivalent to

$$N_n^{\#} := \sum_{i=1}^{\infty} \epsilon_{(\frac{i}{n}, \boldsymbol{Z}_i/b_n)} \Rightarrow N_0^{\#} := \sum_i \epsilon_{(t_i, \boldsymbol{j}_i)} = \mathrm{PRM}(\mathbb{LEB} \times \nu), \tag{7.3}$$

in $M_p([0,\infty) \times \mathbb{E})$.

**Proposition 7.2 (weak convergence to extremal processes).** *Suppose* $\{\boldsymbol{Z}_n, n \ge 1\}$ *are iid nonnegative random vectors. The multivariate regular variation condition is equivalent to*

$$Y_n(\cdot) := \bigvee_{i=1}^{[n\cdot]} \frac{\boldsymbol{Z}_i}{b_n} \Rightarrow Y_0(\cdot) = \bigvee_{t_k \le \cdot} \boldsymbol{j}_k \tag{7.4}$$

*in* $D([0,\infty), [\boldsymbol{0}, \boldsymbol{\infty}))$, *the space of functions whose domain is* $[0,\infty)$ *and range is* $[\boldsymbol{0}, \boldsymbol{\infty})$, *that are right continuous, and that possess finite left limits on* $(0,\infty)$.

*Proof.* Define the almost surely continuous functional

$$\chi^{\#} : M_p([0,\infty) \times \mathbb{E}) \mapsto D([0,\infty), [\boldsymbol{0}, \boldsymbol{\infty}))$$

by

$$\chi\left(\sum_k \epsilon_{(t_k, \boldsymbol{x}_k)}\right)(t) = \bigvee_{t_k \le t} \boldsymbol{x}_k.$$

Apply this and the continuous mapping theorem to (7.3). □

The unchecked claim is the statement that $\chi^{\#}$ is almost surely continuous. We will prove something similar in connection with weak convergence of partial sums to Lévy processes, so we defer to that section. The impatient may wish to consult [260, p. 214]. Proposition 7.2 was originally proved for $d = 1$ by Lamperti [195] with a traditional finite-dimensional convergence plus tightness proof. The connection to weak convergence of point processes is in [258, 296]. Applications of the result in one dimension using the structure of extremal processes and records are in [258, 260].

## 7.2 Partial sums

We now explore ideas which lead to weak convergence of partial sum processes to the limiting jump Lévy processes constructed in Section 5.5 (p. 146). This will also allow approximation of partial sums of heavy-tailed iid random variables by the stable Lévy motions discussed in Section 5.5.2 (p. 154).

For this section, we need the definitions of $D([0,\infty),\mathbb{R}^d)$, the space of functions on $[0,\infty)$ with range $\mathbb{R}^d$ that are right continuous and with finite left limits. Review the Skorohod metric from Section 3.3.4 (p. 46). In particular, we need that if $x_n, n \ge 0$ are functions in $D([0,\infty),\mathbb{R}^d)$, then $x_n \to x_0$ in the Skorohod metric on $D([0,\infty),\mathbb{R}^d)$, if $x_n \to x_0$ in the Skorohod metric on $D([0,T],\mathbb{R}^d)$ for any $T$ which is a continuity point of the limit. Recall the Skorohod metric on $[0,T]$ is a uniform metric after small deformations of the time scale and that local uniform convergence implies Skorohod convergence (see p. 47).

### 7.2.1 Weak onvergence of partial sum processes to Lévy processes

Theorem 7.1 is flexible enough for many purposes. For this $d$-dimensional result, set $\mathbb{E} = [-\infty,\infty]\setminus\{\mathbf{0}\}$. The argument is adapted from [241]. As usual, we denote random vectors by $\boldsymbol{X} = (X^{(1)},\dots,X^{(d)})$.

**Theorem 7.1.** *Suppose for each $n \ge 1$ that $\{\boldsymbol{X}_{n,j}, j \ge 1\}$ are iid random vectors such that*

$$n\mathbb{P}[\boldsymbol{X}_{n,1} \in \cdot\,] \stackrel{v}{\to} \nu(\cdot) \tag{7.5}$$

*in $M_+(\mathbb{E})$, where $\nu$ is a Lévy measure (see Section 5.5.1 (p. 146)), and suppose further that for each $j = 1,\dots,d$,*

$$\lim_{\varepsilon\downarrow 0}\limsup_{n\to\infty} n\mathbf{E}((X_{n,1}^{(j)})^2 1_{[|X_{n,1}^{(j)}|\le\varepsilon]}) = 0. \tag{7.6}$$

*Define the partial sum stochastic process based on the nth row of the array by*

$$\boldsymbol{X}_n(t) := \sum_{k=1}^{[nt]}(\boldsymbol{X}_{nk} - \mathbf{E}(\boldsymbol{X}_{n,k}1_{[\|\boldsymbol{X}_{n,k}\|\le 1]})), \quad t \ge 0.$$

*Then (7.5) and (7.6) imply that*

$$\boldsymbol{X}_n \Rightarrow \boldsymbol{X}_0,$$

*in $D([0,\infty),\mathbb{R}^d)$, where $\boldsymbol{X}_0(\cdot)$ is a Lévy jump process with Lévy measure $\nu$, as constructed in Section 5.5 (p. 146).*

*Proof.* We break the proof into several steps. The idea of each step is simple. Several of the steps involve assertions that certain functionals are continuous, and to promote flow, we delay proofs of continuity.

STEP 1. From the variant of the basic convergence, Corollary 6.1 (p. 183), we know that (7.5) implies that

$$\sum_{k=1}^{\infty} \epsilon_{(\frac{k}{n}, \boldsymbol{X}_{n,k})} \Rightarrow \sum_{k} \epsilon_{(t_k, \boldsymbol{j}_k)} = \text{PRM}(\mathbb{LEB} \times \nu) \tag{7.7}$$

in $M_p([0,\infty) \times \mathbb{E})$. Here and in the rest of the discussion, we always assume $\varepsilon$ is chosen so that $\varepsilon$ is not a jump of the function

$$\tau(t) = \nu\{\boldsymbol{x} : \|\boldsymbol{x}\| > t\}.$$

Later, when $\varepsilon \downarrow 0$, we assume convergence to 0 is through a sequence of values $\{\varepsilon_\ell\}$ that are also not jumps of this function. Since $\tau(\cdot)$ has only a countable number of jumps, this can be arranged. For convenience, we assume 1 is not a jump of $\tau(\cdot)$.

STEP 2. Two continuity assertions:

(i) The restriction map defined by

$$m \mapsto m|_{[0,\infty) \times \{\boldsymbol{x} : \|\boldsymbol{x}\| > \varepsilon\}}$$

is almost surely continuous from $M_p([0,\infty) \times \mathbb{E}) \mapsto M_p([0,\infty) \times \{\boldsymbol{x} : \|\boldsymbol{x}\| > \varepsilon\})$ with respect to the distribution of $\text{PRM}(\mathbb{LEB} \times \nu)$. (See Problem 6.3 (p. 206).)

(ii) The summation functional defined by

$$\sum_{k} \epsilon_{(\tau_k, \boldsymbol{J}_k)} \to \sum_{\tau_k \le (\cdot)} \boldsymbol{J}_k$$

is almost surely continuous from $M_p([0,\infty) \times \{\boldsymbol{x} : \|\boldsymbol{x}\| > \varepsilon\}) \mapsto D([0,T], \mathbb{R}^d)$ (see Section 7.2.3 (p. 221)) with respect to the distribution of $\text{PRM}(\mathbb{LEB} \times \nu)$.

STEP 3. From the first continuity assertion in Step 2, the convergence statement in Step 1, and the continuous mapping theorem, Theorem 3.1 (p. 42), we get the restricted convergence

$$\sum_{k} 1_{[\|\boldsymbol{X}_{n,k}\| > \varepsilon]} \epsilon_{(\frac{k}{n}, \boldsymbol{X}_{n,k})} \Rightarrow \sum_{k} 1_{[\|\boldsymbol{j}_k\| > \varepsilon]} \epsilon_{(t_k, \boldsymbol{j}_k)} \tag{7.8}$$

in $M_p([0,\infty) \times \{\boldsymbol{x} : \|\boldsymbol{x}\| > \varepsilon\})$. From the second continuity assertion in Step 2, we get from (7.8) that

$$\sum_{k=1}^{[n\cdot]} X_{n,k} 1_{[\|X_{n,k}\|>\varepsilon]} \Rightarrow \sum_{t_k \le (\cdot)} j_k 1_{[\|j_k\|>\varepsilon]} \tag{7.9}$$

in $D([0,T],\mathbb{R}^d)$. Similarly, we get

$$\sum_{k=1}^{[n\cdot]} X_{n,k} 1_{[\varepsilon<\|X_{n,k}\|\le 1]} \Rightarrow \sum_{t_k \le (\cdot)} j_k 1_{[\varepsilon<\|j_k\|\le 1]}. \tag{7.10}$$

STEP 4. In (7.10), take expectations and apply (7.5) to get

$$[n\cdot]\mathbf{E}(X_{n,1} 1_{[\varepsilon<\|X_{n,1}\|\le 1]}) \to (\cdot) \int_{\{x:\varepsilon<\|x\|\le 1\}} x\nu(dx) \tag{7.11}$$

in $D([0,T],\mathbb{R}^d)$. To justify this, observe first for any $t > 0$ that

$$\begin{aligned}[nt]\mathbf{E}(X_{n,1} 1_{[\varepsilon<\|X_{n,1}\|\le 1]}) &= \frac{[nt]}{n} \int_{\{x:\|x\|\in(\varepsilon,1]\}} x n\mathbb{P}[X_{n,1} \in dx] \\ &\to t \int_{\{x:\|x\|\in(\varepsilon,1]\}} x\nu(dx)\end{aligned}$$

since $n\mathbb{P}[X_{n,1} \in \cdot] \xrightarrow{v} \nu(\cdot)$ and $\varepsilon$ and 1 are not jumps of $\tau(\cdot)$. Convergence is locally uniform in $t$ and hence convergence takes place in $D([0,T],\mathbb{R}^d)$.

STEP 5. Difference (7.9)–(7.11). The result is

$$\begin{aligned}X_n^{(\varepsilon)}(\cdot) &= \sum_{k=1}^{[n\cdot]} X_{n,k} 1_{[\|X_{n,k}\|>\varepsilon]} - [n\cdot]\mathbf{E}(X_{n,1} 1_{[\varepsilon<\|X_{n,1}\|\le 1]}) \\ &\Rightarrow X_0^{(\varepsilon)}(\cdot) := \sum_{t_k\le\cdot} j_k 1_{[\|j_k\|>\varepsilon]} - (\cdot)\int_{\{x:\|x\|\in(\varepsilon,1]\}} x\nu(dx).\end{aligned} \tag{7.12}$$

(One must check that in $D([0,\infty),\mathbb{R}^d)$, differencing is almost surely continuous.) From the Itô representation of a Lévy process (see Section 5.5.3 (p. 155)), for almost all $\omega$, as $\varepsilon \downarrow 0$,

$$X_0^{(\varepsilon)}(\cdot) \to X_0(\cdot)$$

locally uniformly in $t$. Let $d(\cdot,\cdot)$ be the Skorohod metric on $D[0,\infty)$. Since local uniform convergence implies Skorohod convergence, we get

$$d\big(X_0^{(\varepsilon)}(\cdot), X_0(\cdot)\big) \to 0$$

almost surely as $\varepsilon \downarrow 0$, and hence, since almost sure convergence implies weak convergence,

$$\boldsymbol{X}_0^{(\varepsilon)}(\cdot) \Rightarrow \boldsymbol{X}_0(\cdot).$$

in $D([0,\infty),\mathbb{R}^d)$.

STEP 6. By the second converging together theorem, Theorem 3.5 (p. 56), it suffices to show that

$$\lim_{\varepsilon\downarrow 0}\limsup_{n\to\infty}\mathbb{P}[d(\boldsymbol{X}_n^{(\varepsilon)},\boldsymbol{X}_n)>\delta]=0.$$

To prove convergence in $D([0,\infty),\mathbb{R}^d)$, it is sufficient to prove Skorohod convergence in $D([0,T],\mathbb{R}^d)$ for any $T$, and since the Skorohod metric on $D([0,T],\mathbb{R}^d)$ is bounded above by the uniform metric on $D([0,T],\mathbb{R}^d)$, it suffices to show that

$$\lim_{\varepsilon\downarrow 0}\limsup_{n\to\infty}\mathbb{P}\left[\sup_{0\le t\le T}\|\boldsymbol{X}_n^{(\varepsilon)}(t)-\boldsymbol{X}_n(t)\|>\delta\right]=0$$

for any $\delta>0$. Recalling the definitions, we have

$$\begin{aligned}\|\boldsymbol{X}_n^{(\varepsilon)}(t)-\boldsymbol{X}_n(t)\| &= \left\|\sum_{k=1}^{[nt]}\boldsymbol{X}_{n,k}1_{[\|\boldsymbol{X}_{n,k}\|\le\varepsilon]}-[nt]\mathbf{E}(\boldsymbol{X}_{n,1}1_{[\|\boldsymbol{X}_{n,1}\|\le\varepsilon]})\right\|\\ &= \left\|\sum_{k=1}^{[nt]}\left(\boldsymbol{X}_{n,k}1_{[\|\boldsymbol{X}_{n,k}\|\le\varepsilon]}-\mathbf{E}(\boldsymbol{X}_{n,k}1_{[\|\boldsymbol{X}_{n,k}\|\le\varepsilon]})\right)\right\|,\end{aligned}$$

so

$$\begin{aligned}&\mathbb{P}\left[\sup_{0\le t\le T}\|\boldsymbol{X}_n^{(\varepsilon)}(t)-\boldsymbol{X}_n(t)\|>\delta\right]\\ &\quad\le\mathbb{P}\left[\sup_{0\le t\le T}\left\|\sum_{k=1}^{[nt]}\left(\boldsymbol{X}_{n,k}1_{[\|\boldsymbol{X}_{n,k}\|\le\varepsilon]}-\mathbf{E}(\boldsymbol{X}_{n,k}1_{[\|\boldsymbol{X}_{n,k}\|\le\varepsilon]})\right)\right\|>\delta\right]\\ &\quad=\mathbb{P}\left[\sup_{0\le j\le nT}\left\|\sum_{k=1}^{j}\left(\boldsymbol{X}_{n,k}1_{[\|\boldsymbol{X}_{n,k}\|\le\varepsilon]}-\mathbf{E}(\boldsymbol{X}_{n,k}1_{[\|\boldsymbol{X}_{n,k}\|\le\varepsilon]})\right)\right\|>\delta\right].\end{aligned}$$

Now using the fact that $\|\boldsymbol{x}\|\le d\bigvee_{i=1}^d|x^{(i)}|$, we get the bound

$$\le\sum_{i=1}^d\mathbb{P}\left[\sup_{0\le j\le nT}\left|\sum_{k=1}^{j}\left(X_{n,k}^{(i)}1_{[\|\boldsymbol{X}_{n,k}\|\le\varepsilon]}-\mathbf{E}(X_{n,k}^{(i)}1_{[\|\boldsymbol{X}_{n,k}\|\le\varepsilon]})\right)\right|>\frac{\delta}{d}\right];$$

and by Kolmogorov's inequality (see (5.43) (p. 157), Lemma 5.3 (p. 157), or [24]), this has upper bound

$$
\begin{aligned}
&\leq (\delta/d)^{-2} \sum_{i=1}^{d} \operatorname{Var}\left( \sum_{k=1}^{[nT]} X_{n,k}^{(i)} 1_{[\|X_{n,k}\| \leq \varepsilon]} \right) \\
&= (\delta/d)^{-2} \sum_{i=1}^{d} [nT] \operatorname{Var}(X_{n,1}^{(i)} 1_{[\|X_{n,1}\| \leq \varepsilon]}) \\
&\leq (\delta/d)^{-2} \sum_{i=1}^{d} [nT] \mathbf{E}\big( (X_{n,1}^{(i)})^2 1_{[|X_{n,1}^{(i)}| \leq \varepsilon]} \big).
\end{aligned}
$$

Taking $\lim_{\varepsilon \downarrow 0} \limsup_{n \to \infty}$, we easily get 0 by (7.6). □

### 7.2.2 Weak convergence to stable Lévy motion

Although we focus on $d = 1$ for the following functional limit theorem, versions of this result in $D([0, \infty), \mathbb{R}^d)$ are easy based on the work of the previous section. Conditions for convergence of sums of iid heavy-tailed random vectors were first formulated in [269]. See also [241] and Problem 7.6 (p. 248).

For this result

$$\mathbb{E} = [-\infty, \infty] \setminus \{0\}.$$

Recall from Section 6.5.5 (p. 201) that (7.15), regular variation of the tail probabilities on the cone $\mathbb{R} \setminus \{0\}$, is equivalent to

$$\lim_{x \to \infty} \frac{\mathbb{P}[Z_1 > x]}{\mathbb{P}[|Z_1| > x]} = p, \qquad \lim_{x \to \infty} \frac{\mathbb{P}[Z_1 \leq -x]}{\mathbb{P}[|Z_1| > x]} = q,$$

and

$$\mathbb{P}[|Z_1| > x] \in RV_{-\alpha}.$$

**Corollary 7.1.** *Consider the special case where $\{Z_n, n \geq 1\}$ are iid random variables on $\mathbb{R}$, and set $X_{n,j} = Z_j / b_n$ for some $b_n \to \infty$. Define $\nu$ for $x > 0$ and $0 < \alpha < 2$ by*

$$\nu((x, \infty]) = p x^{-\alpha}, \qquad \nu([-\infty, -x]) = q x^{-\alpha}, \tag{7.13}$$

*where $0 \leq p \leq 1$ and $q = 1 - p$. Then*

$$\sum_{j=1}^{[n\cdot]} \frac{Z_j}{b(n)} - [n\cdot] \mathbf{E}\left( \frac{Z_1}{b(n)} 1_{[|\frac{Z_1}{b(n)}| \leq 1]} \right) \Rightarrow X_\alpha(\cdot), \tag{7.14}$$

*in $D[0,\infty)$, where the limit is $\alpha$-stable Lévy motion with Lévy measure $\nu$, iff*

$$n\mathbb{P}\left[\frac{Z_1}{b_n} \in \cdot\right] \stackrel{v}{\to} \nu \tag{7.15}$$

*in $M_+(\mathbb{E})$.*

*Proof.*

SUFFICIENCY. Given the regular variation of the $Z$s, it is clear that (7.5) is satisfied with the limit measure given in (7.13). For the truncated second moment condition (7.6), we have

$$n\mathbf{E}\left(\left(\frac{Z_1}{b(n)}\right)^2 1_{[|\frac{Z_1}{b(n)}|\leq\varepsilon]}\right) \to \int_{[|x|\leq\varepsilon]} x^2\nu(dx) \quad (n\to\infty)$$
$$= \frac{p\alpha\varepsilon^{2-\alpha}}{2-\alpha} + \frac{q\alpha\varepsilon^{2-\alpha}}{2-\alpha} = (\text{const})\varepsilon^{2-\alpha}$$

by Karamata's theorem, and as $\varepsilon\to 0$, we have $\varepsilon^{2-\alpha}\to 0$, as required for the partial sum process to converge to the Lévy process.

NECESSITY. Conversely, suppose (7.14) holds.

We begin by observing that if (7.14) holds, then it is also true that

$$\sum_{j=1}^{[n\cdot]} \frac{Z_j}{b(n)} - (n\cdot)\mathbf{E}\left(\frac{Z_1}{b(n)} 1_{[|\frac{Z_1}{b(n)}|\leq 1]}\right) \Rightarrow X_\alpha(\cdot), \tag{7.16}$$

where the centering is now a continuous function. To verify this, we take the difference between the centering in (7.16) and the one in (7.14) and show that this goes to zero in $D[0,\infty)$. It suffices to show that the difference converges to zero locally uniformly. For any $T>0$, we observe that

$$\sup_{0\leq s\leq T} |[ns]-ns| \leq 1,$$

and hence it suffices to show that

$$\mathbf{E}\left(\frac{Z_1}{b_n} 1_{[|Z_1|\leq b_n]}\right) \to 0.$$

We have

$$\mathbf{E}\left(\frac{|Z_1|}{b_n} 1_{[|Z_1|\leq b_n]}\right) = \int_{x=0}^{1}\left(\int_{y=0}^{x} dy\right)\mathbb{P}\left[\left|\frac{Z_1}{b_n}\right| \in dx\right]$$

$$= \int_{y=0}^{1} \left( \int_{x=y}^{1} \mathbb{P}\left[ \left| \frac{Z_1}{b_n} \right| \in dx \right] \right) dy$$

$$= \int_{0}^{1} \left( \mathbb{P}\left[ \left| \frac{Z_1}{b_n} \right| > y \right] - \mathbb{P}\left[ \left| \frac{Z_1}{b_n} \right| > 1 \right] \right) dy.$$

The last integrand is bounded by 2 and goes to zero as $n \to \infty$ since $b_n \to \infty$. Hence by dominated convergence, the integral converges to 0.

Define the functionals $T^{\pm}$, $T^{\text{abs}}$ from $D[0, \infty) \mapsto \mathbb{R}$ by

$$\begin{aligned} T^{+}(x) &= \sup_{0 \le t \le 1} \{(x(t) - x(t-))1_{(0,\infty)}(x(t) - x(t-))\}, \\ T^{-}(x) &= \sup_{0 \le t \le 1} \{|x(t) - x(t-)|1_{(-\infty,0)}(x(t) - x(t-))\}, \\ T^{\text{abs}}(x) &= \sup_{0 \le t \le 1} \{|x(t) - x(t-)|\}. \end{aligned}$$

So $T^{+}(x)$ is the maximal positive jump of the function $x$ in $[0, 1]$ and $T^{\text{abs}}$ is the maximal absolute value of the jumps in $[0, 1]$. These are almost surely continuous functionals with respect to the distribution of the stable Lévy motion. Applying these functionals to the convergence in (7.16) yields

$$\begin{aligned} T^{+}\left( \sum_{j=1}^{[n\cdot]} \frac{Z_j}{b(n)} \right) &= \frac{1}{b_n} \bigvee_{j=1}^{n} Z_j 1_{(0,\infty)}(Z_j) \Rightarrow T^{+}(X_\alpha), \\ T^{-}\left( \sum_{j=1}^{[n\cdot]} \frac{Z_j}{b(n)} \right) &= \frac{1}{b_n} \bigvee_{j=1}^{n} |Z_j| 1_{(-\infty,0)}(Z_j) \Rightarrow T^{-}(X_\alpha), \\ T^{\text{abs}}\left( \sum_{j=1}^{[n\cdot]} \frac{Z_j}{b(n)} \right) &= \frac{1}{b_n} \bigvee_{j=1}^{n} |Z_j| \Rightarrow T^{\text{abs}}(X_\alpha). \end{aligned}$$

Now

$$T^{+}(X_\alpha) = \bigvee_{t_k \le 1} j_k^{+},$$

and hence for $x > 0$,

$$\mathbb{P}[T^{+}(X_\alpha) \le x] = \mathbb{P}[N([0, 1] \times (x, \infty]) = 0] = e^{-px^{-\alpha}};$$

similarly for the other two functionals. If $0 < p < 1$, use the equivalence of regular variation and weak convergence of normalized maxima of iid random variables (see Proposition 7.1 (p. 211)) to get that

$$\mathbb{P}[|Z_1| > x], \quad \mathbb{P}[Z_1 > x], \quad \text{and} \quad \mathbb{P}[Z_1 \le -x]$$

are all regularly varying with equivalent tails. If $p = 1$, apply this argument using $T^+$ and $T^{\text{abs}}$. If $p = 0$, apply the argument using $T^-$ and $T^{\text{abs}}$. This proves the converse. □

### 7.2.3 Continuity of the summation functional

The proof of Theorem 7.1 (p. 214) about weak convergence of partial sums to Lévy processes was dependent on the continuity of the summation functional (see p. 215), and we discuss this point in more detail here. We restrict attention to the case $d = 1$ and so assume $\mathbb{E} = [-\infty, \infty] \setminus \{0\}$.

Let $N$ be PRM($\mathbb{LEB} \times \nu$) and assume $\nu \in M_+(\mathbb{E})$ and $\nu\{\pm\infty\} = \nu\{\pm\varepsilon\} = 0$. Set

$$\mathbb{E}^{>\varepsilon} = \{x \in \mathbb{E} : |x| > \varepsilon\},$$

and define the map

$$\chi : M_p([0, \infty) \times \mathbb{E}) \mapsto D[0, \infty)$$

by

$$\begin{aligned} \chi\left(\sum_i \epsilon_{(\tau_i, y_i)}\right)(t) &= \chi\left(\sum_i \epsilon_{(\tau_i, y_i)}(([0, \infty) \times \mathbb{E}^{>\varepsilon}) \cap \cdot)\right)(t) \\ &= \sum_{\tau_i \le t} y_i 1_{[|y_i| > \varepsilon]}. \end{aligned} \tag{7.17}$$

Note this is a finite sum since $[0, t] \times \mathbb{E}^{>\varepsilon}$ is a relatively compact subset of $[0, \infty) \times \mathbb{E}$. Equivalently, one can regard $\chi$ as a mapping with domain $M_p([0, \infty) \times \mathbb{E}^{>\epsilon})$.

Fix some $T > 0$. We will show that if $m_1, m_2 \in M_p([0, \infty) \times \mathbb{E})$ are close, then $\chi(m_1)$ and $\chi(m_2)$ are close as functions in $D[0, T]$. Define the subset of $M_p([0, \infty) \times \mathbb{E})$ by (refer to Figure 7.1)

$$\begin{aligned} \Lambda := \{m \in M_p([0, \infty) \times \mathbb{E}) : &\, m([0, \infty) \times \{\pm\varepsilon\}) = 0, \\ &m([0, \infty) \times \{\pm\infty\}) = 0, \\ &m(\{0\} \times \mathbb{E}^{>\varepsilon}) = m(\{T\} \times \mathbb{E}^{>\varepsilon}) = 0, \\ &m\{[0, T] \times \mathbb{E}^{>\varepsilon}\} < \infty, \text{ and} \\ &\text{no vertical line contains two points of } m(([0, T] \times \mathbb{E}^{>\varepsilon}) \cap \cdot)\}. \end{aligned}$$

We make two claims.

Claim (1). $\mathbb{P}[N \in \Lambda] = 1$.

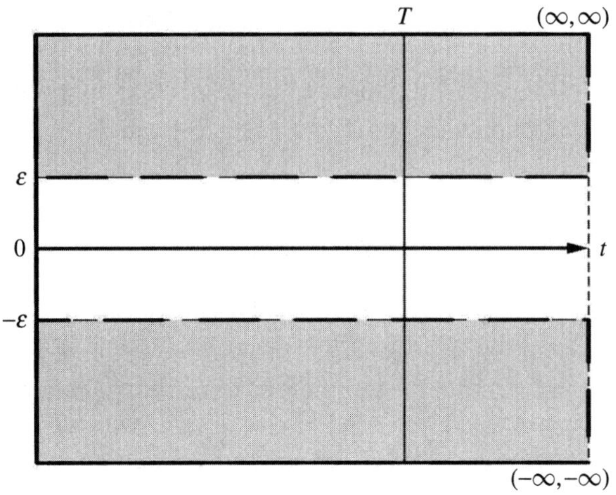


**Fig. 7.1.** The region $[0, T] \times \{x : |x| > \epsilon\}$. Dotted lines indicate an open boundary.

CLAIM (2). If $m \in \Lambda$, then $\chi$ is continuous at $m$ as a function into $D[0, T]$, and therefore $\chi$ is almost surely continuous with respect to $\mathbb{P} \circ N^{-1}$, the distribution of $N$.

Why is Claim (1) true? We analyze $\Lambda$ as an intersection of several sets and show that each of the intersecting sets has probability 1.

(a) First, we have that

$$\mathbf{E}(N[0, T] \times \mathbb{E}^{>\varepsilon}) = T\nu(\mathbb{E}^{>\varepsilon}) < \infty,$$

so

$$\mathbb{P}[N([0, T] \times \mathbb{E}^{>\varepsilon}) < \infty] = 1.$$

(b) Second, we have that

$$\mathbb{LEB} \times \nu(\{0\} \times \mathbb{E}^{>\varepsilon}) = \mathbb{LEB}\{0\} \cdot \nu(\mathbb{E}^{>\varepsilon}) = 0,$$

so

$$\mathbf{E}(N(\{0\} \times \mathbb{E}^{>\varepsilon})) = 0,$$

and therefore

$$\mathbb{P}[N(\{0\} \times \mathbb{E}^{>\varepsilon}) = 0] = 1.$$

Similarly, we have

$$\mathbb{P}[N(\{T\} \times \mathbb{E}^{>\varepsilon}) = 0] = 1.$$

(c) Next, we show that

$$\mathbb{P}\{\text{no vertical line contains two points of } N(([0,T]\times\mathbb{E}^{>\varepsilon})\cap\cdot)\}=1. \tag{7.18}$$

Pick any $M>0$. We can represent

$$N(([0,T]\times\mathbb{E}^{>\varepsilon})\cap\cdot)=\sum_{1}^{\xi}\epsilon_{(U_i,V_i)}(\cdot),$$

where $\xi$ is a Poisson random variable with parameter $T\nu(\mathbb{E}^{>\varepsilon})$, $\{U_i, i\geq 1\}$ are iid uniformly distributed on $(0,T)$, and $\{V_i, i\geq 1\}$ are iid with distribution $\nu(\mathbb{E}^{>\varepsilon}\cap\cdot)/\nu(\mathbb{E}^{>\varepsilon})$. Then

$$\begin{aligned}
&\mathbb{P}\{\text{some vertical line contains two points of } N([0,M]\times\mathbb{E}^{>\varepsilon}\cap\cdot)\}\\
&=\mathbb{P}\left\{\bigcup_{1\leq i<j\leq\xi}[U_i=U_j]\right\}\\
&=\sum_{n=0}^{\infty}\mathbb{P}\left\{\bigcup_{1\leq i<j\leq n}[U_i=U_j]\right\}\mathbb{P}[\xi=n]\\
&\leq\sum_{n=0}^{\infty}\binom{n}{2}\mathbb{P}[U_1=U_2]\mathbb{P}[\xi=n]\\
&=\sum_{n=0}^{\infty}\binom{n}{2}\cdot 0\cdot\mathbb{P}[\xi=n]=0.
\end{aligned}$$

This gives (7.18).

Why is Claim (2) true on p. 222? Suppose $m\in\Lambda$ and $m_n\stackrel{v}{\to}m$. We show, for any $T>0$, that $\chi_{m_n}\to\chi_m$ in $D[0,T]$. This argument is based on the following lemma [230], which describes what it means for two point measures $m_1$ and $m_2$ to be close: In any compact region of the state space, the finite number of points of $m_1$ are close in location to the finite number of points of $m_2$. In Lemma 7.1, $\mathbb{X}$ is any nice state space.

**Lemma 7.1.** *Suppose $m_n$, $n\geq 0$, are point measures in $M_p(\mathbb{X})$ and $m_n\stackrel{v}{\to}m_0$. For $K\in\mathcal{K}(\mathbb{X})$, such that $m_0(\partial K)=0$, we have for $n\geq n(K)$ a labeling of the points of $m_n$ and $m_0$ in $K$ such that*

$$m_n(\cdot\cap K)=\sum_{i=1}^{P}\epsilon_{x_i^{(n)}}(\cdot),\qquad m_0(\cdot\cap K)=\sum_{i=1}^{P}\epsilon_{x_i^{(0)}}(\cdot),$$

*and in* $\mathbb{E}^P$,

$$(x_i^{(n)}, 1 \le i \le P) \to (x_i^{(0)}, 1 \le i \le P)$$

*as* $n \to \infty$.

*Proof.* Write

$$m_0(\cdot \cap K) = \sum_{r=1}^{s} c_r \epsilon_{y_r}(\cdot),$$

where $y_1, \dots, y_s$ are the atoms of $m$ in $K^0$ and $c_1, \dots, c_s$ are integers giving multiplicities.

For each $y_r$, choose a neighborhood $G_r \subset K^0$ and with $G_1, \dots, G_s$ disjoint and satisfying $m_0(\partial G_r) = 0$, so that the neighborhoods do not intersect and do not go outside $K^0$, and so boundaries of the neighborhoods contain no points of $m_0$. Then $m_n(G_r) \to m_0(G_r)$ (see Theorem 3.2 (p. 52)). Because the counting measures are integer valued, a converging sequence actually equals its limit from some point on. So for $n$ sufficiently large, say, $n \ge n(K)$,

$$m_n(G_r) = m_0(G_r), \quad 1 \le r \le s,$$

and also $m_n(K) = m_0(K)$. Labeling points properly now gives the result. □

Back to the discussion of Claim (2): There exist nonnegative integers $k_0$ and $n_0$ such that for $n \ge n_0$,

$$m_n([0, T] \times \mathbb{E}^{>\varepsilon}) = m_0([0, T] \times \mathbb{E}^{>\varepsilon}) = k_0.$$

For $n \ge n_0$, write

$$m_n(([0, T] \times \mathbb{E}^{>\varepsilon}) \cap \cdot) = \sum_{i=1}^{k_0} \epsilon_{(\tau_i^{(n)}, y_i^{(n)})}(\cdot),$$

where, since $m_0 \in \Lambda$,

$$0 = \tau_0 < \tau_1 < \cdots < \tau_{k_0} < T < \tau_{k_0+1}.$$

Pick $\delta$ so small that

$$\tau_1 - \delta > \delta, \qquad \tau_i + \delta < \tau_{i+1} - \delta, \quad i = 1, \dots, k_0 - 1, \qquad \tau_{k_0} + \delta < T - \delta.$$

There exists $n_1 \ge n_0$ such that for $n \ge n_1$,

$$(\tau_i^{(n)}, y_i^{(n)}) \in (\tau_i^{(0)} - \delta, \tau_i^{(0)} + \delta) \times (y_i^{(0)} - \delta, y_i^{(0)} + \delta), \quad 1 \le i \le k_0.$$

Define homeomorphisms $\lambda_n : [0, T] \mapsto \lambda_n[0, T]$ by

$$\lambda_n(0) = 0, \qquad \lambda_n(T) = T,$$
$$\lambda_n(\tau_i^{(n)}) = \tau_i^{(0)}, \quad i = 1, \ldots, k_0;$$

between these points, $\lambda_n(\cdot)$ is defined by linear interpolation. Then writing $\|x(\cdot)\|_{[0,T]}$ for the sup-norm of the function $x(\cdot)$ on $[0, T]$, we have

$$\begin{aligned}
\|\chi(m_n) \circ \lambda_n^{-1} - \chi(m_0)\|_{[0,T]} &= \sup_{0 \le t \le T} \left| \sum_{\tau_k^{(n)} \le \lambda_n^{-1}(t)} y_k^{(n)} - \sum_{\tau_k^{(0)} \le t} y_k^{(0)} \right| \\
&= \sup_{0 \le t \le T} \left| \sum_{\lambda_n(\tau_k^{(n)}) \le t} y_k^{(n)} - \sum_{\tau_k^{(0)} \le t} y_k^{(0)} \right| \\
&= \sup_{0 \le t \le T} \left| \sum_{\tau_k^{(0)} \le t} y_k^{(n)} - \sum_{\tau_k^{(0)} \le t} y_k^{(0)} \right| \\
&\le \sum_{\tau_k^{(0)} \le T} |y_k^{(n)} - y_k^{(0)}| \\
&\le k_0 \delta. \qquad (7.19)
\end{aligned}$$

Also, we claim (recall that $e(t) = t$)

$$\|\lambda_n - e\|_{[0,T]} \le 3\delta. \qquad (7.20)$$

To see this, write

$$\|\lambda_n - e\|_{[0,T]} = \bigvee_{i=0}^{k_0} \bigvee_{s \in [\tau_i^{(n)}, \tau_{i+1}^{(n)}]} |\lambda_n(s) - s|.$$

For the first interval,

$$\begin{aligned}
\sup_{0 \le s \le \tau_1^{(n)}} |\lambda_n(s) - s| &= \sup_{0 \le s \le \tau_1^{(n)}} \left| \frac{\tau_1^{(0)}}{\tau_1^{(n)}} s - s \right| = \sup_{s \le \tau_1^{(n)}} \left| \frac{\tau_1^{(0)}}{\tau_1^{(n)}} - 1 \right| s \\
&\le \left| \frac{\tau_1^{(0)}}{\tau_1^{(n)}} - 1 \right| \tau_1^{(n)} = |\tau_1^{(0)} - \tau_1^{(n)}| \le \delta.
\end{aligned}$$

On $[\tau_i^{(n)}, \tau_{i+1}^{(n)}]$, $i = 1, \dots, k_0$, we have (with obvious abbreviations)

$$\lambda_n(s) = \frac{\tau_{i+1}^{(0)} - \tau_i^{(0)}}{\tau_i^{(n)} - \tau_i^{(n)}} s + \tau_i^{(0)} - \left( \frac{\tau_{i+1}^{(0)} - \tau_i^{(0)}}{\tau_{i+1}^{(n)} - \tau_i^{(n)}} \right) \cdot \tau_i^{(n)}$$
$$= \frac{\Delta\tau^{(0)}}{\Delta\tau^{(n)}} s + \tau_i^{(0)} - \frac{\Delta\tau^{(0)}}{\Delta\tau^{(n)}} \cdot \tau_i^{(n)}$$

and

$$\sup_{\tau_i^{(n)} \le s \le \tau_{i+1}^{(n)}} \left| \frac{\Delta\tau^{(0)}}{\Delta\tau^{(n)}} s + \tau_i^{(0)} - \frac{\Delta\tau^{(0)}}{\Delta\tau^{(n)}} \cdot \tau_i^{(n)} - s \right|.$$

Setting $y = s - \tau_i^{(n)}$, we have

$$= \sup_{0 \le y \le \Delta\tau^{(n)}} \left| \frac{\Delta\tau^{(0)}}{\Delta\tau^{(n)}} (y + \tau_i^{(n)}) + \tau_i^{(0)} - \frac{\Delta\tau^{(0)}}{\Delta\tau^{(n)}} \cdot \tau_i^{(n)} - (y + \tau_i^{(n)}) \right|$$
$$= \sup_{0 \le y \le \Delta\tau^{(n)}} \left| \left( \frac{\Delta\tau^{(0)}}{\Delta\tau^{(n)}} - 1 \right) y + (\tau_i^{(0)} - \tau_i^{(n)}) \right|$$
$$\le \sup_{0 \le y \le \Delta\tau^{(n)}} \left( \left| \frac{\Delta\tau^{(0)}}{\Delta\tau^{(n)}} - 1 \right| \Delta\tau^{(n)} + |\tau_i^{(0)} - \tau_i^{(n)}| \right)$$
$$\le |\Delta\tau^{(0)} - \Delta\tau^{(n)}| + \delta = |\tau_{i+1}^{(0)} - \tau_i^{(0)} - (\tau_{i+1}^{(n)} - \tau_i^{(n)})| + \delta$$
$$\le \delta + |\tau_{i+1}^{(0)} - \tau_{i+1}^{(n)}| + |\tau_i^{(0)} - \tau_i^{(n)}| \le 3\delta.$$

If $d_{[0,T]}$ is the Skorohod metric on $D[0, T]$, we get from (7.19) and (7.20) that

$$d_{[0,T]}(\chi_{m_n}, \chi_{m_0}) \le k_0\delta \vee 3\delta = (k_0 \vee 3)\delta$$

if $n \ge n_1$. Since $\delta$ is arbitrarily chosen we get

$$d_{[0,T]}(\chi_{m_n}, \chi_{m_0}) \to 0,$$

so $\chi$ is continuous at $m$. □

## 7.3 Transformations

There are several useful results describing how regular variation of a distribution of a random vector is affected by various transformations of that random vector. This section presents a sample of results with the goal of illustrating a variety of techniques.

### 7.3.1 Addition

Here are some results about how regular variation of the distribution tail of a vector is affected by addition of components.

#### Linear combinations of components of a random vector

Given a random vector $\boldsymbol{Z}$ with a regularly vary tail, what happens if we sum the components of $\boldsymbol{Z}$ or, more generally, take linear combinations? For $\boldsymbol{t} \in \mathbb{R}^d$, define

$$\boldsymbol{t} \cdot \boldsymbol{Z} = \sum_{i=1}^{d} t^{(i)} Z^{(i)}.$$

**Proposition 7.3.** *Suppose, as usual, that* $\mathbb{E} = [\mathbf{0}, \boldsymbol{\infty}] \setminus \{\mathbf{0}\}$ *and that* $\boldsymbol{Z}$ *satisfies the regular variation condition*

$$n\mathbb{P}\left[\frac{\boldsymbol{Z}}{b_n} \in \cdot\right] \xrightarrow{v} \nu$$

*in* $M_+(\mathbb{E})$. *Then for any* $\boldsymbol{t} > \mathbf{0}$, *we have*

$$n\mathbb{P}[\boldsymbol{t} \cdot \boldsymbol{Z} > b_n x] \to c x^{-\alpha} \quad (n \to \infty),$$

*where* $c = \nu\{\boldsymbol{y} \in \mathbb{E} : \boldsymbol{t} \cdot \boldsymbol{y} > 1\}$.

*In particular, suppose the components of* $\boldsymbol{Z}$ *are iid, and* $b_n$ *is chosen so that for each* $i \in \{1, \ldots, d\}$, *we have*

$$n\mathbb{P}[Z^{(i)} > b_n x] \to x^{-\alpha} = \nu_\alpha(x, \infty] \quad (n \to \infty).$$

*Then for* $\boldsymbol{t} > \mathbf{0}$, *as* $x \to \infty$,

$$\mathbb{P}[\boldsymbol{t} \cdot \boldsymbol{Z} > x] \sim \sum_{i=1}^{d} (t^{(i)})^\alpha \mathbb{P}[Z^{(1)} > x].$$

*Proof.* This is the polar coordinate transformation (Section 6.1.2 (p. 168)) in disguise. Fix $\boldsymbol{t} > \mathbf{0}$, and define the norm on $\mathbb{R}^d$ by

$$\|\boldsymbol{x}\| = \sum_{i=1}^{d} t^{(i)} |x^{(i)}|.$$

From the polar coordinate transformation (see (6.15) (p. 179)), we know that

$$n\mathbb{P}[\boldsymbol{t} \cdot \boldsymbol{Z} > b_n x] = n\mathbb{P}[\|\boldsymbol{Z}\| > b_n x] \to c x^{-\alpha}$$

as $n \to \infty$, where

$$c = \nu\{\boldsymbol{y} : \|\boldsymbol{y}\| > 1\} = \nu\{\boldsymbol{y} : \boldsymbol{t} \cdot \boldsymbol{y} > 1\}.$$

For the iid case, $\nu$ concentrates mass on each axis according to the measure $\nu_\alpha$, and we get

$$\begin{aligned}\nu\{\boldsymbol{y} : \boldsymbol{t} \cdot \boldsymbol{y} > 1\} &= \sum_{i=1}^{d} \nu_\alpha\{w : t^{(i)} w > 1\} \\ &= \sum_{i=1}^{d} (t^{(i)})^\alpha.\end{aligned}$$

□

The converse to this problem is still not completely settled, although important progress has been achieved: If $\boldsymbol{t} \cdot \boldsymbol{Z}$ is regularly varying for every $\boldsymbol{t}$, does this imply regular variation of the distribution tail of $\boldsymbol{Z}$? In general, the answer is negative for $\alpha$s that are integers. See [13, 170, 184, 220].

### Adding independent vectors

If $\boldsymbol{X}$ and $\boldsymbol{Y}$ are independent random vectors each of whose distributions have regularly varying tails, does the sum $\boldsymbol{X} + \boldsymbol{Y}$ preserve the regular variation? A qualified *yes* can be given to this question. See [168, 171, 207, 259, 260].

For this investigation, it is convenient to have the notation

$$\mathbb{E}_d = [0, \infty]^d \setminus \{\boldsymbol{0}\}$$

for the $d$-dimensional compactified nonnegative orthant punctured by removal of the origin.

**Lemma 7.2 (binding).** *Suppose $\boldsymbol{X} \in \mathbb{R}_+^{d_1}$ and $\boldsymbol{Y} \in \mathbb{R}_+^{d_2}$ are defined on the same probability space, independent, and satisfy the regular variation conditions ($n \to \infty$)*

$$n\mathbb{P}\left[\frac{\boldsymbol{X}}{b_n} \in \cdot\right] \xrightarrow{v} \nu_{\boldsymbol{X}}(\cdot) \quad \textit{in } M_+(\mathbb{E}_{d_1}), \tag{7.21}$$

$$n\mathbb{P}\left[\frac{\boldsymbol{Y}}{b_n} \in \cdot\right] \xrightarrow{v} \nu_{\boldsymbol{Y}}(\cdot) \quad \textit{in } M_+(\mathbb{E}_{d_2}), \tag{7.22}$$

*with the same sequence $b_n \to \infty$. Then the distribution tail of $(\boldsymbol{X}, \boldsymbol{Y})$ is also regularly varying:*

$$n\mathbb{P}\left[\left(\frac{\boldsymbol{X}}{b_n}, \frac{\boldsymbol{Y}}{b_n}\right) \in \cdot\right] \xrightarrow{v} \nu \quad \textit{in } M_+(\mathbb{E}_{d_1+d_2}), \tag{7.23}$$

*where*

$$\nu(d\boldsymbol{x}, d\boldsymbol{y}) = \nu_{\boldsymbol{X}}(d\boldsymbol{x})\epsilon_{\boldsymbol{0}}(d\boldsymbol{y}) + \epsilon_{\boldsymbol{0}}(d\boldsymbol{x})\nu_{\boldsymbol{Y}}(d\boldsymbol{y}). \tag{7.24}$$

*Proof.* Suppose $f \in C_K^+(\mathbb{E}_{d_1+d_2})$. We first prove that

$$n\mathbf{E}f\left(\frac{\boldsymbol{X}}{b_n}, \frac{\boldsymbol{Y}}{b_n}\right) - n\mathbf{E}f\left(\frac{\boldsymbol{X}}{b_n}, \mathbf{0}\right) - n\mathbf{E}f\left(\mathbf{0}, \frac{\boldsymbol{Y}}{b_n}\right) \to 0. \tag{7.25}$$

Since (7.21) implies that

$$n\mathbf{E}f\left(\frac{\boldsymbol{X}}{b_n}, \mathbf{0}\right) \to \int_{\mathbb{E}_{d_1}} f(\boldsymbol{x}, \mathbf{0})\nu_{\boldsymbol{X}}(d\boldsymbol{x})$$

with a similar statement for $\boldsymbol{Y}$, (7.25) implies (7.23).

It is convenient to use the $L_\infty$-norm. For fixed $d$, set

$$\|\boldsymbol{x}\|_\infty = \bigvee_{i=1}^{d} |x^{(i)}|, \quad \boldsymbol{x} \in \mathbb{R}^d.$$

(Note that this notation will be used even when the dimension of the vector changes. This abuse of notation should not cause confusion.) There exists a fixed $\delta > 0$ such that the support of $f$ is contained in $\{\boldsymbol{x} \in \mathbb{E}_{d_1+d_2} : \|\boldsymbol{x}\|_\infty > \delta\}$. So

$$\begin{aligned}
n\mathbf{E}f\left(\frac{\boldsymbol{X}}{b_n}, \frac{\boldsymbol{Y}}{b_n}\right) &= n\mathbf{E}f\left(\frac{\boldsymbol{X}}{b_n}, \frac{\boldsymbol{Y}}{b_n}\right) 1_{[\|\boldsymbol{X}\|_\infty \vee \|\boldsymbol{Y}\|_\infty > b_n\delta]} \\
&= n\mathbf{E}f\left(\frac{\boldsymbol{X}}{b_n}, \frac{\boldsymbol{Y}}{b_n}\right) \big(1_{[\|\boldsymbol{X}\|_\infty > b_n\delta, \|\boldsymbol{Y}\|_\infty \le b_n\delta]} \\
&\quad + 1_{[\|\boldsymbol{X}\|_\infty \le b_n\delta, \|\boldsymbol{Y}\|_\infty > b_n\delta]} + 1_{[\|\boldsymbol{X}\|_\infty > b_n\delta, \|\boldsymbol{Y}\|_\infty > b_n\delta]}\big) \\
&= A + B + C.
\end{aligned}$$

Now

$$\begin{aligned}
|C| &\le (\text{const})n\mathbb{P}[\|\boldsymbol{X}\|_\infty > b_n\delta, \|\boldsymbol{Y}\|_\infty > b_n\delta] \\
&= (\text{const})(n\mathbb{P}[\|\boldsymbol{X}\|_\infty > b_n\delta])\mathbb{P}[\|\boldsymbol{Y}\|_\infty > b_n\delta] \\
&\to (\text{const})'\delta^{-\alpha} \cdot 0 = 0
\end{aligned}$$

since $b_n \to \infty$.

Pick any $\eta < \delta$. The same argument used for $C$ gives for $A$,

$$\begin{aligned}
A &= n\mathbf{E}f\left(\frac{\boldsymbol{X}}{b_n}, \frac{\boldsymbol{Y}}{b_n}\right) 1_{[\|\boldsymbol{X}\|_\infty > b_n\delta, \|\boldsymbol{Y}\|_\infty \le b_n\delta]} \\
&= n\mathbf{E}f\left(\frac{\boldsymbol{X}}{b_n}, \frac{\boldsymbol{Y}}{b_n}\right) 1_{[\|\boldsymbol{X}\|_\infty > b_n\delta, \|\boldsymbol{Y}\|_\infty \le b_n\eta]} + o(1).
\end{aligned}$$

Inspecting the right side, apart from $o(1)$, gives $\boldsymbol{Y}$ close to $\boldsymbol{0}$, so it makes sense to compare this with $n\mathbf{E}f(b_n^{-1}\boldsymbol{X}, \boldsymbol{0})$. We have

$$\begin{aligned}
&\left| n\mathbf{E}f\left(\frac{\boldsymbol{X}}{b_n}, \frac{\boldsymbol{Y}}{b_n}\right) 1_{[\|\boldsymbol{X}\|_\infty > b_n\delta, \|\boldsymbol{Y}\|_\infty \le b_n\eta]} - n\mathbf{E}f\left(\frac{\boldsymbol{X}}{b_n}, \boldsymbol{0}\right)\right| \\
&= \left| n\mathbf{E}f\left(\frac{\boldsymbol{X}}{b_n}, \frac{\boldsymbol{Y}}{b_n}\right) 1_{[\|\boldsymbol{X}\|_\infty > b_n\delta, \|\boldsymbol{Y}\|_\infty \le b_n\eta]} - n\mathbf{E}f\left(\frac{\boldsymbol{X}}{b_n}, \boldsymbol{0}\right) 1_{[\|\boldsymbol{X}\|_\infty > b_n\delta]}\right| \\
&= \left| n\mathbf{E}f\left(\frac{\boldsymbol{X}}{b_n}, \frac{\boldsymbol{Y}}{b_n}\right) 1_{[\|\boldsymbol{X}\|_\infty > b_n\delta, \|\boldsymbol{Y}\|_\infty \le b_n\eta]} \right. \\
&\qquad \left. - n\mathbf{E}f\left(\frac{\boldsymbol{X}}{b_n}, \boldsymbol{0}\right) 1_{[\|\boldsymbol{X}\|_\infty > b_n\delta, \|\boldsymbol{Y}\|_\infty \le b_n\eta]}\right| + o(1) \\
&\le \omega_f(\eta) n\mathbb{P}\left[\left\|\frac{\boldsymbol{X}}{b_n}\right\|_\infty > \delta\right] \mathbb{P}\left[\left\|\frac{\boldsymbol{Y}}{b_n}\right\|_\infty \le \eta\right] + o(1) \\
&\to \omega_f(\eta)\delta^{-\alpha}(\text{const}),
\end{aligned}$$

where recall that $\omega_f(\eta)$ is the modulus of continuity of $f$. We can manipulate the free parameter $\eta$. If $\eta \to 0$, then $\omega_f(\eta) \to 0$, which finishes showing that the difference converges to 0. The term $B$ is handled similarly. □

We now consider addition of independent random vectors, and for this the only reasonable assumption is $d_1 = d_2 = d$.

**Proposition 7.4.** *Suppose* $\boldsymbol{X}$ *and* $\boldsymbol{Y}$ *satisfy the assumptions of Lemma 7.2 with the restriction that* $d_1 = d_2 = d$. *Then*

$$n\mathbb{P}\left[\frac{\boldsymbol{X}+\boldsymbol{Y}}{b_n} \in \cdot\right] \stackrel{v}{\to} \nu_{\boldsymbol{X}} + \nu_{\boldsymbol{Y}} \tag{7.26}$$

*in* $M_+(\mathbb{E}_d)$.

*Proof.* Define the map SUM : $\mathbb{E}_d \times \mathbb{E}_d \to \mathbb{E}_d$ by

$$\text{SUM}(\boldsymbol{x}, \boldsymbol{y}) = \boldsymbol{x} + \boldsymbol{y}.$$

Provided the compactness condition (5.19) (p. 141) is satisfied, the result follows from Proposition 5.5 (p. 141) since applying SUM to (7.23) gives

$$n\mathbb{P} \circ \text{SUM}^{-1}\left[\left(\frac{\boldsymbol{X}}{b_n}, \frac{\boldsymbol{Y}}{b_n}\right) \in \cdot\right] = n\mathbb{P}\left[\frac{\boldsymbol{X}+\boldsymbol{Y}}{b_n} \in \cdot\right] \stackrel{v}{\to} \nu \circ \text{SUM}^{-1} = \nu_{\boldsymbol{X}} + \nu_{\boldsymbol{Y}}.$$

Thus it remains to show (5.19) in the form

$$\mathrm{SUM}^{-1}(\mathcal{K}(\mathbb{E}_d)) \subset \mathcal{K}(\mathbb{E}_{2d}).$$

Suppose $K_d \in \mathcal{K}(\mathbb{E}_d)$. Since SUM is continuous, $\mathrm{SUM}^{-1}(K_d)$ is closed in $\mathbb{E}_{2d}$. Also, there exists $\delta > 0$ such that

$$K_d \subset \{\boldsymbol{x} \in \mathbb{E}_d : \|\boldsymbol{x}\|_\infty \geq \delta\}.$$

Now

$$\begin{aligned}\mathrm{SUM}^{-1}(\{\boldsymbol{z} \in \mathbb{E}_d : \|\boldsymbol{z}\|_\infty \geq \delta\}) &= \{(\boldsymbol{x}, \boldsymbol{y}) \in \mathbb{E}_{2d} : \|\boldsymbol{x} + \boldsymbol{y}\|_\infty \geq \delta\} \\ &\subset \{(\boldsymbol{x}, \boldsymbol{y}) \in \mathbb{E}_{2d} : \|\boldsymbol{x}\|_\infty \vee \|\boldsymbol{y}\|_\infty \geq \delta/2\}.\end{aligned}$$

The last set is closed and, being bounded away from the origin, is also compact in $\mathbb{E}_{2d}$. So $\{(\boldsymbol{x}, \boldsymbol{y}) \in \mathbb{E}_{2d} : \|\boldsymbol{x} + \boldsymbol{y}\|_\infty \geq \delta\}$ is compact, and so is $\mathrm{SUM}^{-1}(K_d)$, being a closed subset of a compact set. □

### 7.3.2 Products

Here we take a random vector $\boldsymbol{Z}$ with a regularly varying tail, multiply by a scalar random variable, and examine the tail of the product. We consider two cases, (i) where the multiplier has a relatively thin tail and (ii) where the multiplier is jointly regularly varying with $\boldsymbol{Z}$ but not asymptotically independent of $\boldsymbol{Z}$. The second result receives a direct analytic treatment, and to illustrate alternative probabilistic methods, we prove the first result, Breiman's theorem, by the point process method, which parallels the analytic proof given in [30].

We revert to our notation $\mathbb{E} = [\mathbf{0}, \boldsymbol{\infty}] \setminus \{\mathbf{0}\}$.

#### Breiman's theorem: A factor has a relatively thin tail

**Proposition 7.5 (Breiman's theorem).** *Suppose $\boldsymbol{Z}$ is a nonnegative random vector satisfying the usual multivariate regular variation condition with exponent $-\alpha$:*

$$n\mathbb{P}\left[\frac{\boldsymbol{Z}}{b_n} \in \cdot\right] \xrightarrow{v} \nu.$$

*Suppose further that $Y \geq 0$ is a random variable with a moment greater than $\alpha$. This is equivalent to the existence of $\epsilon > 0$, such that*

$$\mathbf{E}(Y^{\alpha(1+2\epsilon)}) < \infty. \tag{7.27}$$

*Then*

$$n\mathbb{P}\left[\frac{Y\mathbf{Z}}{b_n} \in \cdot\right] \xrightarrow{v} \mathbf{E}(Y^\alpha)\nu.$$

*In particular, if $d = 1$, we have that*

$$\lim_{x\to\infty} \frac{\mathbb{P}[YZ > x]}{\mathbb{P}[Z > x]} = \mathbf{E}(Y^\alpha).$$

*Remark* 7.1. The result for $d = 1$ was first proved by Breiman [30]. A result requiring asymptotic independence of $Y$ and $\mathbf{Z}$ instead of independence is in [215]. The case where $\mathbf{Z}$ is a $d$-dimensional vector is from [259]. A result where $Y$ is a matrix independent of $\mathbf{Z}$ is considered in [14, see Proposition A.1 (p. 113) and Corollary A.2 (p. 114)]. For a nice application of the multivariate Breiman result to solutions of stochastic differential equations, see [169]; see also [5]. For $d = 1$, a refinement which drops (7.27) in favor of a condition that $P[Y > x] = o(P[Z > x])$ is given in [128]. A product result in [44], quoted in [75, p. 542], of a slightly different character describes the case in $d = 1$ of $Y \stackrel{d}{=} Z$, $Y, Z$ independent, $P[Z > x] \in \text{RV}_{-\alpha}$, $\alpha > 0$, and $EZ^\alpha = \infty$.

Breiman's theorem has a straightforward analytic proof using dominated convergence outlined in Problem 7.10 (p. 251); see [30]. The Breiman proof requires judicious carving up of the region of integration arising from the distribution of the product of $Y$ and $Z$. Our proof, based on the Poisson transform, offers an indication of how to decompose the region of integration as the decomposition is guided by the necessity to truncate the state space to get a compact set.

*Proof.* Suppose $\{\mathbf{Z}_n, n \geq 1\}$ are iid copies of $\mathbf{Z}$. The regular variation condition is equivalent to (Theorem 6.2 (p. 179))

$$\sum_{i=1}^n \epsilon_{\mathbf{Z}_i/b_n} \Rightarrow \sum_k \epsilon_{\boldsymbol{j}_k} = \text{PRM}(\nu)$$

in $M_p(\mathbb{E})$. Now let $\{Y_j\}$ be iid copies of $Y$ that are independent of $\{\mathbf{Z}_n\}$ as well as independent of $\{\boldsymbol{j}_k\}$. It follows that in $M_p(\mathbb{E} \times (0, \infty))$,

$$\sum_{i=1}^n \epsilon_{(\mathbf{Z}_i/b_n, Y_i)} \Rightarrow \sum_k \epsilon_{(\boldsymbol{j}_k, Y_k)} = \text{PRM}(\nu \times \mathbb{P}[Y \in \cdot]). \tag{7.28}$$

See Problem 6.7 (p. 208) or the argument leading to (6.20) (p. 181). Note infinities are included in $\mathbb{E}$ and so 0 is excluded from $(0, \infty)$ to avoid the potential problem of $0 \cdot \infty$ when we take products.

Define the product map $\text{PROD} : \mathbb{E} \times (0, \infty) \mapsto \mathbb{E}$ by $\text{PROD}(z, y) = yz$. The compactness condition (5.19) (p. 141) fails (see Figure 7.2), so truncation of the state

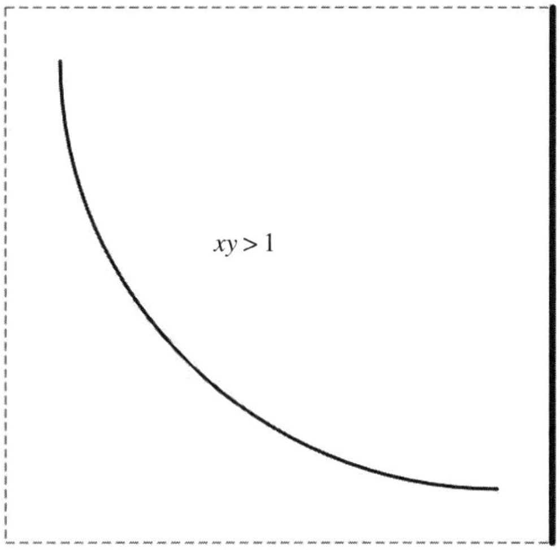


**Fig. 7.2.** The region $\{(x, y) : xy \geq 1\}$ is not compact in $\mathbb{E} \times (0, \infty)$, pictured for the case $d = 1$. Dotted lines indicate an open boundary.

space is necessary, followed by an application of the second converging together method of Theorem 3.5 (p. 56). We proceed in steps.

STEP 1: *Restrict the state space to a compact set.* Consider the compact subset of $\mathbb{E} \times (0, \infty)$

$$\Lambda_\delta := \{(z, y) \in \mathbb{E} \times (0, \infty) : \|z\| \geq \delta^{1+\epsilon}, y \in [\delta, \delta^{-1}]\}.$$

Applying the restriction functional (Problem 6.3 (p. 206)) to (7.28) yields

$$\sum_{i=1}^{n} \epsilon_{(\mathbf{Z}_i/b_n, Y_i)}(\Lambda_\delta \cap \cdot) \Rightarrow \sum_k \epsilon_{(\mathbf{j}_k, Y_k)}(\Lambda_\delta \cap \cdot) \tag{7.29}$$

in $M_p(\Lambda_\delta)$.

STEP 2: *Apply the functional* PROD. From (7.29) we get by applying PROD that

$$\sum_{i=1}^{n} 1_{\Lambda_\delta}\left(\frac{\mathbf{Z}_i}{b_n}, Y_i\right) \epsilon_{Y_i \mathbf{Z}_i/b_n} \Rightarrow \sum_k 1_{\Lambda_\delta}(\mathbf{j}_k, Y_k) \epsilon_{Y_k \mathbf{j}_k} \tag{7.30}$$

in $M_p(\mathbb{E})$.

STEP 3: *The limit in the restricted convergence converges to the desired limit when the restriction evaporates.* Take the limit point process in (7.30) and let $\delta \downarrow 0$ to get pointwise

$$\sum_k 1_{\Lambda_\delta}(\boldsymbol{j}_k, Y_k)\epsilon_{Y_k \boldsymbol{j}_k} \overset{\delta \to 0}{\to} \sum_k \epsilon_{Y_k \boldsymbol{j}_k},$$

vaguely $M_p(\mathbb{E})$.

Step 4: *Show what you want to converge is close to what you know converges.* With $\rho(\cdot, \cdot)$ the vague metric on $M_p(\mathbb{E})$, we need to show that for any $\eta > 0$,

$$\lim_{\delta \downarrow 0} \limsup_{n \to \infty} \mathbb{P}\left[\rho\left(\sum_{i=1}^n 1_{\Lambda_\delta}\left(\frac{\boldsymbol{Z}_i}{b_n}, Y_i\right)\epsilon_{Y_i \boldsymbol{Z}_i/b_n}, \sum_{i=1}^n \epsilon_{Y_i \boldsymbol{Z}_i/b_n}\right) > \eta\right] = 0. \tag{7.31}$$

The expression (7.31) is true, provided that for any $f \in C_K^+(\mathbb{E})$, we have

$$\begin{aligned}\lim_{\delta \downarrow 0} \limsup_{n \to \infty} \mathbb{P}&\left[\left|\sum_{i=1}^n 1_{\Lambda_\delta}\left(\frac{\boldsymbol{Z}_i}{b_n}, Y_i\right) f(Y_i \boldsymbol{Z}_i/b_n) - \sum_{i=1}^n f(Y_i \boldsymbol{Z}_i/b_n)\right| > \eta\right] \\ &= \lim_{\delta \downarrow 0} \limsup_{n \to \infty} \mathbb{P}\left[\sum_{i=1}^n 1_{\Lambda_\delta^c}\left(\frac{\boldsymbol{Z}_i}{b_n}, Y_i\right) f(Y_i \boldsymbol{Z}_i/b_n) > \eta\right] = 0. \end{aligned} \tag{7.32}$$

Unpacking the definition of $\Lambda_\delta^c$, it is enough to verify the following:

$$\lim_{\delta \downarrow 0} \limsup_{n \to \infty} \mathbb{P}\left[\sum_{i=1}^n f(Y_i \boldsymbol{Z}_i/b_n) 1_{[Y_i \in [\delta, \delta^{-1}], \|b_n^{-1} \boldsymbol{Z}_i\| < \delta^{1+\epsilon}]} > \eta_1\right] = 0, \tag{7.33}$$

$$\lim_{\delta \downarrow 0} \limsup_{n \to \infty} \mathbb{P}\left[\sum_{i=1}^n f(Y_i \boldsymbol{Z}_i/b_n) 1_{[Y_i \notin [\delta, \delta^{-1}] \cap [\|b_n^{-1} \boldsymbol{Z}_i\| \geq \delta^{1+\epsilon}]} > \eta_2\right] = 0, \tag{7.34}$$

$$\lim_{\delta \downarrow 0} \limsup_{n \to \infty} \mathbb{P}\left[\sum_{i=1}^n f(Y_i \boldsymbol{Z}_i/b_n) 1_{[Y_i \notin [\delta, \delta^{-1}] \cap [\|b_n^{-1} \boldsymbol{Z}_i\| < \delta^{1+\epsilon}]} > \eta_3\right] = 0 \tag{7.35}$$

for $\eta_i > 0, i = 1, 2, 3.$

Suppose the support of $f$ is in $\{\boldsymbol{x} \in \mathbb{E} : \|\boldsymbol{x}\| \geq \xi\}$ for some $\xi > 0$. The conditions in the indicator in (7.33) imply that

$$\|Y_i \boldsymbol{Z}_i/b_n\| \leq \delta^{-1}\delta^{1+\epsilon} = \delta^\epsilon < \xi,$$

if $\delta$ is sufficiently small. Thus, for small $\delta$, we have the probability in (7.33) equal to 0.

For (7.34), it is enough to show that

$$\lim_{\delta \downarrow 0} \limsup_{n \to \infty} \mathbb{P}\left[\sum_{i=1}^n f(Y_i \boldsymbol{Z}_i/b_n) 1_{[Y_i < \delta] \cap [\|b_n^{-1} \boldsymbol{Z}_i\| \geq \delta^{1+\epsilon}]} > \eta_{21}\right] = 0, \tag{7.36}$$

$$\lim_{\delta \downarrow 0} \limsup_{n\to\infty} \mathbb{P}\left[\sum_{i=1}^{n} f(Y_i \mathbf{Z}_i/b_n) 1_{[Y_i > \delta^{-1}] \cap [\|b_n^{-1}\mathbf{Z}_i\| \geq \delta^{1+\epsilon}]} > \eta_{22}\right] = 0 \tag{7.37}$$

for $\eta_{21} > 0$, $\eta_{22} > 0$. The probability in (7.36) is bounded by

$$\begin{aligned}
&P\left\{\bigcup_{i=1}^{n}\left[\|Y_i\mathbf{Z}_i/b_n\| \geq \xi, \frac{\|\mathbf{Z}_i\|}{b_n} \geq \delta^{1+\epsilon}, Y_i < \delta\right]\right\} \\
&\quad \leq n\mathbb{P}\left[\|Y_1\mathbf{Z}_1/b_n\| \geq \xi, \frac{\|\mathbf{Z}_1\|}{b_n} \geq \delta^{1+\epsilon}, Y_1 < \delta\right] \\
&\quad \leq n\mathbb{P}\left[\|\mathbf{Z}_1/b_n\| \geq \xi\delta^{-1} \bigvee \delta^{1+\epsilon}\right] \\
&\quad \to \left(\xi\delta^{-1} \bigvee \delta^{1+\epsilon}\right)^{-\alpha} \qquad (n \to \infty), \\
&\quad = \xi^{-\alpha}\delta^{\alpha} \to 0 \qquad (\delta \downarrow 0).
\end{aligned}$$

The probability in (7.37) is bounded by

$$\begin{aligned}
&\mathbb{P}\left\{\bigcup_{i=1}^{n}[\|Y_i\mathbf{Z}_i/b_n\| \geq \xi, Y_i > \delta^{-1}, \|b_n^{-1}\mathbf{Z}_i\| \geq \delta^{1+\epsilon}]\right\} \\
&\quad \leq n\mathbb{P}[\|Y_1\mathbf{Z}_i/b_n\| \geq \xi, Y_1 > \delta^{-1}, \|b_n^{-1}\mathbf{Z}_1\| \geq \delta^{1+\epsilon}] \\
&\quad \leq n\mathbb{P}[\|b_n^{-1}\mathbf{Z}_1\| \geq \delta^{1+\epsilon}, Y_1 > \delta^{-1}] \\
&\quad = n\mathbb{P}[\|b_n^{-1}\mathbf{Z}_1\| \geq \delta^{1+\epsilon}]\mathbb{P}[Y_1 > \delta^{-1}] \\
&\quad \sim \delta^{-(1+\epsilon)\alpha}\mathbb{P}[Y_1 > \delta^{-1}] \qquad (n \to \infty),
\end{aligned}$$

and applying the Chebychev inequality to the tail probability for $Y_1$ gives the bound

$$\begin{aligned}
&\leq \delta^{-(1+\epsilon)\alpha}\mathbf{E}Y_1^{(1+2\epsilon)\alpha}\delta^{(1+2\epsilon)\alpha} \\
&\to 0 \qquad (\delta \to 0).
\end{aligned}$$

Finally, we show (7.35). First, for small $\delta > 0$,

$$n\mathbb{P}[Y_1\|\mathbf{Z}_1/b_n\| > \xi, Y_1 \leq \delta, \|b_n^{-1}\mathbf{Z}_1\| < \delta^{1+\epsilon}] = 0,$$

as for (7.33). For the other case, we have, using Karamata's theorem (p. 25),

$$\begin{aligned}
&n\mathbb{P}[Y_1\|\mathbf{Z}_n/b_n\| > \xi, Y_1 > \delta^{-1}, \|\mathbf{Z}_1/b_n\| < \delta^{1+\epsilon}] \\
&\qquad = n\mathbb{P}[Y_1 1_{[Y_1 > \delta^{-1}]} \cdot \|\mathbf{Z}_n/b_n\| 1_{[\|\mathbf{Z}_1/b_n\| < \delta^{1+\epsilon}]} > \xi] \\
&\qquad \leq \frac{n}{\xi^{\alpha+\epsilon}}\mathbf{E}(Y_1^{\alpha+\epsilon})\mathbf{E}(\|Z_1/b_n\|^{\alpha+\epsilon} 1_{[\|\mathbf{Z}_1/b_n\| < \delta^{1+\epsilon}]})
\end{aligned}$$

$$\rightarrow (\text{const}) \int_0^{\delta^{1+\epsilon}} z^{\alpha+\epsilon} \alpha z^{-\alpha-1} dz \qquad (n \to \infty)$$

$$= (\text{const}) \int_0^{\delta^{1+\epsilon}} z^{\epsilon-1} dz = (\text{const})(\delta^{1+\epsilon})^{\epsilon} \to 0 \quad (\delta \to 0).$$

STEP 5: *Wrap-up.* We conclude that

$$\sum_{i=1}^{n} \epsilon_{Y_i \mathbf{Z}_i / b_n} \Rightarrow \sum_k \epsilon_{Y_k \mathbf{j}_k}$$

in $M_p(\mathbb{E})$. The result follows by another application of Theorem 6.2 (p. 179). □

## Products of heavy-tailed random variables which are jointly regularly varying

The product of two random variables that are not asymptotically independent, but whose tails satisfy multivariate regular variation, offers contrasting behavior to the case just considered. We consider $(Y, \mathbf{Z})$, which satisfy the nonstandard form of regular variation (p. 204).

**Proposition 7.6.** *Suppose $Y$ is a nonnegative random variable satisfying, for some $\alpha_Y > 0$,*

$$\mathbb{P}[Y > \cdot] \in RV_{-\alpha_Y} \tag{7.38}$$

*with quantile function $b_Y(\cdot) \in RV_{1/\alpha_Y}$. Let $\mathbf{Z}$ be a $\mathbb{R}_+^d$-valued random vector, defined on the same probability space as $Y$, whose distribution tail is regularly varying with index $\alpha_Z > 0$,*

$$t\mathbb{P}\left[\frac{\mathbf{Z}}{b_Z(t)} \in \cdot\right] \stackrel{v}{\to} \nu_Z(\cdot) \quad (t \to \infty) \tag{7.39}$$

*in $M_+(\mathbb{E}_d)$, where $\mathbb{E}_d = [0, \infty]^d \setminus \{\mathbf{0}\}$. This means that $b_Z(t)$ is regularly varying with index $1/\alpha_Z$. Suppose further that $(Y, \mathbf{Z})$ is multivariate regularly varying in the sense that*

$$t\mathbb{P}\left[\left(\frac{Y}{b_Y(t)}, \frac{\mathbf{Z}}{b_Z(t)}\right) \in \cdot\right] \stackrel{v}{\to} \nu(\cdot) \not\equiv 0 \quad (t \to \infty) \tag{7.40}$$

*on $\mathbb{E}_{d+1}$, $\nu$ concentrates on $[0, \infty)^{d+1} \setminus \{\mathbf{0}\}$, and there exists $\delta > 0$ such that*

$$\nu\{(y, z) : |y| \wedge \|z\| > \delta\} > 0. \tag{7.41}$$

*Then $Y\mathbf{Z}$ has a regularly varying distribution tail with index $-\frac{\alpha_Y \alpha_Z}{\alpha_Y + \alpha_Z}$, a scaling function $b_Y(\cdot) b_Z(\cdot)$, and limit measure*

$$\nu\{(y, z) : yz \in \cdot\}. \tag{7.42}$$

*Remark* 7.2. The condition (7.41) prevents the limit measure in (7.42) from being trivial.

*Proof.* We first give the proof assuming $d = 1$. Fix $x > 0$, and define for any positive number $K$,

$$A_{K,x} := \{(y, z) : yz > x, y \leq K, z \leq K\}.$$

Then

$$t\mathbb{P}\left[\frac{YZ}{b_Y(t)b_Z(t)} > x\right] \geq t\mathbb{P}\left[\left(\frac{Y}{b_Y(t)}, \frac{Z}{b_Z(t)}\right) \in A_{K,x}\right].$$

Next, let $t$ converge to $\infty$ first, and then let $K$ go to $\infty$ through a sequence so that $A_{K,x}$ is a $\nu$-continuity set. This results in

$$\liminf_{t\to\infty} t\mathbb{P}\left[\frac{YZ}{b_Y(t)b_Z(t)} > x\right] \geqslant \nu(\{(y, z) : yz > x\}).$$

On the other hand, we have

$$t\mathbb{P}\left[\frac{YZ}{b_Y(t)b_Z(t)} > x\right] \leqslant t\mathbb{P}\left[\left(\frac{Y}{b_Y(t)}, \frac{Z}{b_Y(t)}\right) \in A_{K,x}\right] + t\mathbb{P}\left[\frac{Y}{b_Y(t)} > K\right] + t\mathbb{P}\left[\frac{Z}{b_Z(t)} > K\right].$$

Now, from the regular variation of the tails of $Y$ and $Z$, the last two terms converge to $K^{-\alpha_Y}$ and $K^{-\alpha_Z}$, respectively, as $t \to \infty$. Then letting $K$ go to $\infty$ through a sequence so that $A_{K,x}$ is a $\nu$-continuity set, both terms go to zero. Hence

$$\limsup_{t\to\infty} t\mathbb{P}\left[\frac{YZ}{b_Y(t)b_Z(t)} > x\right] \leqslant \nu(\{(y, z) : yz > x\}).$$

Thus

$$t\mathbb{P}\left[\frac{YZ}{b_Y(t)b_Y(t)} > z\right] \to \nu(\{(x, y) : xy > z\}).$$

Then since $b_Y b_Z$ is a regularly varying function of index $\frac{\alpha_Y+\alpha_Z}{\alpha_Y\alpha_Z}$ and since $\nu(\{(y, z) : yz > x\}) > 0$ for some $x > 0$, we have

$$\mathbb{P}[YZ > \cdot] \in \mathrm{RV}_{-\frac{\alpha_Y\alpha_Z}{\alpha_Y+\alpha_Z}}.$$

For $d > 1$, define the map

$$\text{IDPOLAR} : (y, z) = \left(y, \|z\|, \frac{z}{\|z\|}\right).$$

Using the method that showed the equivalences in Theorem 6.1 (p. 173), we get from (7.40) that

$$t\mathbb{P}\left[\left(\frac{Y}{b_Y(t)}, \frac{\|\boldsymbol{Z}\|}{b_Z(t)}, \frac{\boldsymbol{Z}}{\|\boldsymbol{Z}\|}\right) \in \cdot\right] \xrightarrow{v} \nu \circ \text{IDPOLAR}^{-1}(\cdot). \tag{7.43}$$

Define

$$\text{PRODID}(y, r, \boldsymbol{a}) = (yr, \boldsymbol{a});$$

applying this to (7.43) yields

$$t\mathbb{P}\left[\left(\frac{Y\|\boldsymbol{Z}\|}{b_Y(t)b_Z(t)}, \frac{\boldsymbol{Z}}{\|\boldsymbol{Z}\|}\right) \in \cdot\right] = t\mathbb{P}\left[\left(\frac{Y\|\boldsymbol{Z}\|}{b_Y(t)b_Z(t)}, \frac{Y\boldsymbol{Z}}{\|Y\boldsymbol{Z}\|}\right) \in \cdot\right]$$
$$\xrightarrow{v} \nu \circ \text{IDPOLAR}^{-1} \circ \text{PRODID}^{-1}(\cdot).$$

This gives regular variation of $Y\boldsymbol{Z}$ in polar coordinate form. □

In Proposition 7.6, if $\alpha_Y$ and $\alpha_Z$ are between 1 and 2, i.e., $Y$ and $\|\boldsymbol{Z}\|$ have finite mean but infinite variance, then the product $Y\boldsymbol{Z}$ has a regularly varying tail of index $-\frac{\alpha_Y\alpha_Z}{\alpha_Y+\alpha_Z} \in (\frac{1}{2}, 1)$; i.e., the product has a much heavier tail with infinite mean. This result contrasts with Breiman's theorem (Proposition 7.5), where the product of asymptotically independent random variables has tail behavior similar to the factor with the heavier tail.

## Internet data

Recall the Boston University study, mentioned in Example 1.1 (p. 4) and Section 5.2.2 (p. 125). This was a study of World Wide Web downloads in sessions initiated by logins at a Boston University computer laboratory. The study kept track of

$$F = \text{the file size of the requested document,}$$
$$L = \text{the duration of the download,}$$
$$R = \text{throughput of the request} = F/L.$$

Empirical evidence indicates all three quantities have heavy tails. Table 7.1 gives empirical estimates for the tail parameters for $F$, $R$, and $L$ for the BU measurements arrived at by a combination of QQ plotting and Hill plotting.

| $\alpha$ | $\hat{\alpha}_F$ | $\hat{\alpha}_R$ | $\hat{\alpha}_L$ |
|---|---|---|---|
| estimated value | 1.15 | 1.13 | 1.4 |

**Table 7.1.** Tail parameter estimates.

What conclusions can we make about the dependence structure of $(F, R, L)$? Since $F = LR$, the tail parameters $(\alpha_F, \alpha_R, \alpha_L)$ cannot be arbitrary. Consider the following two possibilities:

- Proposition 7.6 (p. 236) is applicable for $L$, $R$. This means that $(L, R)$ possess a jointly regularly varying tail but are *not* asymptotically independent. The conclusion from Proposition 7.6 is that

$$\hat{\alpha}_F = \frac{\hat{\alpha}_L \hat{\alpha}_R}{\hat{\alpha}_L + \hat{\alpha}_R} = .625 \neq 1.15.$$

Unfortinately, the empirical estimates do not match the theoretical predictions, indicating that the model posed by Proposition 7.6 is unlikely to be correct.

- Proposition 7.5 (p. 231) applies. This would be the case if $(L, R)$ were independent or if [215] $(L, R)$ obey some form of asymptotic independence. In this case, Proposition 7.5 predicts that

$$\alpha_F = \alpha_R \wedge \alpha_L,$$

assuming that $\alpha_R \neq \alpha_L$. In our example,

$$1.15 \approx 1.13 \wedge 1.4.$$

So for the BU data, evidence seems to support some form of independence for $(R, L)$. Interestingly, for other data sets (see [38]), large values of $R$ and $F$ are independent. Input models taking account of $F$, $R$, $L$ will differ in their predictions depending on what is assumed about the dependence of these three quantities. See [68].

### 7.3.3 Laplace transforms

Suppose $U$ is a Radon measure on $[\mathbf{0}, \infty) = [0, \infty)^d$, written $U \in M_+[\mathbf{0}, \infty)$, whose Laplace transform $\hat{U}$ exists:

$$\hat{U}(\boldsymbol{\lambda}) = \hat{U}(\lambda^{(1)}, \ldots, \lambda^{(d)}) = \int_{[\mathbf{0},\infty)} \exp\left\{-\sum_{i=1}^{d} \lambda^{(i)} x^{(i)}\right\} U(d\boldsymbol{x}) \tag{7.44}$$

$$= \int_{[\mathbf{0},\infty)} e^{-\boldsymbol{\lambda}\cdot\boldsymbol{x}} U(d\boldsymbol{x}) < \infty \quad (\boldsymbol{\lambda} > \mathbf{0}). \tag{7.45}$$

Let $U(\boldsymbol{x}) = U[\mathbf{0}, \boldsymbol{x}]$ be the distribution function of the measure $U$, and assume that $U(\boldsymbol{x})$ satisfies the regular variation condition (6.1) (p. 167). We assume that $U(\boldsymbol{x})$ is regularly varying on the cone $(\mathbf{0}, \infty)$ and that there exists a function $g(t) \in \mathrm{RV}_\rho$, $\rho > 0$, and a limit measure $V \in M_+[\mathbf{0}, \infty)$ with distribution function $V(\boldsymbol{x}) = V[\mathbf{0}, \boldsymbol{x}]$,

$$\lim_{t\to\infty} U_t(\boldsymbol{x}) := \lim_{t\to\infty} \frac{U(t\boldsymbol{x})}{g(t)} = V(\boldsymbol{x}), \quad \boldsymbol{x} \in (\mathbf{0}, \infty), \tag{7.46}$$

at points of continuity of the limit. This means that

$$U_t \xrightarrow{v} V \quad (t \to \infty) \tag{7.47}$$

in $M_+[\mathbf{0}, \infty)$.

Assume that $V \not\equiv 0$, and as a normalization, suppose $V(\mathbf{1}) = 1$. The argument given on p. 167 shows that

$$V(c\boldsymbol{x}) = c^\rho V(\boldsymbol{x}), \quad c > 0, \quad \boldsymbol{x} > \mathbf{0},$$

and the argument following (6.13) (p. 177) gives

$$V\left\{\boldsymbol{x} \in [\mathbf{0}, \infty) : \|\boldsymbol{x}\| \leq r, \frac{\boldsymbol{x}}{\|\boldsymbol{x}\|} \in \Lambda\right\} = r^\rho S(\Lambda), \quad r > 0, \quad \Lambda \subset \aleph_+,$$

where $\Lambda$ is a Borel set.

The Laplace transform of $V$ exists since, if $\boldsymbol{\lambda} > \mathbf{0}$ is fixed, we may define the norm

$$\|\boldsymbol{x}\| = \sum_{i=1}^{d} \lambda^{(i)} |x^{(i)}|, \quad \boldsymbol{x} \in \mathbb{R}^d,$$

so that if $\boldsymbol{x} \geq \mathbf{0}$, then $\|\boldsymbol{x}\| = \lambda \cdot \boldsymbol{x}$. Write

$$\begin{aligned}
\hat{V}(\boldsymbol{\lambda}) &= \int_{[\mathbf{0},\infty)} e^{-\boldsymbol{\lambda}\cdot\boldsymbol{u}} V(d\boldsymbol{u}) = \int_{[\mathbf{0},\infty)} e^{-\|\boldsymbol{u}\|} V(d\boldsymbol{u}) \\
&= \iint_{\boldsymbol{a}\in\aleph_+, s>0} e^{-s} \rho s^{\rho-1} ds\, S(d\boldsymbol{a}) \\
&= S(\aleph_+) \int_0^\infty e^{-s} \rho s^{\rho-1} ds < \infty
\end{aligned}$$

for $\rho > 0$.

Recall our convention from Appendix 10 (p. 359) that operations should be interpreted componentwise, so that, for instance,

$$\frac{1}{\boldsymbol{x}} = \left(\frac{1}{x^{(1)}}, \ldots, \frac{1}{x^{(d)}}\right),$$

but keep in mind that $\boldsymbol{\lambda} \cdot \boldsymbol{x} = \sum_{i=1}^{d} \lambda^{(i)} x^{(i)}$. Similarly, we recall that

$$\boldsymbol{x}/\boldsymbol{y} = (x^{(1)}/y^{(1)}, \ldots, x^{(d)}/y^{(d)}).$$

Now suppose that

$$\sum_k \epsilon_{\boldsymbol{u}_k^{(t)}} = \mathrm{PRM}(U_t), \qquad \sum_k \epsilon_{\boldsymbol{v}_k} = \mathrm{PRM}(V)$$

with state space $[\mathbf{0}, \infty)$. Then from Problem 5.3 (p. 163), (7.47) is equivalent to

$$\sum_k \epsilon_{\boldsymbol{u}_k^{(t)}} \Rightarrow \sum_k \epsilon_{\boldsymbol{v}_k}$$

in $M_p[\mathbf{0}, \infty)$. Define

$$\{\boldsymbol{E}_n = (E_n^{(1)}, \dots, E_n^{(d)}); n \geq 1\}$$

to be iid $d$-dimensional random vectors each of whose components are iid unit exponential random variables. Set

$$\mathbb{EXP}(\cdot) = \mathbb{P}[\boldsymbol{E}_1 \in \cdot]$$

for the joint distribution of a $d$-dimensional vector of iid unit exponential random variables. Assuming $\{\boldsymbol{E}_k\}$ independent of both $\{\boldsymbol{u}_k^{(t)}\}$ and $\{\boldsymbol{v}_k\}$, we get from augmentation (Proposition 5.3 (p. 123)) that

$$N_t := \sum_k \epsilon_{(\boldsymbol{u}_k^{(t)}, \boldsymbol{E}_k)} = \mathrm{PRM}(U_t \times \mathbb{EXP})$$

and

$$N_\infty := \sum_k \epsilon_{(\boldsymbol{v}_k, \boldsymbol{E}_k)} = \mathrm{PRM}(V \times \mathbb{EXP}),$$

each with state space $[\mathbf{0}, \infty) \times (\mathbf{0}, \infty]$ and

$$N_t \Rightarrow N_\infty \tag{7.48}$$

in $M_p([\mathbf{0}, \infty) \times (\mathbf{0}, \infty])$. Apply the map

$$\mathrm{RATIO} : [\mathbf{0}, \infty) \times (\mathbf{0}, \infty] \mapsto [\mathbf{0}, \infty)$$

defined by

$$\mathrm{RATIO}(\boldsymbol{u}, \boldsymbol{e}) = (\boldsymbol{u}/\boldsymbol{e}).$$

We hope Proposition 5.2 (p. 121) is applicable (almost—we have not checked the compactness condition in Proposition 5.2) so that $N_t \circ \mathrm{RATIO}^{-1}$ is a Poisson process. The mean measure ($\boldsymbol{z} \in (\mathbf{0}, \infty)$) is

$$\mathbf{E}(N_t \circ \mathrm{RATIO}^{-1}([\mathbf{0}, z]))$$

$$= \iint_{\{(\boldsymbol{x},\boldsymbol{y}):\boldsymbol{x}/\boldsymbol{y}\le \boldsymbol{z}\}} U_t(d\boldsymbol{x})\,\mathbb{EXP}(d\boldsymbol{y}) = \int_{[\mathbf{0},\boldsymbol{\infty})} \prod_{i=1}^{d} e^{-x^{(i)}/z^{(i)}} U_t(d\boldsymbol{x})$$

$$= \int_{[\mathbf{0},\boldsymbol{\infty})} \prod_{i=1}^{d} \exp\left\{\frac{-x^{(i)}}{z^{(i)}}\right\} \frac{1}{g(t)} U(t d\boldsymbol{x}) = \frac{1}{g(t)} \hat{U}\left(\frac{\mathbf{1}}{t z}\right). \tag{7.49}$$

Condition (7.45) forces a finite expectation, and (7.49) shows that $\hat{U}(\frac{\mathbf{1}}{z})$, $z \in (\mathbf{0}, \boldsymbol{\infty})$ is the distribution function of a measure in $M_+[\mathbf{0}, \boldsymbol{\infty})$.

We now state the multidimensional Tauberian theorem. The result is from an unpublished technical report by Stam [284]; the approach is from [261]. See also [281–283].

**Proposition 7.7.** *If* $U \in M_+[\mathbf{0}, \boldsymbol{\infty})$ *has a finite Laplace transform given in* (7.45) *and its distribution function* $U(\boldsymbol{x})$ *satisfies* (7.46), *then the distribution function* $\hat{U}(\mathbf{1}/\boldsymbol{x})$ *is also regularly varying on the cone* $(\mathbf{0}, \boldsymbol{\infty})$,

$$\frac{1}{g(t)} \hat{U}\left(\frac{\mathbf{1}}{t\boldsymbol{x}}\right) \to \hat{V}\left(\frac{\mathbf{1}}{\boldsymbol{x}}\right), \quad \boldsymbol{x} \in (\mathbf{0}, \boldsymbol{\infty}). \tag{7.50}$$

*The limit function of* $\hat{U}(\mathbf{1}/\boldsymbol{x})$ *is* $\hat{V}(\mathbf{1}/\boldsymbol{x})$.

For discussion of the case in which $d = 1$, see, for instance, [26, 135]. The converse is also true, but it would take us a bit further afield. See [261, 284].

*Proof.* Given (7.46), if we did not have to worry about whether the map RATIO satisfies the compactness condition (5.19) (p. 141), then (7.48) would imply convergence of Poisson processes

$$N_t \circ \mathrm{RATIO}^{-1} \Rightarrow N_\infty \circ \mathrm{RATIO}^{-1}$$

in $M_p[\mathbf{0}, \boldsymbol{\infty})$, which would imply the mean measures converge. Hence, from Problem 5.3 (p. 163), we could conclude that for $z > \mathbf{0}$,

$$\mathbf{E}(N_t \circ \mathrm{RATIO}^{-1}([\mathbf{0}, z])) = \frac{1}{g(t)} \hat{U}\left(\frac{\mathbf{1}}{tz}\right)$$

$$\to \mathbf{E}(N_\infty \circ \mathrm{RATIO}^{-1}([\mathbf{0}, z])) = \hat{V}\left(\frac{\mathbf{1}}{z}\right), \tag{7.51}$$

which gives the result. How do we fill the gap in this outline?

Reviewing Remark 5.2 (p. 142) suggests truncating the state space $[\mathbf{0}, \boldsymbol{\infty}) \times (\mathbf{0}, \boldsymbol{\infty})$ to the compact set

$$K_M := \{(\boldsymbol{u}, \boldsymbol{y}) \in [\mathbf{0}, \boldsymbol{\infty}) \times (\mathbf{0}, \boldsymbol{\infty}) : \|\boldsymbol{u}\| \le M, M^{-1}\mathbf{1} \le \boldsymbol{y} \le M\mathbf{1}\}.$$

Applying the restriction functional (Problem 6.3 (p. 206)) to (7.48), we get

$$N_t^M := N_t(\cdot \cap K_M) \Rightarrow N_\infty(\cdot \cap K_M) =: N_\infty^M$$

in $M_p([\mathbf{0}, \infty) \times (\mathbf{0}, \infty))$; then, using the fact that RATIO is continuous, we get from Remark 5.2 (p. 142) that

$$N_t^{M,R} := N_t(\cdot \cap K_M) \circ \text{RATIO}^{-1} \Rightarrow N_\infty(\cdot \cap K_M) \circ \text{RATIO}^{-1} =: N_\infty^{M,R}.$$

Unpacking this result gives, as $t \to \infty$,

$$\sum_k 1_{[(\boldsymbol{u}_k^{(t)}, E_k) \in K_M]} \epsilon_{\boldsymbol{u}_k^{(t)}/E_k} \Rightarrow \sum_k 1_{[(\boldsymbol{v}_k, E_k) \in K_M]} \epsilon_{\boldsymbol{v}_k/E_k}.$$

As $M \to \infty$, we have vague convergence in $M_p(\mathbf{0}, \infty)$,

$$N_\infty^{M,R} := \sum_k 1_{[(\boldsymbol{v}_k, E_k) \in K_M]} \epsilon_{\boldsymbol{v}_k/E_k} \to N_\infty^R := \sum_k \epsilon_{\boldsymbol{v}_k/E_k}.$$

We are now prepared for an application of the second converging together theorem, Theorem 3.5 (p. 56). With

$$N_t^R = \sum_k \epsilon_{\boldsymbol{u}_k^{(t)}/E_k},$$

it suffices to show that for any $\eta > 0$ and $d(\cdot, \cdot)$ the vague metric,

$$\lim_{M \to \infty} \limsup_{t \to \infty} \mathbb{P}[d(N_t^{M,R}, N_t^R) > \eta] = 0, \tag{7.52}$$

since then the second converging together theorem plus the fact that PRMs converge iff their mean measures converge (Problem 5.3 (p. 163)) justifies the desired result (7.51).

The proof of (7.52) follows the usual pattern. Suppose $h \in C_K^+[\mathbf{0}, \infty)$; it then suffices to show for any such $h$ that for $\delta > 0$,

$$\lim_{M \to \infty} \limsup_{t \to \infty} \mathbb{P}[|N_t^{M,R}(h) - N_t^R(h)| > \delta] = 0. \tag{7.53}$$

Now $h$ has compact support, so suppose for convenience that the support of $h$ is contained in $[\mathbf{0}, c\mathbf{1}]$ for some $c > 0$; for typographical ease, just set $c = 1$. Observe

$$\begin{aligned} |N_t^{M,R}(h) - N_t^R(h)| &= \left| \sum_k 1_{[(\boldsymbol{u}_k^{(t)}, E_k) \in K_M]} h\left(\frac{\boldsymbol{u}_k^{(t)}}{E_k}\right) - \sum_k h\left(\frac{\boldsymbol{u}_k^{(t)}}{E_k}\right) \right| \\ &= \left| \sum_k 1_{[(\boldsymbol{u}_k^{(t)}, E_k) \in K_M^c]} h\left(\frac{\boldsymbol{u}_k^{(t)}}{E_k}\right) \right|, \end{aligned}$$

and therefore, by Chebychev's inequality,

$$\begin{aligned}
&\mathbb{P}[|N_t^{M,R}(h) - N_t^R(h)| > \delta] \\
&\quad \le \delta^{-1}\mathbf{E}\left(\sum_k 1_{[(\boldsymbol{u}_k^{(t)},\boldsymbol{E}_k)\in K_M^c]} h\left(\frac{\boldsymbol{u}_k^{(t)}}{\boldsymbol{E}_k}\right)\right) \\
&\quad = \delta^{-1}\iint_{\{(\boldsymbol{u},\boldsymbol{y}):(\boldsymbol{u},\boldsymbol{y})\in K_M^c\}} h(\boldsymbol{u}/\boldsymbol{y})U_t(d\boldsymbol{u})\,\mathbb{EXP}(d\boldsymbol{y}) \\
&\quad \le \delta^{-1}\sup_{\boldsymbol{x}} h(\boldsymbol{x}) \iint_{\{(\boldsymbol{u},\boldsymbol{y}):(\boldsymbol{u},\boldsymbol{y})\in K_M^c,\boldsymbol{u}/\boldsymbol{y}\le\mathbf{1}\}} U_t(d\boldsymbol{u})\,\mathbb{EXP}(d\boldsymbol{y}),
\end{aligned}$$

where the last inequality takes into account the compact support of $h$.

Now we deal with the double integral. Note first that

$$\begin{aligned}
K_M^c = \{(\boldsymbol{u},\boldsymbol{y}) : \|\boldsymbol{u}\| > M\} &\cup \{(\boldsymbol{u},\boldsymbol{y}) : \|\boldsymbol{u}\| \le M, \boldsymbol{y} > M\mathbf{1}\} \\
&\cup \{(\boldsymbol{u},\boldsymbol{y}) : \|\boldsymbol{u}\| \le M, \boldsymbol{y} \le M^{-1}\mathbf{1}\}.
\end{aligned} \tag{7.54}$$

Because the region of integration is compact,

$$\begin{aligned}
&\iint_{\{\boldsymbol{u}/\boldsymbol{y}\le 1,\|\boldsymbol{u}\|>M\}} U_t(d\boldsymbol{u})\,\mathbb{EXP}(d\boldsymbol{y}) \\
&\quad \to \iint_{\{\boldsymbol{u}\le\boldsymbol{y},\|\boldsymbol{u}\|>M\}} V(d\boldsymbol{u})\,\mathbb{EXP}(d\boldsymbol{y}) \qquad (t\to\infty) \\
&\quad = \int_{\{\|\boldsymbol{u}\|>M\}} \prod_{i=1}^d e^{-u^{(i)}} V(d\boldsymbol{u}) \\
&\quad = \int_{\{\|\boldsymbol{u}\|>M\}} e^{-\mathbf{1}\cdot\boldsymbol{u}} V(d\boldsymbol{u}) \\
&\quad \to 0 \qquad (M\to\infty),
\end{aligned}$$

since the Laplace function $\hat{V}(\boldsymbol{\lambda})$ exists.

Now consider the last set in the decomposition of $K_M$ given in (7.54). We have

$$\begin{aligned}
&\iint_{\{\|\boldsymbol{u}\|\le M,\boldsymbol{y}\le M^{-1}\mathbf{1}\}} U_t(d\boldsymbol{u})\,\mathbb{EXP}(d\boldsymbol{y}) \\
&\quad \le (1-e^{-M^{-1}})^d U_t(\{\boldsymbol{u} : \|\boldsymbol{u}\| \le M\}) \\
&\quad \to (1-e^{-M^{-1}})^d V(\{\boldsymbol{u} : \|\boldsymbol{u}\| \le M\}) \qquad (t\to\infty) \\
&\quad = M^{\rho}(1-e^{-M^{-1}})^d V(\{\boldsymbol{u} : \|\boldsymbol{u}\| \le 1\}) \\
&\quad \to 0 \qquad (M\to\infty).
\end{aligned}$$

The rest is very similar. □

### Special case for $d = 1$: Karamata's Tauberian theorem

Suppose $d = 1$ in Proposition 7.7. Then if $U \in \mathrm{RV}_\rho$, $\rho > 0$, we have

$$\frac{U(tx)}{U(t)} \to x^\rho =: V(x), \quad x > 0, \quad t \to \infty,$$

and $g(t) = U(t)$. It follows that, as $t \to \infty$, (7.50) becomes

$$\frac{\hat{U}\left(\frac{1}{tx}\right)}{U(t)} \to \int_0^\infty e^{-x^{-1}s} \rho s^{\rho-1} ds = \Gamma(\rho+1)x^\rho. \tag{7.55}$$

When we set $x = 1$, we get

$$\hat{U}\left(\frac{1}{t}\right) \sim U(t)\Gamma(\rho+1) \quad (t \to \infty). \tag{7.56}$$

We have not proved converses here, but they hold as well. See [26, 135, 182].

### Renewal theory

Consider an ordinary renewal process $\{S_n, n \geq 0\}$ such that

$$S_0 = 0, \qquad S_n = \sum_{i=1}^{n} X_i, \quad n \geq 1,$$

and $\{X_n, n \geq 1\}$ is a sequence of iid nonnegative random variables with common distribution $F$. The function that counts renewals is

$$N = \sum_{n=0}^{\infty} \epsilon_{S_n}, \tag{7.57}$$

so that

$$N(t) := N([0,t]) = \sum_{n=0}^{\infty} 1_{[S_n \leq t]}, \quad t > 0. \tag{7.58}$$

The renewal function [135, 262] is

$$U(t) := EN(t) = \sum_{n=0}^{\infty} \mathbb{P}[S_n \leq t] = \sum_{n=0}^{\infty} F^{n*}(t), \tag{7.59}$$

where $F^{n*}$ is the $n$th convolution power of $F$.

Suppose

$$1 - F(t) = t^{-\alpha} L(t) \in RV_{-\alpha}, \quad t \to \infty, \quad 0 < \alpha < 1. \tag{7.60}$$

What is the asymptotic form of $U$?

Set

$$H(x) = \int_0^x (1 - F(s))ds,$$

and by Karamata's theorem (Theorem 2.1 (p. 25)),

$$H(x) \sim \frac{x\bar{F}(x)}{1-\alpha} \in RV_{1-\alpha} \quad (x \to \infty). \tag{7.61}$$

From (7.56), we conclude that

$$\hat{H}\left(\frac{1}{t}\right) \sim H(t)\Gamma(2-\alpha). \tag{7.62}$$

However, by integrating by parts, one quickly sees that

$$\hat{H}(\lambda) = \frac{1 - \hat{F}(\lambda)}{\lambda}, \quad \lambda > 0. \tag{7.63}$$

Put (7.61)–(7.63) into the blender, and out comes

$$1 - \hat{F}\left(\frac{1}{t}\right) \sim \bar{F}(t)\Gamma(1-\alpha). \tag{7.64}$$

From the definition of $U(t)$, the transform satisfies

$$\hat{U}(\lambda) = \frac{1}{1 - \hat{F}(\lambda)}, \quad \lambda > 0.$$

Again from (7.56),

$$\hat{U}\left(\frac{1}{t}\right) = \frac{1}{1 - \hat{F}\left(\frac{1}{t}\right)} \sim U(t)\Gamma(1+\alpha);$$

(7.64) gives the final alchemy:

$$U(t) \sim \frac{1}{1 - F(t)} \frac{1}{\Gamma(1-\alpha)\Gamma(1+\alpha)} =: \frac{c(\alpha)}{1 - F(t)}, \quad t \to \infty. \tag{7.65}$$

## 7.4 Problems

**7.1 ([42, 259]).** In the context of Corollary 7.1 (p. 218), prove that

$$\left( \bigvee_{i=1}^{[n\cdot]} \frac{Z_i}{b_n}, \frac{S_{[n\cdot]}}{b_n} - [n\cdot]\mathbf{E}\left( \frac{Z_1}{b_n} 1_{[|Z_1/b_n|\le 1]} \right) \right) \Rightarrow (Y_\alpha(\cdot), X_\alpha(\cdot))$$

in $D([0,\infty), \mathbb{R}^2)$, where $\{Y_\alpha(t), t \ge 0\}$ is an extremal process and $\{X_\alpha(t), t \ge 0\}$ is a stable Lévy motion.

**7.2.** In Theorem 7.1 (p. 214) and Corollary 7.1 (p. 218), the partial sums are centered by *truncated* first moments since no assumption is made about existence of first moments. What if you knew the first moments were finite. Could these be used for centering?

**7.3 (Convergence to stable subordinators).** Suppose $\{X_i, i \ge 1\}$ are iid, nonnegative random variables and

$$\bar{F}(x) = \mathbb{P}[X_1 > x] \in \mathrm{RV}_{-\alpha}$$

for $0 < \alpha < 1$. As usual, let $b(t)$ be the quantile function

$$b(t) = \frac{1}{1-F}^{\leftarrow}(t).$$

Show that

$$\frac{1}{b(n)} \sum_{i=1}^{[nt]} X_i \Rightarrow X_\alpha(t)$$

in $D[0,\infty)$, where $X_\alpha(\cdot)$ is a stable subordinator.

**7.4 (Bootstrap the sample mean [10]).** Review Corollary 7.1 (p. 218) and Propositions 6.2 (p. 188) and 6.3 (p. 189). For $1 < \alpha < 2$, assume that $\{Z_i, i \ge 1\}$ are iid with a common distribution in the domain of attraction of a stable law of index $\alpha$; that is, the global regular variation of Corollary 7.1 holds.

Suppose $Z_1, \dots, Z_n$ are observed with sample mean $\bar{Z}_n$, and then a bootstrap sample $Z_1^*, \dots, Z_n^*$ is drawn, which has sample mean $\bar{Z}_n^*$. Prove that for $x_1, \dots, x_l$ fixed, that as $n \to \infty$,

$$\left( \mathbb{P}\left[ \frac{n}{b_n}(\bar{Z}_n^* - \bar{Z}_n) \le x_i \mid Z_1, \dots, Z_n \right], i = 1, \dots, l \right)$$

converges to a random distribution limit evaluated at $x_1, \dots, x_l$. To eliminate the unknown $b_n$, prove that the same conclusion holds for

$$\left(\mathbb{P}\left[\frac{n}{\bigvee_{j=1}^{n} Z_j}(\bar{Z}_n^* - \bar{Z}_n) \le x_i \mid Z_1, \dots, Z_n\right], i = 1, \dots, l\right).$$

What if the bootstrap sample size is reduced to $m$, where $m = m(n) = o(n)$?

**7.5 (Karamata's theorem).** Suppose $U : [0, \infty) \mapsto [0, \infty)$ is nondecreasing. Suppose

$$\sum_k \epsilon_{(t_k, u_k)} = \mathrm{PRM}(\mathbb{LEB} \times U)$$

is Poisson on $M_p([0, \infty) \times [0, \infty))$.

1. $U \in \mathrm{RV}_\alpha$, $\alpha > 0$, iff there exists a sequence of constants $b_n \to \infty$ such that

$$N_n := \sum_k \epsilon_{(nt_k, u_k/b_n)} \Rightarrow \mathrm{PRM}(\mathbb{LEB} \times \mu_\alpha),$$

where $\mu_\alpha[0, x] = x^\alpha$.

2. Consider the map $T : (0, \infty) \times [0, \infty) \mapsto (0, \infty) \times [0, \infty)$ defined by

$$T(t, x) = (t, x/t).$$

Check that $T^{-1}([a, b] \times [0, y])$ is compact for $0 < a < b < \infty$ and $y > 0$. From this, conclude that $T^{-1}(K)$ is compact whenever $K \subset (0, \infty) \times [0, \infty)$ is compact.

3. From these facts, prove Karamata's theorem, that

$$\lim_{x \to \infty} \frac{\int_0^x U(s)ds}{xU(x)} = \frac{1}{\alpha + 1}.$$

**7.6 (Convergence of sums in the nonstandard case [241]).** Suppose we have iid vectors $\{\mathbf{Z}_n = (Z_n^{(1)}, Z_n^{(2)}), n \ge 1\}$ in $\mathbb{R}^2$ such that for $i = 1, 2$,

$$n\mathbb{P}\left[\frac{Z_1^{(i)}}{b_n^{(i)}} \in \cdot\right] \xrightarrow{v} \nu_i$$

in $M_+([-\infty, \infty] \setminus \{0\})$; that is, marginally we have regular variation. Suppose that $b_n^{(i)}$ is the restriction to the integers of a regularly varying function with index $1/\alpha_i$, where

$$0 < \alpha_i < 2, \quad i = 1, 2,$$

but that it is not necessarily the case that $\alpha_1 = \alpha_2$. Assume the two-dimensional global condition

$$n\mathbb{P}\left[\left(\frac{Z_1^{(1)}}{b_n^{(1)}}, \frac{Z_1^{(2)}}{b_n^{(2)}}\right) \in \cdot\right] \xrightarrow{v} \nu,$$

in $M_+([-\infty, \infty] \setminus \{\mathbf{0}\})$, where $\nu$ is a Lévy measure, that is, a Radon measure on $[-\infty, \infty] \setminus \{\mathbf{0}\}$ satisfying

$$\int_{\{\boldsymbol{x}:\|\boldsymbol{x}\|\leq 1\}} \|\boldsymbol{x}\|^2 \nu(d\boldsymbol{x}) < \infty.$$

Show that the sequence of processes

$$\left\{\left(\sum_{i=1}^{[nt]} \frac{Z_i^{(1)}}{b_n^{(1)}}, \sum_{i=1}^{[nt]} \frac{Z_i^{(2)}}{b_n^{(2)}}\right), t \geq 0\right\}$$

converges weakly as $n \to \infty$, after suitable centering, in $D([0, \infty), \mathbb{R}^2)$, the space of right-continuous functions with domain $[0, \infty)$ and range $\mathbb{R}^2$. Describe the limit process.

**7.7 (Sample variance [259]).** Suppose $\{Z_n, n \geq 1\}$ are iid with a distribution $F$. Suppose for simplicity that $Z_i \geq 0$ and $1 - F \in \mathrm{RV}_{-\alpha}$ and suppose $0 < \alpha < 1$.

1. Show that the sequence of processes

$$\left\{\left\{\left(\sum_{i=1}^{[nt]} Z_i, \sum_{i=1}^{[nt]} Z_i^2\right), t \geq 0\right\}, n \geq 1\right\}$$

converges weakly in $D([0, \infty), \mathbb{R}^2)$ after centering and scaling. Describe the limit process.

2. Set

$$\bar{Z}_n := \frac{1}{n}\sum_{i=1}^{n} Z_i, \qquad \mathcal{S}_n = \frac{1}{n}\sum_{i=1}^{n}(Z_i - \bar{Z}_n)^2.$$

Show that the sequence of processes

$$\{\{(\bar{Z}_{[nt]}, \mathcal{S}_{[nt]}), t \geq 0\}, n \geq 1\}$$

converges weakly in $D([0, \infty), \mathbb{R}^2)$ after suitable centering and scaling of the components. Describe the limit process. (The normalization by $\frac{1}{n}$ is traditional but inappropriate in the heavy-tailed case.)

**7.8 (More products [44, 75]).** Suppose $Z_1$, $Z_2$ are iid, nonnegative random variables with $P[Z_i > x] \in \mathrm{RV}_{-\alpha}$, $\alpha > 0$, and $EZ_i^\alpha = \infty$. Show that $P[Z_1 Z_2 > x] \in \mathrm{RV}_{-\alpha}$ and

$$\lim_{x\to\infty} \frac{\mathbb{P}[Z_1 Z_2 > x]}{\mathbb{P}[Z_1 > x]} = \infty.$$

**7.9 (Partial converse of Breiman's theorem [217]).** Suppose $\xi$ and $\eta$ are two independent, nonnegative random variables, and $\xi$ has a Pareto distribution with parameter 1:

$$\mathbb{P}[\xi > x] = x^{-1}, \quad x \geq 1.$$

(a) We have

$$\mathbb{P}[\xi\eta > x] \in \mathrm{RV}_{-\alpha}, \quad \alpha < 1,$$

iff

$$\mathbb{P}[\eta > x] \in \mathrm{RV}_{-\alpha},$$

and then

$$\frac{\mathbb{P}[\xi\eta > x]}{\mathbb{P}[\eta > x]} \to \frac{1}{1-\alpha}.$$

(b) If $\mathbb{P}[\xi\eta > x] \in \mathrm{RV}_{-1}$ and $\xi\eta$ has a heavier tail than $\xi$, meaning that

$$\frac{\mathbb{P}[\xi\eta > x]}{\mathbb{P}[\xi > x]} = \int_0^x \mathbb{P}[\eta > y]dy \to \infty,$$

i.e., $\mathbb{E}[\eta] = \infty$, then

$$\int_0^x \mathbb{P}[\eta > s]ds =: L(x) \uparrow \infty$$

is slowly varying. If, in addition, $L(x) \in \Pi$, the de Haan function class $\Pi$ (see Problems 2.10 (p. 37) and 2.11 and [26, 90, 102, 144, 260]), then

$$\mathbb{P}[\eta > x] \in \mathrm{RV}_{-1}$$

and

$$\frac{L(x)}{x\mathbb{P}[\eta > x]} = \frac{\mathbb{P}[\xi\eta > x]}{\mathbb{P}[\eta > x]} \to \infty.$$

As an example, consider

$$\mathbb{P}[\eta > x] = \frac{e \log x}{x}, \quad x > e$$

and show that

$$\mathbb{P}[\xi\eta > x] \sim \frac{1}{2} e x^{-1} (\log x)^2, \quad x \to \infty.$$

**7.10 (Analytic proof of Breiman's theorem).** Review the statement of Breiman's theorem in Proposition 7.5. Proceed as follows to construct an analytic proof:

1. Assume that $d = 1$. Write $\mathbb{P}[YZ > u]$ as an integral on $[0, \infty)$ with respect to the distribution of $Y$. Divide through by $\mathbb{P}[Z > u]$.
2. Split the region of integration $[0, \infty) = [0, u/M] \cup [u/M, \infty]$. On $[0, u/M]$, bound the ratio integrand with a uniform bound using Potter's bounds.
3. On $[u/M, \infty)$, bound the ratio integrand by $\mathbb{P}[Y > u/M]/\mathbb{P}[Z > u]$. The asymptotic behavior of this ratio is controlled by (7.27).
4. Apply dominated convergence to get $\mathbb{P}[YZ > u]/\mathbb{P}[Z > u]$ to converge to the desired limit.
5. For $d > 1$, let $K \in \mathcal{K}(\mathbb{E})$ be compact in $\mathbb{E}$. Then for some $\delta > 0$, $K \subset \{z : \|z\| > \delta\}$. Bound $n\mathbb{P}[Y\boldsymbol{Z}/b_n \in K] \leq \int n\mathbb{P}[y\|\boldsymbol{Z}\|/b_n > \delta]\mathbb{P}[Y \in dy]$, and apply Fatou to get a $\mathbf{E}(Y^\alpha)\nu(K)$ is an upper bound to the lim sup of $n\mathbb{P}[Y\boldsymbol{Z}/b_n \in K]$. Construct a lower bound to the lim inf similarly after changing $K$ to a relatively compact open set. Apply Theorem 3.2 (p. 52).

**7.11 (Choice theory [243, 244]).** Suppose $\boldsymbol{Y}_\infty$ is the limit random vector given in (7.1) (p. 212). The limit measure is $\nu$ and the angular probability measure is $S$. Define

$$\mathbb{E}^{1>} = \left\{ \boldsymbol{x} \in \mathbb{E} : x^{(1)} > \bigvee_{i=2}^{d} x^{(i)} \right\}$$

and

$$\aleph^{1>} = \aleph \cap \mathbb{E}^{1>}.$$

Prove that

$$\mathbb{P}\left[ Y^{(1)} > \bigvee_{i=2}^{d} Y^{(i)} \right] = S(\aleph^{1>}).$$

Furthermore, for $y > 0$,

$$\mathbb{P}\left[ Y^{(1)} > \bigvee_{i=2}^{d} Y^{(i)}, \bigvee_{i=1}^{d} Y^{(i)} \leq y \right] = S(\aleph^{1>})e^{-y^{-1}}.$$

**7.12 (Convex hulls [78, 228]).** Let $\mathcal{K}[\boldsymbol{0}, \infty)$ be the compact sets of $[\boldsymbol{0}, \infty)$ metrized by the Hausdorff metric [214, 228]. Suppose $\{\boldsymbol{Z}_1, \ldots, \boldsymbol{Z}_n\}$ are iid random vectors in $[\boldsymbol{0}, \infty)$ with common distribution $F$ satisfying the regular variation condition with scaling function $b_n = b(n)$. Prove the convex hull of $\{\boldsymbol{Z}_1/b_n, \ldots, \boldsymbol{Z}_n/b_n\}$ converges weakly in $\mathcal{K}[\boldsymbol{0}, \infty)$ to a limit which is the convex hull of the points of the limiting Poisson point process associated with (7.2) (p. 212).

# 8

# Applied Probability Models and Heavy Tails

This chapter uses the heavy-tail machinery in service of various applied probability models of networks and queuing systems.

## 8.1 A network model for cumulative traffic on large time scales

The simple infinite-node Poisson based model discussed in Section 5.2.2 (p. 125) and formalized in Section 5.2.4 (p. 127) offers a compelling explanation of how heavy-tailed file sizes induce long-range dependence in the traffic rates. To decide if our model is an accurate enough reflection of reality, however, we need to see how well data measurements fit the model. So we require a partial catalogue of features of the model to see if such features are found in data measurements. In this section, based on [222, 267], we analyze what the model predicts about the cumulative traffic process over large time scales. An alternate approach [68, pp. 373–404], based on small time scales more consistent with empirical observations of *burstiness*, examines cumulative traffic in small time slots as slot length goes to zero.

### 8.1.1 Model review

The infinite-node Poisson model with heavy-tailed file sizes allows cumulative traffic at large time scales to look either heavy tailed or Gaussian, depending on whether the rate at which transmissions are initiated (crudely referred to as the *connection rate*) is moderate or quite large. Here we discuss why stable Lévy motion is a possible approximation.

The process describing offered traffic is $A(t)$, the cumulative input in $[0, t]$ by all sources. Recall from (5.3) and (5.4) (p. 129) that the model assumes unit rate transmissions, and $A(t)$ is the integral of $M(s)$ over $[0, t]$. For large $T$, we think of $(A(Tt), t \geq 0)$ as the process on large time scales. The results show that if the connection rate $\lambda(\cdot)$ is allowed to depend on $T$ in such a way that it has a growth rate in $T$ that is moderate (in a manner to be made precise), then $A(T\cdot)$ looks like an $\alpha$-stable Lévy motion, while if the connection rate grows faster than a critical value, $A(T\cdot)$ looks like a fractional Brownian motion. These statements can be made precise by adopting a heavy-traffic outlook. We imagine a family of models indexed by $T$, where the $T$th model has connection rate $\lambda(T)$ and file size distribution $F_{\text{on}}$. Depending on growth rates, the $T$th model is approximated by either Lévy stable motion or fractional Brownian motion [178, 222].

As in Section 5.2.5 (p. 130), let $(\Gamma_k, -\infty < k < \infty)$ be the points of the rate $\lambda$ homogeneous Poisson process on $\mathbb{R}$, and now label the points so that $\Gamma_0 < 0 < \Gamma_1$, and hence $\{-\Gamma_0, \Gamma_1, (\Gamma_{k+1} - \Gamma_k, k \neq 0)\}$ are iid exponentially distributed random variables with parameter $\lambda$. The random measure that counts the points is denoted by $\sum_{k=-\infty}^{\infty} \epsilon_{\Gamma_k}$ and is a Poisson random measure with mean measure $\lambda\,\mathbb{LEB}$, where $\mathbb{LEB}$ is Lebesgue measure. The network has an infinite number of nodes or *sources*, and at time $\Gamma_k$ a connection is made and some node begins a transmission at constant rate to the server. As a normalization, this constant rate is taken to be unity. The lengths of transmissions are random variables $L_k$. Assume that $L_{\text{on}}, L_1, L_2, \ldots$ are iid and independent of $\{\Gamma_k\}$, and

$$\mathbb{P}(L_{\text{on}} > x) = \bar{F}_{\text{on}}(x) = x^{-\alpha}L(x), \quad x > 0, \quad 1 < \alpha < 2, \tag{8.1}$$

where $L$ is a slowly varying function. Since $\alpha \in (1, 2)$, the variance of $L_{\text{on}}$ is infinite and its mean $\mu_{\text{on}}$ is finite. We will need the quantile function

$$b(t) = (1/\bar{F}_{\text{on}})^{\leftarrow}(t) =: \inf\left\{x : \frac{1}{1 - F_{\text{on}}(x)} \geq t\right\}, \quad t > 0, \tag{8.2}$$

which is regularly varying with index $1/\alpha$. Recall the two-dimensional Poisson random measure $\xi$ defined by (5.6), which is a counting function on $\mathbb{R} \times [0, \infty]$ corresponding to the points $\{(\Gamma_k, L_k)\}$ and has mean measure $\lambda\,\mathbb{LEB} \times F_{\text{on}}$; cf. [260].

To remind us we consider the $T$th model, we sometimes subscript quantities by $T$. For example, the number of active sources at $t$ or the overall transmission rate at $t$ is denoted by either $M(t)$ or $M_T(t)$. We will consider a family of Poisson processes indexed by the scaling parameter $T > 0$ such that the intensity $\lambda = \lambda(T)$ goes to infinity as $T \to \infty$. The intensity $\lambda = \lambda(T)$ will be referred to as the *connection rate* for the $T$th model.

Recall that heavy-tailed transmission times $L_k$ induce long-range dependence in $M$; the precise expression of this is (5.8) (p. 132). High variability in transmission times causes long-range dependence in the rate at which work is offered to the system.

### 8.1.2 The critical input rate

Recall that $\lambda = \lambda(T)$ is the parameter governing the connection rate in the $T$th model, and suppose $\lambda = \lambda(T)$ is a nondecreasing function of $T$. We phrase our condition first in terms of the quantile function $b$ defined in (8.2). The asymptotic behavior of $A_T(\cdot)$ depends on whether

$$\textit{Condition 1 (slow-growth condition):} \qquad \lim_{T\to\infty} \frac{b(\lambda T)}{T} = 0$$

or

$$\textit{Condition 2 (fast-growth condition):} \qquad \lim_{T\to\infty} \frac{b(\lambda T)}{T} = \infty$$

holds. Notice that $b(\cdot)$ is regularly varying with index $1/\alpha$.

There is an alternative, more intuitive, way to express the conditions.

**Lemma 8.1.** *Assume that $F_{\text{on}}$ satisfies (8.1). In the $T$th model, assume that the Poisson process of session initiations is constructed on $\mathbb{R}$ and $M_T(\cdot)$ is a stationary process on $\mathbb{R}$. Note that $M_T(t)$ represents the number of active sources at time $t$ in the $T$th model.*

1. *The slow-growth condition (Condition 1) is equivalent to either of the following two conditions:*

$$\lim_{T\to\infty} \lambda T \bar{F}_{\text{on}}(T) = 0 \quad \textit{or} \quad \lim_{T\to\infty} \text{Cov}(M_T(0), M_T(T)) = 0. \tag{8.3}$$

2. *The fast-growth condition (Condition 2) is equivalent to either of the following two conditions:*

$$\lim_{T\to\infty} \lambda T \bar{F}_{\text{on}}(T) = \infty \quad \textit{or} \quad \lim_{T\to\infty} \text{Cov}(M_T(0), M_T(T)) = \infty. \tag{8.4}$$

If we think of the model with time scaled by $T$, the covariance appearing in (8.3) and (8.4) is the lag 1 covariance. As we proceed through our family of models indexed by $T$, under slow growth, the lag 1 covariance is diminishing at large scales, and under fast growth, the lag 1 covariance is getting very strong.

*Proof.* In the case of Condition 1, there exists a function $0 < \epsilon(T) \to 0$ such that $T\epsilon(T) \to \infty$ and $b(\lambda T) = T\epsilon(T)$. Thus, by inversion,

$$\lambda T \sim 1/\bar{F}_{\text{on}}(T\epsilon(T)). \tag{8.5}$$

Therefore, Condition 1 implies that

$$\lambda T \bar{F}_{\text{on}}(T) \sim \bar{F}_{\text{on}}(T)/\bar{F}_{\text{on}}(T\epsilon(T)) \to 0. \tag{8.6}$$

Conversely, if $\delta(T) := \lambda T \bar{F}_{\text{on}}(T) \to 0$, then using $b^{\leftarrow}(T) \sim 1/\bar{F}_{\text{on}}(T)$, we get

$$\frac{b(\lambda T)}{T} \sim \frac{b(\delta(T) b^{\leftarrow}(T))}{b(b^{\leftarrow}(T))} \to 0,$$

and so Condition 1 and (8.6) are equivalent. Similarly, Condition 2 is the same as

$$\lambda T \bar{F}_{\text{on}}(T) \to \infty. \tag{8.7}$$

To get the equivalence in terms of the covariances, use (5.8) (p. 132). □

The following fact expedites proofs in subsequent sections.

**Lemma 8.2.** *If Condition* 1 *holds, then*

$$\lim_{T\to\infty} \frac{\lambda T^2 \bar{F}_{\text{on}}(T)}{b(\lambda T)} = 0, \tag{8.8}$$

*and if Condition* 2 *holds, this limit is infinite.*

*Proof.* Assume that Condition 1 holds. As with (8.5), set $\epsilon(T) = b(\lambda T)/T \to 0$, so that $\epsilon(T)T \to \infty$. Denoting the ratio in (8.8) by $r(T)$, we see that

$$r(T) \sim \frac{\bar{F}_{\text{on}}(T)}{\epsilon(T)\bar{F}_{\text{on}}(T\epsilon(T))},$$

and using the Karamata representation of a regularly varying function (see Section 2.3.3 (p. 29) and (2.24) (p. 29)), we obtain

$$r(T) \sim [\epsilon(T)]^{-1} \exp\left\{-\int_{T\epsilon(T)}^{T} u^{-1}\alpha(u)du\right\} \tag{8.9}$$

for some function $\alpha(u) \to \alpha$ as $u \to \infty$. Since $1 < \alpha < 2$, we may pick $\delta$ so small that $\alpha - \delta > 1$. Since $T\epsilon(T) \to \infty$, we have, for $T$ sufficiently large, that the right-hand side in (8.9) is bounded from above by

$$[\epsilon(T)]^{-1} \exp\{-(\alpha-\delta)\log(1/\epsilon(T))\} = [\epsilon(T)]^{\alpha-\delta-1},$$

and the right-hand side converges to zero as $T \to \infty$. The proof of an infinite limit under Condition 2 is similar. □

### 8.1.3 Why stable Lévy motion can approximate cumulative input under slow growth

We now assume that Condition 1 holds and show why at large time scales, the process $A$ is approximately an $\alpha$-stable Lévy motion. The following is the result under the slow-growth condition. We will not discuss the fractional Brownian motion limit obtained under fast growth or the intermediate cases. A skillful overview is given in [178]. See also [139, 179].

**Theorem 8.1.** *If Condition* 1 *holds, then the process* $(A(Tt), t \geq 0)$ *describing the total cumulative input in* $[0, Tt]$, $t \geq 0$, *satisfies the limit relation*

$$X^{(T)}(\cdot) := \frac{A(T\cdot) - T\lambda\mu_{\text{on}}(\cdot)}{b(\lambda T)} \overset{\text{fidi}}{\to} X_\alpha(\cdot), \tag{8.10}$$

*where* $X_\alpha(\cdot)$ *is an* $\alpha$*-stable Lévy motion. Here* $\overset{\text{fidi}}{\to}$ *denotes convergence of the finite-dimensional distributions.*

*Remark* 8.1. The mode of convergence cannot be extended to $J_1$ convergence in the Skorohod space $D[0, \infty)$. This follows, for example, from Konstantopoulos and Lin [190], who show that a sequence of processes with a.s. continuous sample paths cannot converge in distribution in $(D[0, \infty), J_1)$ to a process with a.s. discontinuous sample paths. A thorough discussion of this phenomena is in [301]; see also [254] and Problem 3.22 (p. 69).

Here is a discussion of the proof.

#### The basic decomposition

We start by giving a useful decomposition of the random variable $A(T)$ corresponding to a decomposition of $(-\infty, T] \times [0, \infty)$:

$$\begin{aligned} R_1 &:= \{(s, y) : 0 < s \leq T, 0 < y, s + y \leq T\}, \\ R_2 &:= \{(s, y) : 0 < s \leq T, T < s + y\}, \\ R_3 &:= \{(s, y) : s \leq 0, 0 < s + y \leq T\}, \\ R_4 &:= \{(s, y) : s \leq 0, T < s + y\} \end{aligned} \tag{8.11}$$

(see Figure 8.1). Compare this decomposition to the one used in Figure 5.3 (p. 131). Rewrite $A(T)$ using (5.4) as

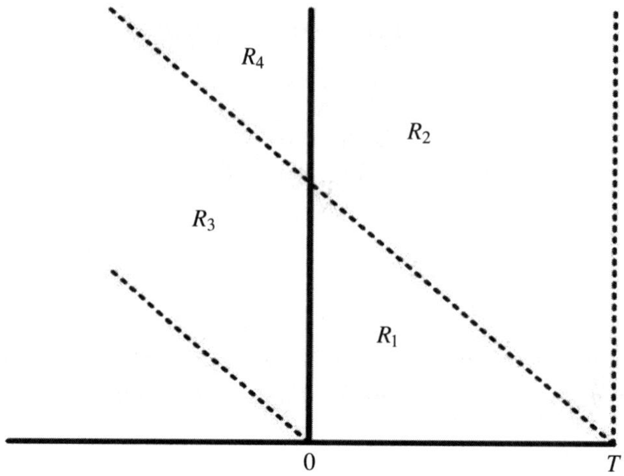


**Fig. 8.1.** The regions $R_1$, $R_2$, $R_3$, $R_4$.

$$
\begin{aligned}
A(T) &= \sum_k L_k 1_{[(\Gamma_k, L_k)\in R_1]} + \sum_k (T - \Gamma_k) 1_{[(\Gamma_k, L_k)\in R_2]} \\
&\quad + \sum_k (L_k + \Gamma_k) 1_{[(\Gamma_k, L_k)\in R_3]} + \sum_k T 1_{[(\Gamma_k, L_k)\in R_4]} \qquad (8.12)\\
&=: A_1 + A_2 + A_3 + A_4.
\end{aligned}
$$

Recall the definition of the PRM $\xi$ from (5.6) (p. 130) with mean measure $\lambda\,\mathbb{LEB}\times F_{\text{on}}$. Note that $A_i$ is a function of the points of $\xi$ in region $R_i$, and since the $R_i$s are disjoint, $A_i$, $i = 1, \dots, 4$, are mutually independent. Calculating as in (5.7) (p. 130) and using Karamata's theorem, we get that as $T \to \infty$,

$$
\begin{aligned}
\lambda m_1 &:= E\xi(R_1) = \lambda \int_0^T F_{\text{on}}(T-s)ds \sim \lambda T, \\
\lambda m_2 &:= E\xi(R_2) = \lambda \int_0^T \bar{F}_{\text{on}}(T-s)ds \sim \lambda \mu_{\text{on}}, \qquad (8.13)\\
\lambda m_3 &:= E\xi(R_3) = \lambda \int_{s=-\infty}^{0} (F_{\text{on}}(T+|s|) - F_{\text{on}}(|s|))ds \sim \lambda\mu_{\text{on}}, \\
\lambda m_4 &:= E\xi(R_4) = \lambda \int_{s=-\infty}^{0}\int_{y=-s+T}^{\infty} F_{\text{on}}(dy)ds = \lambda \int_T^\infty \bar{F}_{\text{on}}(u)du, \\
&\sim \lambda T \bar{F}_{\text{on}}(T)/(\alpha - 1) \to 0.
\end{aligned}
$$

So the mean measure $E\xi(\cdot)$ restricted to $R_i$ is finite for $i = 1, \dots, 4$, which implies that the points of $\xi\big|_{R_i}$ can be represented as a Poisson number of iid random vectors (see Section 5.4.2 (p. 143)):

$$\xi\Big|_{R_i} \stackrel{d}{=} \sum_{k=1}^{P_i} \epsilon_{(t_{k,i},\, j_{k,i})}, \quad i = 1, \ldots, 4, \tag{8.14}$$

where $P_i$ is a Poisson random variable with mean $\lambda m_i$, which is independent of the iid pairs $(t_{k,i}, j_{k,i})$, $k \geq 1$, with common distribution

$$\left.\frac{\lambda\, \mathbb{LEB}(ds) F_{\text{on}}(dy)}{\lambda m_i}\right|_{R_i} = \left.\frac{\mathbb{LEB}(ds) F_{\text{on}}(dy)}{m_i}\right|_{R_i}, \tag{8.15}$$

for $i = 1, \ldots, 4$. Notice that the distributions of $((t_{k,i}, j_{k,i}))$ are independent of $\lambda$, which only enters into the specification of the mean of $P_i$, $i = 1, \ldots, 4$. This means that for fixed $T$, we can represent the $A_i$s as sums of a Poisson number of iid random variables,

$$\begin{aligned} A_1 &\stackrel{d}{=} \sum_{k=1}^{P_1} j_{k,1}, & A_2 &\stackrel{d}{=} \sum_{k=1}^{P_2} (T - t_{k,2}), \\ A_3 &\stackrel{d}{=} \sum_{k=1}^{P_3} (j_{k,3} + t_{k,3}), & A_4 &\stackrel{d}{=} \sum_{k=1}^{P_4} T = T P_4. \end{aligned} \tag{8.16}$$

### One-dimensional convergence

We show under Condition 1 or (8.3) that $A(T)$ is asymptotically an $\alpha$-stable random variable by showing that $A_1(T) = A_1$ is asymptotically stable and $A_i(T) = A_i$, $i = 2, 3, 4$, are asymptotically negligible.

It is relatively easy to see that

$$A_i / b(\lambda T) \stackrel{P}{\to} 0, \quad i = 2, 3, 4. \tag{8.17}$$

Here is a sample calculation for the case $i = 2$; a similar argument works for $i = 3, 4$. We write

$$\mathbf{E}(A_2) = \mathbf{E}(P_2)\mathbf{E}(T - t_{k,2}) = [\lambda m_2]\mathbf{E}(T - t_{k,2}),$$

and from (8.15), this is

$$\begin{aligned} &= \lambda m_2 \iint_{\substack{0 \leq x \leq T \\ s+y > T}} (T - s) ds \frac{F_{\text{on}}(dy)}{m_2} \\ &= \lambda \int_0^T \bar{F}_{\text{on}}(T - s)(T - s) ds \end{aligned}$$

$$= \lambda \int_0^T s \bar{F}_{\text{on}}(s) ds.$$

Therefore, from Karamata's theorem (p. 25),

$$\frac{\lambda m_2 \mathbf{E}(T - t_{k,2})}{\lambda T^2 \bar{F}_{\text{on}}(T)} = \frac{\int_0^T s \bar{F}_{\text{on}}(s) ds}{T^2 \bar{F}_{\text{on}}(t)}$$
$$\to \int_0^1 s \cdot s^{-\alpha} ds = \frac{1}{2 - \alpha}.$$

Then Lemma 8.2 (p. 256) and, in particular, (8.8) give

$$\mathbf{E}(A_2) = o(b(\lambda T)),$$

as desired.

Thus it remains to consider $A_1$. The representation of $A_1$ given in (8.16) yields the decomposition

$$A_1 - \lambda \mu_{\text{on}} T = \sum_{k=1}^{P_1} (j_{k,1} - \mathbf{E}(j_{k,1})) + \mathbf{E}(j_{k,1})[P_1 - \mathbf{E}(P_1)] + [\mathbf{E}(A_1) - \lambda \mu_{\text{on}} T]$$
$$= A_{11} + A_{12} + A_{13}.$$

It is readily checked that $Ej_{k,1} \sim \mu_{\text{on}}$ since

$$\lim_{T\to\infty} \mathbf{E}(j_{k,1}) \sim \lim_{T\to\infty} \iint_{\substack{0 \le s \le T \\ 0 \le s+y \le T}} y ds \frac{F_{\text{on}}(dy)}{T}$$
$$= \lim_{T\to\infty} \frac{1}{T} \int_{s=0}^{T} \left( \int_{y=0}^{T-s} y F_{\text{on}}(dy) \right) ds$$
$$= \lim_{T\to\infty} \frac{1}{T} \int_{s=0}^{T} \left( \int_{y=0}^{s} y F_{\text{on}}(dy) \right) ds$$
$$= \lim_{T\to\infty} \int_0^T y F_{\text{on}}(dy) = \mu_{\text{on}}.$$

Furthermore, $P_1$ is Poisson with mean $\lambda m_1 \to \infty$, so it satisfies the central limit theorem, i.e.,

$$[\lambda m_1]^{-1/2} [P_1 - \lambda m_1] \Rightarrow N(0, 1). \tag{8.18}$$

Since $\lambda m_1 \sim \lambda T$, we conclude that

$$A_{12} = O_P([\lambda T]^{1/2}) = o_P(b(\lambda T)), \tag{8.19}$$

since

$$\lim_{T\to\infty} \frac{\sqrt{\lambda T}}{b(\lambda T)} = \lim_{s\to\infty} \frac{s^{1/2}}{b(s)} = \lim_{s\to\infty} s^{1/2-1/\alpha}/L(s),$$

and $1 < \alpha < 2$ implies that $\frac{1}{2} < \frac{1}{\alpha} < 1$.

By (8.16) and (8.18), $A_{11}$ is a sum of approximately $\lambda m_1 \sim \lambda T$ iid summands. Under Condition 1 or (8.3), $b(\lambda T)/T \to 0$, so that for any $x > 0$ fixed, we eventually have $T - b(\lambda T)x > 0$. Therefore, from (8.15), as $x \to \infty$,

$$\begin{aligned}
&\lambda T \mathbb{P}(j_{k,1} > b(\lambda T)x) \\
&\quad = \lambda T \iint_{\substack{0\le s\le T\\ 0\le s+y\le T\\ y>b(\lambda T)x}} \frac{ds\,F_{\text{on}}(dy)}{m_1} \\
&\quad = \lambda T \int_{s=0}^{T-b(\lambda T)x} \left( \int_{y=b(\lambda T)x}^{T-s} \frac{F_{\text{on}}(dy)}{m_1} \right) ds \\
&\quad = \lambda T \left[ \frac{1}{m_1} \bar{F}_{\text{on}}(b(\lambda T)x)(T - b(\lambda T)x) - \frac{1}{m_1} \int_{s=0}^{T-b(\lambda T)x} \bar{F}_{\text{on}}(T-s)ds \right] \\
&\quad \sim \left(1 - \frac{b(\lambda T)x}{T}\right) \lambda T \bar{F}_{\text{on}}(b(\lambda T)x) - \frac{b(\lambda T)}{T} \int_x^{T/b(\lambda T)} \lambda T \bar{F}_{\text{on}}(b(\lambda T)s)ds \\
&\quad \sim x^{-\alpha}.
\end{aligned} \tag{8.20}$$

From this, we would like to conclude by Theorem 7.1 (p. 214) that

$$Y^{(T)}(\cdot) := (b(\lambda T))^{-1} \sum_{k=1}^{[\lambda T\cdot]} (j_{k,1} - \mathbf{E}(j_{k,1})) \Rightarrow X_\alpha(\cdot) \tag{8.21}$$

in $D[0,\infty)$, where the limit is a totally skewed $\alpha$-stable Lévy random motion ($p = 1$, $q = 0$). However, Theorem 7.1 requires us to check (7.6) (p. 214), which controls the truncated second moment. The condition (7.6) becomes, for our case,

$$\lim_{\delta\to 0} \limsup_{T\to\infty} \lambda T \mathbf{E}\left( \left(\frac{j_{k,1}}{b(\lambda T)}\right)^2 1_{[|j_{k,1}|\le b(\lambda T)\delta]} \right) = 0. \tag{8.22}$$

Verifying this is an easy application of Karamata's theorem, as given in Problem 2.5 (p. 36). The left side of (8.22) is asymptotic to

$$\lambda T \iint_{\substack{0\le x\le T\\ 0\le y\le T-s}} \left(\frac{y}{b(\lambda T)}\right)^2 1_{[y\le b(\lambda T)\delta]} ds \frac{F_{\text{on}}(dy)}{T}$$

$$
\begin{aligned}
&= \lambda T \int_{s=0}^{T} \left( \int_{y=0}^{T-s} \left( \frac{y}{b(\lambda T)} \right)^2 1_{[y \le b(\lambda T)\delta]} \frac{F_{\text{on}}(dy)}{T} \right) ds \\
&= \lambda T \int_{s=0}^{1} \left( \int_{y=0}^{Ts} \left( \frac{y}{b(\lambda T)} \right)^2 1_{[y \le b(\lambda T)\delta]} F_{\text{on}}(dy) \right) ds \\
&= \lambda T \int_{s=0}^{1} \left( \int_{y=0}^{Ts/b(\lambda T)} y^2 1_{[y \le \delta]} F_{\text{on}}(b(\lambda T) dy) \right) ds \\
&\le \int_{s=0}^{1} \left( \int_{y=0}^{T/b(\lambda T) \wedge \delta} y^2 \lambda T F_{\text{on}}(b(\lambda T) dy) \right) ds,
\end{aligned}
$$

and because of the slow-growth condition, this is ultimately

$$
\begin{aligned}
&\le \int_0^\delta y^2 \lambda T F_{\text{on}}(b(\lambda T) dy) \\
&\to \int_0^\delta y^2 \alpha y^{-\alpha-1} dy = \frac{\alpha}{2-\alpha} \delta^{2-\alpha} \quad \text{(Karamata's theorem)} \\
&\to 0 \quad (\delta \to 0).
\end{aligned}
$$

This establishes (8.21) in $D[0, \infty)$. By independence, we may couple (8.18) and (8.21) to get joint convergence:

$$
\left( Y^{(T)}(\cdot), \frac{P_1}{\lambda T} \right) \Rightarrow (X_\alpha(\cdot), 1)
$$

in $D[0, \infty) \times \mathbb{R}$. Using composition and the continuous mapping theorem, one obtains

$$
\frac{A_{11}}{b(\lambda T)} = Y^{(T)} \left( \frac{P_1}{\lambda T} \right) = (b(\lambda T))^{-1} \sum_{i=1}^{P_1} (j_{k,1} - \mathbf{E}(j_{k,1})) \Rightarrow X_\alpha(1). \tag{8.23}
$$

Finally, we need to consider $A_{13}$. Write

$$
\begin{aligned}
A_{13} &= \mathbf{E}(A_1) - \lambda \mu_{\text{on}} T = \mathbf{E}(j_{k,1}) \mathbf{E}(P_1) - \lambda T \mu_{\text{on}} \\
&= \lambda \int_0^T \left[ \int_0^s y F_{\text{on}}(dy) - \mu_{\text{on}} \right] ds = -\lambda \int_0^T \int_s^\infty y F_{\text{on}}(dy) ds \\
&\sim -(\text{const}) \lambda T^2 \bar{F}_{\text{on}}(T) = o(b(\lambda T)),
\end{aligned} \tag{8.24}
$$

where we applied Karamata's theorem and (8.8). Combining the limit relations (8.17), (8.19), (8.23), and (8.24) gives the desired $\alpha$-stable limit for $A(T)$.

**Finite-dimensional convergence**

We restrict ourselves to showing convergence of the two-dimensional distributions. The general case is analogous but notationally more cumbersome. First, observe that for $t > 0$,

$$\frac{A(Tt) - \lambda\mu_{\text{on}}Tt}{b(\lambda T)} = \frac{A(Tt) - \lambda\mu_{\text{on}}Tt}{b(\lambda Tt)} \cdot \frac{b(\lambda Tt)}{b(\lambda T)} \Rightarrow X_\alpha(1) \cdot t^{1/\alpha} \stackrel{d}{=} X_\alpha(t).$$

Next, suppose $t_1 < t_2$. The same arguments as for the one-dimensional convergence show that it suffices to consider the joint convergence of

$$[b(\lambda T)]^{-1}(A_1(Tt_i) - \lambda Tt_i\mu_{\text{on}}), \quad i = 1, 2,$$

since the rest will be $o_p(1)$. We can write

$$\begin{aligned} A_1(Tt_2) &= A_1(Tt_1) + \sum_{Tt_1 < \Gamma_k \le Tt_2} L_k 1_{[\Gamma_k + L_k \le Tt_2]} + \sum_{0 < \Gamma_k \le Tt_1} L_k 1_{[Tt_1 < \Gamma_k + X_k \le Tt_2]} \\ &=: A_1(Tt_1) + A_{21} + A_{22}. \end{aligned}$$

The terms $A_1(Tt_1)$ and $A_{21}$ are independent. Also, we have

$$A_{21} \stackrel{d}{=} A_1(T(t_2 - t_1)).$$

To see this, set $\tilde{\Gamma}_k = \Gamma_k - Tt_1$ and note that $\sum_k \epsilon_{\tilde{\Gamma}_k}$ is PRM($\lambda\, \mathbb{LEB}$), so

$$\sum_{Tt_1 < \Gamma_k \le Tt_2} L_k 1_{[\Gamma_k + L_k \le Tt_2]} = \sum_{0 < \tilde{\Gamma}_k \le Tt_2 - Tt_1} L_k 1_{[\tilde{\Gamma}_k + L_k \le Tt_2 - Tt_1]} \stackrel{d}{=} A_{21}.$$

Hence the proof of the convergence of bivariate distributions follows from the one-dimensional convergence if one can show that

$$[b(\lambda T)]^{-1} A_{22} \stackrel{P}{\to} 0.$$

However,

$$\begin{aligned} \mathbf{E}(A_{22}) &= \mathbf{E}\left(\iint_{0 \le s \le Tt_1, Tt_1 < s+u \le Tt_2} u\xi(ds, du)\right) \\ &= \iint_{0 \le s \le Tt_1, Tt_1 < s+u \le Tt_2} u\lambda ds F_{\text{on}}(du) \\ &= \lambda \int_{s=0}^{Tt_1} \left(\int_{u=Tt_1 - s}^{Tt_2 - s} u F_{\text{on}}(du)\right) ds \end{aligned}$$

$$
\begin{aligned}
&= \lambda T^2 \bar{F}_{\text{on}}(T) \int_0^{t_1} \left( \int_{u=t_1-s}^{t_2-s} u \frac{F_{\text{on}}(T du)}{\bar{F}_{\text{on}}(T)} \right) ds \\
&\sim \lambda T^2 \bar{F}_{\text{on}}(T) \left[ \int_0^{t_1} \frac{\alpha}{\alpha - 1} ((t_1 - s)^{-(\alpha-1)} - (t_2 - s)^{-(\alpha-1)}) ds \right] \\
&= o(b(\lambda T))
\end{aligned}
$$

by Lemma 8.2. This concludes the proof of Theorem 8.1. □

## 8.2 A model for network activity rates

This section is based on [223]. Consider an ordinary renewal process $\{S_n, n \geq 0\}$ such that

$$S_0 = 0, \qquad S_n = \sum_{i=1}^{n} X_i, \quad n \geq 1,$$

and $\{X_n, n \geq 1\}$ is a sequence of iid nonnegative random variables with common distribution $F$. At time point $S_n$, an event begins of duration $L_n$, where we assume $\{L_n, n \geq 0\}$ is a sequence of iid nonnegative random variables with common distribution $F_{\text{on}}$ and $\{L_n\}$ is independent of $\{X_n\}$. The event that was initiated at $S_n$ terminates at $S_n + L_n$. In a data network context, $S_n$ would be the time a user initiates a file download and $L_n$ is the download time. In an insurance context, $S_n$ is the time of a disaster or accident and $L_n$ is the length of time during which all insurance claims from this incident are received, so that $S_n + L_n$ is the latest time a claim from the $n$th accident is received. Note that in contrast with the infinite-source Poisson model of Sections 5.2.4 (p. 127) and 8.1 (p. 253), we do not assume that event initiation times are Poisson but only form a renewal sequence.

We focus our attention on

$$M(t) = \sum_{n=1}^{\infty} 1_{[S_n \leq t < S_n + L_n]}, \quad t > 0, \tag{8.25}$$

the *number of active downloads at time* $t$ or the *number of active claims at time* $t$. In Section 5.2.4 (p. 127), the variable $M(t)$ was Poisson distributed for each $t$, but that is not the case here, and the asymptotic behavior of $M(t)$ will vary depending on different heavy-tail assumptions on $\bar{F}$ and $\bar{F}_{\text{on}}$. A fairly complete analysis is in [223]; we give a sample here to illustrate some heavy-tail methodology for applied probability modeling.

Consider the very heavy-tailed cases when for $0 \leq \alpha, \beta \leq 1$,

$$\bar{F}(x) = 1 - F(x) \sim x^{-\alpha} L_F(x), \qquad \bar{F}_{\text{on}}(x) = 1 - F_{\text{on}}(x) \sim x^{-\beta} L_{\text{on}}(x), \quad x \to \infty,$$

for some slowly varying functions $L_F$, $L_{\text{on}}$.

### 8.2.1 Mean value analysis when $\alpha, \beta < 1$

Karamata's Tauberian theorem applied to the renewal function was discussed in Section 7.3.3 (p. 245) and yields the mean-value asymptotic behavior of $M(t)$. The renewal function is given by (7.59), and its asymptotic behavior described by (7.65) (p. 246). Therefore, as $t \to \infty$,

$$\begin{aligned}\mathbf{E}(M(t)) &= \int_0^t \bar{F}_{\text{on}}(t-x)U(dx) = \int_0^1 \frac{\bar{F}_{\text{on}}(t(1-s))}{\bar{F}_{\text{on}}(t)} \frac{U(tds)}{U(t)} (\bar{F}_{\text{on}}(t)U(t)) \\ &\sim c(\alpha) \int_0^1 (1-s)^{-\beta} \alpha s^{\alpha-1} ds \frac{\bar{F}_{\text{on}}(t)}{\bar{F}(t)} = c'(\alpha) \frac{\bar{F}_{\text{on}}(t)}{\bar{F}(t)}. \end{aligned} \tag{8.26}$$

Thus if

$$\begin{aligned} \bar{F}(t)/\bar{F}_{\text{on}}(t) &\to c > 0, \quad &&\text{then} \quad \mathbf{E}(M(t)) \to c'(\alpha)c^{-1}; \\ \bar{F}(t)/\bar{F}_{\text{on}}(t) &\to 0 &&\text{then} \quad \mathbf{E}(M(t)) \to \infty; \\ \bar{F}(t)/\bar{F}_{\text{on}}(t) &\to \infty &&\text{then} \quad \mathbf{E}(M(t)) \to 0. \end{aligned}$$

In the last case, $\mathbf{E}(M(t)) \to 0$ and hence $M(t) \xrightarrow{L_1} 0$, and so it is of lesser interest, corresponding to the case in which renewals are so sparse relative to event durations that at any time there is not likely to be an event in progress.

### 8.2.2 Behavior of $N(t)$, the renewal counting function when $0 < \alpha < 1$

The counting function $N(t) = \sum_{n=0}^{\infty} \epsilon_{S_n}[0, t]$ was defined in Section 7.3.3 (p. 245). Note that $N(x) = S^{\leftarrow}(x)$, where $S(t) = S_{[t]}$ for $t \geq 0$. Let $\sum_k \epsilon_{(t_k, j_k)}$ be PRM($\mathbb{LEB} \times \nu_\alpha$) on

$$\mathbb{E} = [0, \infty) \times (0, \infty].$$

The process $X_\alpha(t) = \sum_{t_k \leq t} j_k$, $t \geq 0$, is $\alpha$-stable Lévy motion with Lévy measure $\nu_\alpha$; see Section 5.5.2 (p. 153) and Problem 7.3 (p. 247). Define, as usual, the quantile function of $F$:

$$b(t) = \left(\frac{1}{1-F}\right)^{\leftarrow}(t).$$

When $\alpha > 0$, we can always choose $b$ as a continuous and strictly increasing function; see Proposition 2.6(vii) (p. 32).

Renewal epochs are asymptotically stable (Section 7.2.2 (p. 218)). If

$$X^{(s)}(t) = \frac{S_{[st]}}{b(s)}, \quad t \geq 0,$$

then in $D[0, \infty)$, we have as $s \to \infty$,

$$X^{(s)} \Rightarrow X_\alpha. \tag{8.27}$$

Furthermore, the inverse processes also converge in $D[0, \infty)$:

$$(X^{(s)})^{\leftarrow} \Rightarrow X_\alpha^{\leftarrow}.$$

Unpacking this last result, we get

$$\frac{N(b(s)\cdot)}{s} \Rightarrow X_\alpha^{\leftarrow}(\cdot) \tag{8.28}$$

in $D[0, \infty)$ or, equivalently, $\bar{F}(s)N(s\cdot) \Rightarrow X_\alpha^{\leftarrow}(\cdot)$ or, equivalently,

$$\frac{1}{s}\sum_{n=0}^{\infty} \epsilon_{\frac{S_n}{b(s)}} \Rightarrow X_\alpha^{\leftarrow}$$

in $M_+[0, \infty)$, where we have used $X_\alpha^{\leftarrow}$ to indicate both the monotone function and the measure. Note that the simple reasoning that gave us Proposition 3.2 (p. 58) does not suffice for justifying the inversion since the limit is not a continuous process. The more sophisticated arguments used in [300] must be employed.

### 8.2.3 Activity rates when $\alpha, \beta < 1$ and tails are comparable

Consider the case $\bar{F}(t) \sim \bar{F}_{\text{on}}(t)$, where the tails of $F$ and $F_{\text{on}}$ are asymptotically equivalent.

#### Counting function of $\{(S_k, T_k), k \geq 0\}$

Suppose $D^{\uparrow}[0, \infty)$ are the nondecreasing functions in $D[0, \infty)$. Define the mapping $T : D^{\uparrow}[0, \infty) \times M_+(\mathbb{E}) \mapsto M_+(\mathbb{E})$ by

$$T(x, m) = \tilde{m}, \tag{8.29}$$

where $\tilde{m}$ is defined by

$$\tilde{m}(f) = \iint f(x(u), v)m(du, dv), \quad f \in C_K^+(\mathbb{E}).$$

This means that $T$ replaces the usual time scale of $m$ by one determined by the function $x$. If $m$ is a point measure with representation $m = \sum_k \epsilon_{(\tau_k, y_k)}$, then

$$T(x, m) = \sum_k \epsilon_{(x(\tau_k), y_k)}.$$

**Proposition 8.1.** *Suppose that as $t \to \infty$,*

$$\bar{F}(t) \sim \bar{F}_{\text{on}}(t) \in \text{RV}_{-\alpha}, \quad 0 < \alpha < 1.$$

*Assume that $N_\infty = \sum_k \epsilon_{(t_k, j_k)}$ is* PRM($\mathbb{LEB} \times \nu_\alpha$). *Set $\mathbb{E} = [0, \infty) \times (0, \infty]$. Then in $M_p(\mathbb{E})$, as $s \to \infty$,*

$$N_s^* = \sum_{k=0}^{\infty} \epsilon_{\left(\frac{S_k}{b(s)}, \frac{L_k}{b(s)}\right)} \Rightarrow N_\infty^* = T(X_\alpha, N_\infty) = \sum_k \epsilon_{(X_\alpha(t_k), j_k)}. \tag{8.30}$$

*Proof.* Begin with the statement (Theorem 6.3 (p. 180))

$$\sum_{k=0}^{\infty} \epsilon_{\left(\frac{k}{s}, \frac{L_k}{b(s)}\right)} \Rightarrow N_\infty \quad (s \to \infty)$$

in $M_p(\mathbb{E})$. Since $\{S_k\}$ is independent of $\{L_k\}$, we use Problem 3.20 (p. 69) (or [24, p. 23]) to get joint convergence in $D[0, \infty) \times M_p(\mathbb{E})$, using (8.27),

$$\left( \frac{S_{[s\cdot]}}{b(s)}, \sum_{k=0}^{\infty} \epsilon_{\left(\frac{k}{s}, \frac{L_k}{b(s)}\right)} \right) \Rightarrow (X_\alpha, N_\infty).$$

The function $T$ is a.s. continuous at $(X_\alpha, N_\infty)$. Hence

$$T\left( \frac{S[s\cdot]}{b(s)}, \sum_{k=0}^{\infty} \epsilon_{\left(\frac{k}{s}, \frac{L_k}{b(s)}\right)} \right) = \sum_{k=0}^{\infty} \epsilon_{\left(\frac{S_{[sk/s]}}{b(s)}, \frac{L_k}{b(s)}\right)} = \sum_{k=0}^{\infty} \epsilon_{\left(\frac{S_k}{b(s)}, \frac{L_k}{b(s)}\right)} \Rightarrow T(X_\alpha, N_\infty). \quad \square$$

## Number of active sources when tails are comparable

Proposition 8.1 leads to the result about $M$, the number of active sources or events.

**Corollary 8.1.** *The finite-dimensional distributions of the counting function $M(t)$ defined in* (8.25) *satisfy, as $s \to \infty$,*

$$M(st) = \sum_{k=0}^{\infty} 1_{\left[\frac{S_k}{s} \le t < \frac{S_k + L_k}{s}\right]} \Rightarrow M_\infty(t) = \sum_k 1_{[X_\alpha(t_k) \le t < X_\alpha(t_k) + j_k]}.$$

*Conditionally on $X_\alpha^{\leftarrow}$, the limit $M_\infty(t)$ is Poisson with mean*

$$\Lambda(t) = \int_0^t (t-u)^{-\alpha} dX_\alpha^{\leftarrow}(u),$$

*and hence the generating function of $M_\infty(t)$ is*

$$\mathbb{E}(\tau^{M_\infty(t)}) = \mathbb{E} \exp\{(\tau - 1)\Lambda(t)\}, \quad \tau \in (0, 1).$$

*Proof.* Fix $t > 0$. An important observation is that $\Lambda(t) < \infty$ almost surely. To see this, note that

$$\mathbb{E}X_\alpha^{\leftarrow}(u) = u^\alpha \mathbb{E}(X_\alpha^{-\alpha}(1)) = d_\alpha u^\alpha.$$

This results from the self-similar scaling of the Lévy process $X_\alpha$:

$$\begin{aligned}\mathbb{E}X_\alpha^{\leftarrow}(u) &= \int_0^\infty \mathbb{P}[X_\alpha^{\leftarrow}(u) > x]dx = \int_0^\infty \mathbb{P}[u > X_\alpha(x)]dx \\ &= \int_0^\infty \mathbb{P}[u > x^{1/\alpha}X_\alpha(1)]dx = u^\alpha \mathbf{E}(X_\alpha^{-\alpha}(1)) = d_\alpha u^\alpha.\end{aligned}$$

The quantity $d_\alpha$ is finite; see [314].

We prove that $\Lambda(t) < \infty$ a.s. for $t = 1$ as an example of the method. Writing $f(u) = (1-u)^{-\alpha}$, $0 < u < 1$, and observing that $f(0) = 1$, we have

$$\begin{aligned}\int_0^1 f(u)dX_\alpha^{\leftarrow}(u) - X_\alpha^{\leftarrow}(1) &= \int_0^1 (f(u) - f(0))dX_\alpha^{\leftarrow}(u) \\ &= \int_0^1 \int_0^u f'(s)ds dX_\alpha^{\leftarrow}(u) \\ &= \int_0^1 \left(\int_s^1 dX_\alpha^{\leftarrow}(u)\right)\alpha(1-s)^{-\alpha-1}ds \\ &= \alpha \int_0^1 (X_\alpha^{\leftarrow}(1) - X_\alpha^{\leftarrow}(s))(1-s)^{-\alpha-1}ds.\end{aligned}$$

Taking expectations, we have

$$\mathbf{E}\left(\int_0^1 f(u)dX_\alpha^{\leftarrow}(u)\right) = d_\alpha + \alpha d_\alpha \int_0^1 (1-s^\alpha)(1-s)^{-\alpha-1}ds.$$

Now, apart from constants, the second term is $\int_0^1 (1-(1-s)^\alpha))s^{-\alpha-1}ds$. The problem for integrability is near 0. However, as $s \downarrow 0$, the integrand is asymptotic $\sim \alpha s^{-\alpha}$, which for $0 < \alpha < 1$ is integrable. This verifies that $\Lambda(1) < \infty$ with probability 1.

Next, we prove that $M(b(s)t) \Rightarrow M_\infty(t)$ for fixed $t > 0$. As before, we choose $t = 1$ in order to demonstrate the method. For positive $\epsilon$, let

$$B_\epsilon = \{(u, v) : u \le 1 < u + v, v > \epsilon\},$$

which is relatively compact in $\mathbb{E}$. By virtue of Proposition 8.1, $N_s^*(B_\epsilon) \Rightarrow N_\infty^*(B_\epsilon)$. Also, by monotone convergence and using $\Lambda(1) < \infty$, with probability 1,

$$N_\infty^*(B_\epsilon) \uparrow N_\infty^*(B_0) = M_\infty(1) < \infty.$$

From the second converging together theorem, Theorem 3.5 (p. 56), it suffices to show that for any $\delta > 0$,

$$\lim_{\epsilon\to 0} \limsup_{s\to\infty} \mathbb{P}[|N_s^*(B_\epsilon) - N_s^*(B_0)| > \delta] = 0. \tag{8.31}$$

Observe that

$$N_s^*(B_0) - N_s^*(B_\epsilon) = \sum_k 1_{[S_k \le b(s) < S_k + L_k, L_k \le \epsilon b(s)]}.$$

By Chebyshev's inequality, it suffices to show that the expectation of this last quantity has a double limit that is zero. We have

$$\begin{aligned}
&\sum_k \mathbb{P}[S_k \le b(s) < S_k + L_k, L_k \le \epsilon b(s)] \\
&= \int_{1-\epsilon}^1 \sum_k F^{k*}(b(s)dx)\mathbb{P}[1 - x < L_k/b(s) \le \epsilon] \\
&= \int_{1-\epsilon}^1 U(b(s)dx)[\bar{F}_{\text{on}}(b(s)(1-x)) - \bar{F}_{\text{on}}(b(s)\epsilon)] \\
&= \int_{1-\epsilon}^1 \frac{\bar{F}_{\text{on}}(b(s)(1-x)) - \bar{F}_{\text{on}}(b(s)\epsilon)}{\bar{F}_{\text{on}}(b(s))} \frac{U(b(s)dx)}{U(b(s))} U(b(s))\bar{F}_{\text{on}}(b(s)) \\
&\to c(\alpha) \int_{1-\epsilon}^1 [(1-x)^{-\alpha} - \epsilon^{-\alpha}]dx^\alpha \quad \text{as } s \to \infty \\
&\to 0 \quad \text{as } \epsilon \downarrow 0.
\end{aligned}$$

Thus we proved that $M(b(s)t) \Rightarrow M_\infty(t)$ for fixed $t > 0$. The convergence of the finite-dimensional distributions follows analogously by an application of Theorem 8.1. Since $b$ can be chosen continuous and strictly increasing, we may rephrase the latter limit relation as $M(st) \Rightarrow M_\infty(t)$. □

### 8.2.4 Activity rates when $0 < \alpha, \beta < 1$, and $F_{\text{on}}$ has a heavier tail

Now we assume $0 < \beta \le \alpha < 1$, and if $0 < \alpha = \beta$, then $\bar{F}(t)/\bar{F}_{\text{on}}(t) \to 0$ as $t \to \infty$. Recall the definition of the measure $\nu_\alpha$ given by $\nu_\alpha(x, \infty] = x^{-\alpha}$ for $x > 0$, some $\alpha > 0$.

As with Proposition 8.1 (p. 267), we first prove a limit result for the point process generated by the scaled points $(b(s))^{-1}(S_k, L_k)$. Then we use this result to derive a distributional limit for $M(s)$ as $s \to \infty$.

**Proposition 8.2.** *Assume that* $0 < \alpha, \beta < 1$, *and* $\bar{F}(t)/\bar{F}_{\text{on}}(t) \to 0$ *as* $t \to \infty$. *Then in* $M_+(\mathbb{E})$, *we have*

$$\frac{\bar{F}(b(s))}{\bar{F}_{\text{on}}(b(s))} \sum_{k=0}^{\infty} \epsilon_{\left(\frac{S_k}{b(s)}, \frac{L_k}{b(s)}\right)} \Rightarrow T(X_\alpha, \mathbb{LEB} \times \nu_\beta), \tag{8.32}$$

*where* $T$ *was defined in* (8.29).

Note that the normalization in (8.32) for both $S_k$ and $L_k$ is by the quantile function $b(s) = (1/\bar{F})^{\leftarrow}(s)$ for the lighter-tailed distribution function. Since this is inappropriate for $L_k$, premultification by the ratio of the tails (which goes to 0) is necessary for convergence.

*Proof.* Begin by observing that

$$\frac{s\bar{F}(b(s))}{\bar{F}_{\text{on}}(b(s))} \bar{F}_{\text{on}}(b(s)\cdot) \xrightarrow{v} \nu_\beta$$

in $M_+(0, \infty]$. Hence, from Theorem 5.3 and especially (5.16) (p. 139), we get

$$\frac{\bar{F}(b(s))}{\bar{F}_{\text{on}}(b(s))} \sum_{k=0}^{[s]} \varepsilon_{\frac{L_k}{b(s)}} \Rightarrow \nu_\beta.$$

This may be extended as we did in Theorem 6.3 (p. 180) (see also Problem 6.17 (p. 210)) to show that in $M_+(E)$,

$$\frac{\bar{F}(b(s))}{\bar{F}_{\text{on}}(b(s)} \sum_{k=0}^{\infty} \varepsilon_{\left(\frac{k}{s}, \frac{L_k}{b(s)}\right)} \Rightarrow \mathbb{LEB} \times \nu_\beta.$$

From independence we get the joint convergence in $D[0, \infty) \times M_+(E)$,

$$\left(\frac{S_{[s\cdot]}}{b(s)}, \frac{\bar{F}(b(s))}{\bar{F}_{\text{on}}(b(s))} \sum_{k=0}^{\infty} \varepsilon_{\left(\frac{k}{s}, \frac{L_k}{b(s)}\right)}\right) \Rightarrow (X_\alpha, \mathbb{LEB} \times \nu_\beta).$$

Now apply the a.s. continuous map $T$ (see (8.29)) to get (8.32). □

**Number of active sources when $\bar{F}_{\text{on}}$ is heavier**

From this result, we get the desired result about $M$, the number of active sources or events.

**Corollary 8.2.** *The finite-dimensional distributions of the counting function $M$ defined in* (8.25) *satisfy, as $s \to \infty$,*

$$\frac{\bar{F}(s)}{\bar{F}_{\text{on}}(s)} M(st) \Rightarrow \int_0^t (t-u)^{-\beta} dX_\alpha^{\leftarrow}(u). \tag{8.33}$$

*For any fixed $t$,*

$$\int_0^t (t-u)^{-\beta} dX_\alpha^{\leftarrow}(u) \stackrel{d}{=} t^{-\beta\alpha} \int_0^1 (1-u)^{-\beta} dX_\alpha^{\leftarrow}(u).$$

*Proof.* We again consider the case of a fixed $t > 0$; the convergence of the finite-dimensional distributions is analogous. We evaluate the convergence in (8.32) on the set $\{(u,v) : 0 \le u \le t < u+v\}$. After a truncation and the Slutsky argument outlined in the proof of Corollary 8.1 (p. 267), we get

$$\frac{\bar{F}(b(s))}{\bar{F}_{\text{on}}(b(s))} M(b(s)t) \Rightarrow T(X_\alpha, \mathbb{LEB} \times \nu_\beta)(f), \tag{8.34}$$

where $T$ is the mapping defined in (8.29) and $f(u,v) = 1_{[u \le t < u+v]}$. Evaluating the right side, we find

$$\begin{aligned} T(X_\alpha, \mathbb{LEB} \times \nu_\beta)(f) &= \int\int f(X_\alpha(v), x) dv d\nu_\beta(x) = \int_0^{X_\alpha^{\leftarrow}(t)} (t - X_\alpha(v))^{-\beta} dv \\ &= \int_0^t (t-v)^{-\beta} dX_\alpha^{\leftarrow}(v), \end{aligned}$$

which is the convolution of the measure $\nu_\beta$ and the nondecreasing function $X_\alpha^{\leftarrow}$. The integral also equals

$$t^{-\beta} \int_0^1 (1-v)^{-\beta} dX_\alpha^{\leftarrow}(tv) \stackrel{d}{=} t^{-\beta\alpha} \int_0^1 (1-v)^{-\beta} dX_\alpha^{\leftarrow}(v).$$

Since $b$ can be chosen continuous and strictly increasing, the $M(b(s)t)$ in (8.34) may be replaced by $M(st)$. This concludes the proof. □

## 8.3 Heavy traffic and heavy tails

Heavy-traffic limit theorems were devised to study the behavior of complex networks that for economic reasons are heavily loaded, so that system parameters are near the boundary of the set of parameters that make the system stable. By studying a sequence of normalized systems when system parameters approach the stability boundary, a heavy-traffic approximation provides various approximations for performance measures.

Original work assumed the component random variables of the model all had finite variance. This research originated with [186, 187] and was nicely surveyed and updated in [298, 299]. See also the summaries in [8] and [262]. This early work served as a foundation for J. M. Harrison, who with coworkers started the subject of diffusion process approximations [157], which is still a subject of active research.

The classical work on heavy-traffic approximations has little methodological relevance to models depending critically on heavy-tailed distributions. The present section, based on [250], was stimulated by investigations by Boxma and Cohen [29, 45–47], who based their attack on Laplace transforms.

Assume we have a sequence of GI/G/1 queuing models, which are sometimes called Lindley queues. The sequence of models is indexed by $k$. For each model, interarrival times of customers form an iid sequence with common distribution, and the service lengths of each customer are iid with a service length distribution. Each model is stable, but as $k \to \infty$, the models become unstable in the sense that the net drift (expected service time minus expected interarrival time) tends to zero. The service length distribution for the $k$th model, as a first approximation, can be thought to be independent of $k$ and heavy tailed. The interarrival time distribution in the $k$th model is lighter than the service time distribution. Since the $k$th model is assumed to be stable, there is a stationary waiting-time distribution. Let $W^{(k)}$ be a random variable with the stationary waiting-time distribution in the $k$th model. For large $k$, $W^{(k)}$ properly scaled has an approximate Mittag–Leffler distribution. This is a distribution with an explicit series representation and a simple Laplace transform. See (8.50), (8.51), and [137, 138].

An instructive way to understand this result is as follows. For the $k$th model, $W^{(k)}$ has a standard interpretation [8, 262] as the maximum of a negative-drift random walk, which has a natural association to the $k$th model. Under the assumptions that make our Mittag–Leffler distribution approximation valid, scaled and time-dilated versions of this sequence of random walks converge weakly in the sense of stochastic processes to a limiting stable Lévy motion with negative drift. This means that a scaled version of $W^{(k)}$, interpreted as the all time maximum of the $k$th random walk, has approximately the same distribution as the all time maximum of the negative-drift stable Lévy motion. The distribution of this maximum is known from the work of [137, 138] and [315].

Roughly stated, the conclusion is that the Mittag–Leffler distribution is an approximation to the equilibrium waiting-time distribution of a heavily loaded GI/G/1 system whose service time distribution has finite mean, infinite second moment and heavy tail. More details follow.

### 8.3.1 Crash course on waiting-time processes

We now give some standard background, abstracted from [262], for the waiting-time process of the G/G/1 queue. The symbols G/G/1 stand for general input (arrivals occur according to a renewal process), general service times (service times of successive customers are iid), and **1** (one) server.

As a convention suppose customer number 0 arrives at time 0. Let $\sigma_{n+1}$ be the interarrival time between the $n$th and the $(n+1)$st arriving customer. Assume $\sigma_n$, $n \geq 1$, are iid with finite mean. Let $t_k$ be the time of arrival of customer $k$, $k \geq 0$, so that $t_0 = 0$, $t_k = \sigma_1 + \cdots + \sigma_k$, $k \geq 1$.

Let $\tau_n$ be the service time of the $n$th arriving customer and suppose $\{\tau_n, n \geq 0\}$ is iid with a finite mean. Define the traffic intensity $\rho$ by

$$\rho = E\tau_0/E\sigma_1 = (E\sigma_1)^{-1}/(E\tau_0)^{-1}, \tag{8.35}$$

so that $\rho$ is the ratio of the arrival rate to the service rate. If $\rho < 1$, then on the average, the server is able to cope with his load. Assume $\{\tau_n\}$ and $\{\sigma_n\}$ are independent. (Sometimes it suffices that $\{(\tau_n, \sigma_{n+1}), n \geq 0\}$ be iid.)

We assume there is one server and that he serves customers on a first-come–first-served basis. A basic process is $W_n$, the waiting time of the $n$th customer until his service commences. This is the elapsed time between the arrival of the $n$th customer and the beginning of his service period. A basic recursion for $W_n$ is

$$W_0 = 0, \qquad W_{n+1} = (W_n + \tau_n - \sigma_{n+1})^+, \quad n \geq 0, \tag{8.36}$$

where $x^+ = x$ if $x \geq 0$ and $= 0$ if $x < 0$. A process satisfying (8.36) is sometimes called a *Lindley process* [206]. Why is the recursion true? There are two possible scenarios. For the first, $W_n + \tau_n > \sigma_{n+1}$, and then the waiting time of the $(n+1)$st customer is positive and equal to $W_n + \tau_n - \sigma_{n+1}$. The second scenario is when $W_n + \tau_n \leq \sigma_{n+1}$. In this case, the $(n+1)$st customer enters service immediately upon arrival and has no wait.

For $n \geq 0$, define

$$X_{n+1} = \tau_n - \sigma_{n+1} \tag{8.37}$$

so that $\{X_n, n \geq 1\}$ is iid. With this notation,

$$W_{n+1} = (W_n + X_{n+1})^+,$$

and $\{W_n\}$ is a random walk with a boundary at 0, that is, a partial sum process prevented from going negative.

Denote the random walk by $\{S_n, n \geq 0\}$, where $S_n = X_1 + \cdots + X_n$. Note that if $\mu = EX_1$, then

$$\begin{aligned}
\mu &< 0 \quad \text{if and only if } \rho < 1, \\
\mu &= 0 \quad \text{if and only if } \rho = 1, \\
\mu &> 0 \quad \text{if and only if } \rho > 1.
\end{aligned}$$

**Proposition 8.3.** *For the waiting time $W_n$ of the* G/G/1 *queuing model, we have*

$$W_n = \max\left\{0, X_n, X_n + X_{n-1}, \ldots, \sum_{i=2}^{n} X_i, S_n\right\} \tag{8.38}$$

$$\stackrel{d}{=} \bigvee_{j=0}^{n} S_j. \tag{8.39}$$

*Proof.* Proceed by induction: The equality (8.38) is trivially true for $n = 0$ and $n = 1$. Assume that it holds for $n$. Then by the induction hypothesis,

$$\begin{aligned}
W_{n+1} &= (W_n + X_{n+1})^+ \\
&= \left(\max\left\{0, X_n, X_n + X_{n-1}, \ldots, \sum_{i=1}^{n} X_i\right\} + X_{n+1}\right)^+ \\
&= \left(\max\left\{X_{n+1}, X_{n+1} + X_n, \ldots, \sum_{i=1}^{n+1} X_i\right\}\right)^+ \\
&= \max\left\{0, X_{n+1}, X_{n+1} + X_n, \ldots, \sum_{i=1}^{n+1} X_i\right\}.
\end{aligned}$$

So if (8.38) holds for $n$, then it holds for $n + 1$.

To prove the equality in distribution $W_n \stackrel{d}{=} \vee_{j=0}^{n} S_j$, we observe that

$$(X_1, \ldots, X_n) \stackrel{d}{=} (X_n, \ldots, X_1)$$

since both vectors consist of iid random variables. Therefore,

$$W_n = \max\left\{0, X_n, X_n + X_{n-1}, \ldots, \sum_{i=2}^{n}, S_n\right\}$$

$$\stackrel{d}{=} \max\left\{0, X_1, X_1 + X_2, \ldots, \sum_{i=1}^{n-1} X_i, S_n\right\} = \bigvee_{j=0}^{n} S_j. \qquad \square$$

This simple result allows us to calculate the asymptotic distribution of $\{W_n\}$. We are interested in the stable case when $\rho < 1$.

**Proposition 8.4.** *For the waiting time $W_n$ of a* G/G/1 *queuing model, the following is true: If $\rho < 1$, then $W_\infty := \vee_{j=1}^{\infty} S_j < \infty$ and*

$$\mathbb{P}[W_n \le x] \to \mathbb{P}[W_\infty \le x].$$

*Proof.* We use the critical fact that $W_n \stackrel{d}{=} \vee_{j=0}^{n} S_j$. Since $\rho < 1$, we have $\mu = \mathbf{E}(X_1) < 0$, so by the strong law of large numbers, $S_n \to -\infty$ almost surely. Thus $W_\infty < \infty$ and

$$\mathbb{P}[W_n \le x] = \mathbb{P}\left[\bigvee_{j=0}^{n} S_j \le x\right] \to \mathbb{P}\left[\bigvee_{j=0}^{\infty} S_j \le x\right]. \qquad \square$$

### 8.3.2 Heavy-traffic approximation for queues with heavy-tailed services

To state the approximation result precisely, we construct a sequence of Lindley queuing models. Suppose $\{\tau_i^{(k)}, i \ge 1\}$ is a nonnegative iid sequence (of service lengths) with common distribution $B^{(k)}(x)$ and $\{\sigma_i^{(k)}, i \ge 1\}$ is an independent sequence of nonnegative iid interarrival times with common distribution $A^{(k)}(x)$. We assume the means of $A^{(k)}(x)$ and $B^{(k)}(x)$ are finite, and that for each $k$, $\{\tau_n^{(k)}, n \ge 0\}$ and $\{\sigma_n^{(k)}, n \ge 1\}$ are independent. The delay or waiting-time process of the $k$th Lindley queue is given by

$$W_0^{(k)} = 0, \qquad W_{n+1}^{(k)} = \left(W_n^{(k)} + \tau_n^{(k)} - \sigma_{n+1}^{(k)}\right)^+, \qquad n \ge 0.$$

The traffic intensity for the $k$th model is

$$\rho^{(k)} = E\left(\tau_1^{(k)}\right) / E\left(\sigma_1^{(k)}\right).$$

For the heavy-traffic approximation to hold, we need the following conditions.

CONDITION (A). Suppose there exists a distribution function $F$ concentrating on $[0, \infty)$ such that $\bar{F} := 1 - F \in \mathrm{RV}_{-\alpha}$, $1 < \alpha < 2$. The quantile function

$$b(t) = \left(\frac{1}{1-F}\right)^{\leftarrow}(t) \tag{8.40}$$

is regularly varying with index $1/\alpha$. Suppose further that $B^{(k)}(x)$ satisfies

$$\lim_{x\to\infty} \frac{\bar{B^{(k)}}(x)}{\bar{F}(x)} = 1 \tag{8.41}$$

uniformly in $k = 1, 2, \ldots$. This means that given $\delta > 0$, there exists $x_0 = x_0(\delta)$ independent of $k$ such that for $x > x_0$ and all $k$, we have

$$1 - \delta < \frac{\bar{B^{(k)}}(x)}{\bar{F}(x)} \le 1 + \delta. \tag{8.42}$$

CONDITION (B). The tails of the distribution of $\sigma_1^{(k)}$ are always lighter than the tail of $F$. A convenient way we ensure this is by assuming that there exists $\eta > \alpha$ such that

$$c^{\vee} := \sup_{k\ge 1} E\left(\sigma_1^{(k)}\right)^{\eta} < \infty. \tag{8.43}$$

CONDITION (C). Assume

$$0 > m(k) := E\left(\tau_1^{(k)}\right) - E\left(\sigma_1^{(k)}\right) \to 0 \tag{8.44}$$

as $k \to \infty$. Set

$$X_i^{(k)} := \frac{\tau_i^{(k)} - \sigma_{i+1}^{(k)}}{b(d(k))}, \quad i \ge 1, \tag{8.45}$$

where the specification of $d(k)$ is given below. We think of $\{X_i^{(k)}, i \ge 1\}$ as steps of the $k$th random walk. The step mean is

$$\mu^{(k)} = \frac{E\left(\tau_1^{(k)}\right) - E\left(\sigma_1^{(k)}\right)}{b(d(k))} = \frac{m(k)}{b(d(k))}.$$

We interpret (8.44) as meaning that the $k$th random walk has negative drift so that the $k$th Lindley queue is stable but that as $k$ increases, the random walk drift becomes more and more negligible so that the associated Lindley models become less and less stable. Hence the need for scaling by $b(d(k))$.

**Definition of $d(k)$.** In order for the random walks with negative drift to be approximated by stable Lévy motion with drift $-1$, the function $d(k)$ must satisfy

$$d(k)\mu^{(k)} := \frac{d(k)m(k)}{b(d(k))} \to -1$$

as $k \to \infty$. The function

$$H(t) := \frac{t}{b(t)} \in \mathrm{RV}_{1-\frac{1}{\alpha}} \tag{8.46}$$

grows like a power function with exponent $1-\alpha^{-1}>0$ and has an asymptotic inverse

$$H^{\leftarrow} \in \mathrm{RV}_{\alpha/(\alpha-1)}.$$

The sequence $d(k)$ must satisfy

$$H(d(k)) \sim \frac{1}{|m(k)|}.$$

Therefore, we choose the sequence $\{d(k)\}$ to be any sequence satisfying

$$d(k) \sim H^{\leftarrow}\left(\frac{1}{|m(k)|}\right), \tag{8.47}$$

where $H$ is specified in (8.46).

We now state the approximation theorem.

**Theorem 8.2.** *Assume Conditions* (A)–(C) *hold. Then with* $\{X_i^{(k)}, i \geq 1\}$ *defined by* (8.45) *and* $\{d(k)\}$ *satisfying* (8.47), *we have in* $D[0,\infty)$,

$$Y^{(k)}(t) := \sum_{i=1}^{[d(k)t]} X_i^{(k)} = \frac{1}{b(d(k))} \sum_{i=1}^{[d(k)t]} \left(\tau_i^{(k)} - \sigma_{i+1}^{(k)}\right) \Rightarrow Y^{(\infty)}(t), \tag{8.48}$$

*where the limit* $Y^{(\infty)}(t) = \xi^{(\infty)}(t) - t$ *and* $\xi^{(\infty)}(t)$ *is a totally skewed to the right, zero mean,* $\alpha$*-stable Lévy motion.*

*Furthermore, the sequence of stationary waiting times indexed by* $k$ *converges in distribution in* $\mathbb{R}$*:*

$$\frac{1}{b(d(k))} W^{(k)} = \bigvee_{n=0}^{\infty} \frac{1}{b(d(k))} \sum_{i=1}^{n} \left(\tau_i^{(k)} - \sigma_{i+1}^{(k)}\right) \Rightarrow W^{(\infty)} = \bigvee_{t=0}^{\infty} Y^{(\infty)}(t). \tag{8.49}$$

*Remark* 8.2. The distribution of the maximum $W^{(\infty)}$ of a negative-drift $\alpha$-stable Lévy motion has been computed in [137, 138] using work of [315]. The limit distribution is a Mittag–Leffler distribution. Thus a queuing system with heavy-tailed service requirements under heavy load has an equilibrium waiting-time distribution which is approximated by the Mittag–Leffler distribution. See, e.g., [137, (3.20)].

We have the following corollary.

**Corollary 8.3.** *Suppose the assumptions of Theorem* 8.2 *hold. Then for every* $t>0$,

$$P(W^{(k)}/b(d(k)) \leq t) \to P(W^{(\infty)} \leq t) = 1 - \sum_{n=0}^{\infty} \frac{(-a)^n}{\Gamma(1+n(\alpha-1))} t^{n(\alpha-1)}, \tag{8.50}$$

*where* $a = (\alpha - 1)/\Gamma(2 - \alpha)$, *and for every* $\lambda \geq 0$,

$$Ee^{-\lambda W^{(k)}} \to Ee^{-\lambda W^{(\infty)}} = \frac{a}{a + \lambda^{\alpha - 1}}. \tag{8.51}$$

*Example* 8.1. Consider a GI/G/1 queue with service times $\{\tau_i, i \geq 1\}$ having the Pareto distribution

$$F(x) = 1 - x^{-3/2}, \quad x \geq 1,$$

and interarrival times $\{\sigma_i, i \geq 1\}$ having the Gamma$(\beta, \lambda)$ distribution. Assume that

$$\rho = E(\tau_1)/E(\sigma_1) < 1,$$

but not by much. One can get approximate values of the probabilities that the stationary waiting time in the system exceeds a given level by thinking, for example, about the scaling setup of Problem 8.3 (or any other setup, e.g., shift setup) as follows. We have

$$b(t) = t^{2/3} \quad \text{for } t \geq 1$$

by (8.40) and

$$m(k) = -(1 - \rho)E(\tau_1) = -2(1 - \rho)$$

by (8.44). The argument $k$ does not make sense here, but we are sticking with the terminology of this section.

The function $H$ in (8.46) is, in this case, given by

$$H(t) = t^{1/3} \quad \text{for } t \geq 1,$$

and its inverse is

$$H^{\leftarrow}(u) = u^3 \quad \text{for } u \geq 1.$$

Suppose that relation (8.47) is an equality; we then have

$$d(k) = \frac{1}{8}(1 - \rho)^{-3}.$$

With $W^{(\infty)}$ having the Mittag–Leffler distribution (8.50), our approximation is then

$$P(W > t) \approx P(W^{(\infty)} > t/b(d(k))) = P(W^{(\infty)} > 4(1 - \rho)^2 t).$$

For example, the approximate values for $P(W > t)$ for $t = 250, 1000$, and 4000 are, correspondingly, .459, .281, and .153 for $\rho = .9$, are .644, .459, and .281 for $\rho = .95$, and are .907, .827, and .697 for $\rho = .99$. □

### 8.3.3 Approximation to a negative-drift random walk

We now proceed to an understanding of the heavy-traffic approximation. Since the stationary waiting-time distribution in a stable Lindley queue can be expressed as the distribution of the supremum of a negative-drift random walk, we begin by studying weak convergence of a sequence of negative-drift random walks. In this section, we assume that for each $k = 1, 2, \ldots$, $\{X_i^{(k)}, i \geq 1\}$ are iid random variables. The $k$th random walk is

$$S_0^{(k)} = 0, \qquad S_n^{(k)} = \sum_{i=1}^{n} X_i^{(k)}, \quad n \geq 1,$$

so that the $k$th random walk has steps $X_i^{(k)}, i = 1, 2, \ldots$.

We need to make the following assumptions.

Assumption 1. There exists a nonnegative sequence of integers $d(k) \to \infty$ such that

$$\nu_k(\cdot) := d(k)\mathbb{P}\left[X_1^{(k)} \in \cdot\right] \xrightarrow{v} \nu(\cdot), \tag{8.52}$$

vaguely in $[-\infty, \infty] \setminus \{0\}$, where $\nu$ is a measure on $[-\infty, \infty] \setminus \{0\}$ satisfying

(a) $\nu$ is a Lévy measure (cf. Section 5.5.1 (p. 146)),
(b) $\nu(-\infty, 0) = 0$,
(c) $\int_1^\infty x\nu(dx) < \infty$.

Assumption 2. How much mass is allowed near 0 is controlled by the condition that for any $M > 0$,

$$\limsup_{k\to\infty} d(k)\mathbf{E}\left(\left(X_1^{(k)}\right)^2 1_{[|X_1^{(k)}|\leq M]}\right) < \infty \tag{8.53}$$

and

$$\lim_{\epsilon\downarrow 0}\limsup_{k\to\infty} d(k)\mathbf{E}\left(\left(X_1^{(k)}\right)^2 1_{[|X_1^{(k)}|\leq \epsilon]}\right) = 0. \tag{8.54}$$

Assumption 3. We assume each $X_i^{(k)}$ has a finite negative mean $\mu^{(k)}$ satisfying

$$\lim_{k\to\infty} d(k)\mu^{(k)} = -1,$$

which implies $0 > \mu^{(k)} \to 0$ as $k \to \infty$.

Assumption 4. Just assuming $\nu$ is a Lévy measure does not provide sufficient control near infinity, so we assume further that

$$\lim_{M\to\infty} \limsup_{k\to\infty} d(k)E\left(\left|X_i^{(k)}\right| 1_{[|X_i^{(k)}|>M]}\right) = 0. \tag{8.55}$$

With these assumptions in place, we state and prove the first result about how a sequence of negative-drift random walks can be approximated by a negative-drift Lévy process.

**Proposition 8.5.** *Assume that Assumptions 1–4 hold. Define the random element of $D[0,\infty)$ as*

$$Y^{(k)}(t) = S^{(k)}_{[d(k)t]}, \quad t \geq 0,$$

*for $k = 1, 2, \ldots$. Let $\{\xi^{(\infty)}(t), t \geq 0\}$ be a totally skewed to the right zero mean Lévy process with Lévy measure $\nu$ and set $Y^{(\infty)}(t) = \xi^{(\infty)}(t) - t$, $t \geq 0$. Then in $D[0,\infty)$,*

$$Y^{(k)}(\cdot) \Rightarrow Y^{(\infty)}(\cdot).$$

*Proof.* Use Theorem 7.1 (p. 214) to conclude that

$$X^{(k)}(\cdot) := \sum_{i=1}^{[d(k)\cdot]} \left(X_i^{(k)} - E\left(X_i^{(k)} 1_{[|X_k^{(k)}|\leq 1]}\right)\right) \Rightarrow X^{(\infty)}(\cdot) \tag{8.56}$$

in $D[0,\infty)$, where

$$\begin{aligned}
X^{(\infty)}(\cdot) &:= \lim_{\delta\downarrow 0}\left(\sum_{t_k\leq(\cdot)} j_k 1_{[j_k>\delta]} - (\cdot)\int_{\delta<x\leq 1} x\nu(dx)\right) \\
&= \lim_{\delta\downarrow 0}\left(\sum_{t_k\leq(\cdot)} j_k 1_{[j_k\in(\delta,1]]} - (\cdot)\int_{\delta<x\leq 1} x\nu(dx)\right. \\
&\qquad \left. + \sum_{t_k\leq(\cdot)} j_k 1_{[j_k>1]} - (\cdot)\int_{x>1} x\nu(dx) + (\cdot)\int_{x>1} x\nu(dx)\right) \\
&= \xi^{(\infty)}(\cdot) + (\cdot)\int_{x>1} x\nu(dx),
\end{aligned}$$

and $\xi^{(\infty)}(t)$ is totally skewed to the right and has Lévy measure $\nu$ and zero mean.

Now center (8.56) to zero expectations. We have

$$d(k)\left(\mu^{(k)} - E\left(X_1^{(k)} 1_{[|X_1^{(k)}|\leq 1]}\right)\right) = d(k)E\left(X_1^{(k)} 1_{[|X_1^{(k)}|>1]}\right)$$

and

$$d(k)E\left(X_1^{(k)}1_{[|X_1^{(k)}|>1]}\right) \to \int_{x>1} x\nu(dx),$$

since the absolute value of the difference can be bounded by

$$\left|d(k)E\left(X_1^{(k)}1_{[1<|X_1^{(k)}|\leq M]}\right) - \int_1^M x\nu(dx)\right| + d(k)E\left(X_1^{(k)}1_{[|X_1^{(k)}|>M]}\right) + \int_M^\infty x\nu(dx)$$
$$= \mathrm{I} + \mathrm{II} + \mathrm{III}$$

for an arbitrary $M$ chosen to avoid the atoms of $\nu$. As $k \to \infty$, $\mathrm{I} \to 0$ by vague convergence (8.52). We can make II as small as desired by (8.55) of Assumption 4, and III is made small by Assumption 1(c). We therefore conclude that

$$\sum_{i=1}^{[d(k)t]} X_i^{(k)} - [d(k)t]\mu^{(k)} \Rightarrow X^{(\infty)}(t) - t\int_1^\infty x\nu(dx)$$

in $D[0, \infty)$ and, furthermore, that

$$Y^{(k)}(t) = S_{[d(k)t]}^{(k)} \Rightarrow X^{(\infty)}(t) - t\int_1^\infty x\nu(dx) - t = Y^{(\infty)}(t)$$

in $D[0, \infty)$, where we have used Assumption 3. This completes the proof. □

### 8.3.4 Approximation to the supremum of a negative-drift random walk

The supremum of a negative-drift random walk is of interest because of its relation to the equilibrium waiting time of GI/G/1 queuing models. In this section we discuss how the approximation of Section 8.3.3 to the negative-drift random walk implies an approximation to the supremum. We continue using the notation defined in Section 8.3.3.

**Proposition 8.6.** *Assume Assumptions* 1–4 *of Section* 8.3.3 *(p.* 279*) hold. Define*

$$W^{(k)} := \bigvee_{t=0}^{\infty} Y^{(k)}(t) = \bigvee_{n=0}^{\infty} S_n^{(k)}.$$

*Then in $\mathbb{R}$, we have the convergence in distribution, as $k \to \infty$,*

$$W^{(k)} \Rightarrow W^{(\infty)} := \bigvee_{t=0}^{\infty} Y^{(\infty)}(t) = \bigvee_{t=0}^{\infty}(\xi^{(\infty)}(t) - t),$$

*where we recall that $\xi^{(\infty)}(\cdot)$ is the zero mean Lévy process of Theorem* 8.5.

*Proof.* We use a method learned from [8] for the finite variance case. For any $T > 0$, the map $x \mapsto \bigvee_{s=0}^{T} x(s)$ from $D[0,\infty) \mapsto \mathbb{R}$ is continuous (cf. Problem 3.21 (p. 69)). So from Theorem 8.5, we have

$$\bigvee_{s=0}^{T} Y^{(k)}(s) \Rightarrow \bigvee_{s=0}^{T} Y^{(\infty)}(s)$$

in $\mathbb{R}$. The desired result will be proven using the second converging together theorem, Theorem 3.5 (p. 56), provided we can show for any $\eta > 0$ that

$$\lim_{T\to\infty} \limsup_{k\to\infty} \mathbb{P}\left[ \bigvee_{j\ge d(k)T} S_j^{(k)} > \eta \right] = 0. \tag{8.57}$$

To prove (8.57), we observe that for any suitably chosen $M > 0$,

$$\begin{aligned}
&\mathbb{P}\left[ \bigvee_{j\ge d(k)T} S_j^{(k)} > \eta \right] \\
&\quad \le \mathbb{P}\left[ \bigvee_{j\ge d(k)T} \frac{S_j^{(k)}}{j} > 0 \right] \\
&\quad \le \mathbb{P}\left[ \bigvee_{j\ge d(k)T} \left( j^{-1}\sum_{i=1}^{j} X_i^{(k)} 1_{[|X_i^{(k)}|\le M]} + j^{-1}\sum_{i=1}^{j} X_i^{(k)} 1_{[|X_i^{(k)}|> M]} \right) > 0 \right] \\
&\quad = \mathbb{P}\left[ \bigvee_{j\ge d(k)T} \left( j^{-1}\sum_{i=1}^{j} \left( X_i^{(k)} 1_{[|X_i^{(k)}|\le M]} - \mathbf{E}\left( X_1^{(k)} 1_{[|X_1^{(k)}|\le M]} \right)\right) \right.\right. \\
&\qquad\qquad \left.\left. + j^{-1}\sum_{i=1}^{j} \left( X_i^{(k)} 1_{[|X_i^{(k)}|> M]} - \mathbf{E}\left( X_1^{(k)} 1_{[|X_1^{(k)}|> M]} \right)\right) \right) > |\mu^{(k)}| \right] \\
&\quad \le \mathbb{P}\left[ \bigvee_{j\ge d(k)T} j^{-1}\sum_{i=1}^{j} \left( X_i^{(k)} 1_{[|X_i^{(k)}|\le M]} - \mathbf{E}\left( X_i^{(k)} 1_{[|X_i^{(k)}|\le M]} \right)\right) > |\mu^{(k)}|/2 \right] \\
&\qquad + \mathbb{P}\left[ \bigvee_{j\ge d(k)T} j^{-1}\sum_{i=1}^{j} \left( X_i^{(k)} 1_{[|X_i^{(k)}|> M]} - \mathbf{E}\left( X_i^{(k)} 1_{[|X_i^{(k)}|> M]} \right)\right) > |\mu^{(k)}|/2 \right] \\
&\quad = \mathrm{I} + \mathrm{II}.
\end{aligned}$$

The centered sample averages are reversed martingales, so we may apply Kolmogorov's inequality [24, 264]. We will use the fact that from Assumption 3, $d(k) \sim 1/|\mu^{(k)}|$ as $k \to \infty$. For I, we have

$$\begin{aligned} \mathrm{I} &\le (\mu^{(k)}/2)^{-2}\,\mathrm{Var}\left(\frac{1}{[d(k)T]}\sum_{i=1}^{[d(k)T]} X_i^{(k)} 1_{[|X_i^{(k)}|\le M]}\right) \\ &\le 4\frac{[d(k)T]}{[d(k)T]^2}\frac{1}{(\mu^{(k)})^2}\,\mathrm{Var}\left(X_1^{(k)}1_{[|X_1^{(k)}|\le M]}\right). \end{aligned}$$

Using $(d(k)\mu^{(k)})^2 \to 1$, as $k \to \infty$, this is

$$\sim \frac{1}{T}d(k)\,\mathrm{Var}\left(X_1^{(k)}1_{[|X_1^{(k)}|\le M]}\right).$$

This converges to 0 as $T \to \infty$ due to (8.53). To kill II, we write, using the martingale maximal inequality [24, 229, 264],

$$\begin{aligned} \mathrm{II} &\le \frac{2}{|\mu^{(k)}|}\mathbf{E}\left|\frac{1}{[d(k)T]}\sum_{i=1}^{[d(k)T]}\left(X_i^{(k)}1_{[|X_i^{(k)}|>M]} - \mathbf{E}\left(X_i^{(k)}1_{[|X_i^{(k)}|>M]}\right)\right)\right| \\ &\le \frac{2}{|\mu^{(k)}|}\mathbf{E}\left|X_1^{(k)}1_{[|X_1^{(k)}|>M]} - \mathbf{E}\left(X_1^{(k)}1_{[|X_1^{(k)}|>M]}\right)\right| \\ &\sim 2d(k)\mathbf{E}\left|X_1^{(k)}1_{[|X_1^{(k)}|>M]} - \mathbf{E}\left(X_1^{(k)}1_{[|X_1^{(k)}|>M]}\right)\right| \\ &\le 4d(k)\mathbf{E}\left(\left|X_1^{(k)}\right|1_{[|X_1^{(k)}|>M]}\right). \end{aligned}$$

From (8.55) of Assumption 4, if we choose $M$ sufficiently large, we can guarantee that $\limsup_{k\to\infty}$ II can be made as small as desired. This completes the proof. □

### 8.3.5 Proof of the heavy-traffic approximation

This section gives the proof of Theorem 8.2 and Corollary 8.3.

*Proof.* Both assertions in the statement of the theorem will be proven if we verify that Conditions (A)–(C) and the definition of $d(k)$ given in (8.47) imply Assumptions 1–4. We begin by showing that Assumption 1 is valid with

$$\nu(dx) = \nu_\alpha(dx) = \alpha x^{-\alpha-1}dx1_{(0,\infty)}(x).$$

On one hand, we have for $x > 0$,

$$d(k)\mathbb{P}\left[X_1^{(k)} > x\right] \leq d(k)\mathbb{P}\left[\tau_1^{(k)} > b(d(k))x\right] \to x^{-\alpha}$$

and on the other, for any $\delta > 0$,

$$\begin{aligned}
d(k)\mathbb{P}\left[X_1^{(k)} > x\right] &\geq d(k)\mathbb{P}\left[\tau_1^{(k)} - \sigma_2^{(k)} > b(d(k))x, \sigma_2^{(k)} \leq \delta b(d(k))\right] \\
&\geq d(k)\mathbb{P}\left[\tau_1^{(k)} > b(d(k))(x+\delta), \sigma_2^{(k)} \leq \delta b(d(k))\right] \\
&= d(k)\mathbb{P}\left[\tau_1^{(k)} > b(d(k))(x+\delta)\right] \\
&\quad - d(k)\mathbb{P}\left[\tau_1^{(k)} > b(d(k))(x+\delta), \sigma_2^{(k)} > \delta b(d(k))\right] \\
&\to (x+\delta)^{-\alpha} - 0 \quad (k \to \infty),
\end{aligned}$$

where the last 0 results from

$$\begin{aligned}
\lim_{k\to\infty} d(k)\mathbb{P}\left[\sigma_2^{(k)} > \delta b(d(k))\right] &\leq \lim_{k\to\infty} \frac{d(k)}{b(d(k))^\eta \delta^\eta} \mathbf{E}\left(\sigma_2^{(k)}\right)^\eta \\
&\leq \lim_{t\to\infty} c^\vee \frac{t}{b(t)^\eta \delta^\eta} = 0 
\end{aligned} \tag{8.58}$$

since $\eta/\alpha > 1$. Thus, since $\delta > 0$ is arbitrary, we conclude that for $x > 0$, $d(k)\mathbb{P}[X_1^{(k)} > x] \to \nu_\alpha(x, \infty]$.

For $x < 0$, note that

$$\begin{aligned}
d(k)\mathbb{P}\left[X_1^{(k)} < x\right] &= d(k)\mathbb{P}\left[\sigma_2^{(k)} - \tau_1^{(k)} > b(d(k))|x|\right] \\
&\leq d(k)\mathbb{P}\left[\sigma_2^{(k)} > b(d(k))x\right] \to 0
\end{aligned}$$

due to (8.58). This verifies that Assumption 1 holds.

To check that Assumption 2 is valid, observe that we have, after integration by parts,

$$\begin{aligned}
&d(k)\mathbf{E}\left(\left(X_1^{(k)}\right)^2 1_{[|X_1^{(k)}|\leq\epsilon]}\right) \leq 2d(k)\int_0^\epsilon x\mathbb{P}\left[\left|X_1^{(k)}\right| > x\right]dx \\
&\leq 2d(k)\int_0^\epsilon x\mathbb{P}\left[\tau_1^{(k)} > b(d(k))x\right]dx + 2d(k)\int_0^\epsilon x\mathbb{P}\left[\sigma_1^{(k)} > b(d(k))x\right]dx \\
&=: \mathrm{I} + \mathrm{II}.
\end{aligned}$$

For I, we have with $x_0$ and $\delta$ as in Condition (A) (see (8.42)),

$$\mathrm{I} = 2d(k)\int_0^{x_0/b(d(k))} x\mathbb{P}\left[\tau_1^{(k)} > b(d(k))x\right]dx$$

$$
\begin{aligned}
&+ 2d(k) \int_{x_0/b(d(k))}^{\epsilon} x\mathbb{P}\left[\tau_1^{(k)} > b(d(k))x\right] dx \\
&\leq d(k) \left(\frac{x_0}{b(d(k))}\right)^2 + 2(1+\delta)d(k) \int_{x_0/b(d(k))}^{\epsilon} x\bar{F}(b(d(k))x)dx \\
&\leq \frac{d(k)}{b(d(k))^2}(\text{const}) + 2(1+\delta) \int_0^{\epsilon} 1_{[x>x_0/b(d(k))]} x d(k)\bar{F}(b(d(k))x)dx \\
&= \text{Ia} + \text{Ib}.
\end{aligned}
$$

Since $t/b^2(t) \to 0$, we have Ia $\to 0$ with $k \to \infty$. By Karamata's theorem, Theorem 2.1 (p. 25), we have Ib $\to 2(1+\delta)(2-\alpha)^{-1}\epsilon^{2-\alpha}$ as $k \to \infty$, which goes to 0 as $\epsilon \to 0$. To verify the limit for Ib, note that for every fixed $x > 0$,

$$
d(k)\bar{F}(b(d(k))x) = d(k)\bar{F}(b(d(k)))\frac{\bar{F}(b(d(k))x)}{\bar{F}(b(d(k)))} \to x^{-\alpha}
$$

as $k \to \infty$ by regular variation, and note that there is are $\alpha < \beta < 2$ and $C > 0$ such that

$$
\frac{\bar{F}(b(d(k))x)}{\bar{F}(b(d(k)))} \leq Cx^{-\beta}
$$

for all $\epsilon > x > x_0/b(d(k))$ and $k$ large enough. (This is a modification of the Potter bounds (2.31) (p. 32); see Problem 2.6 (p. 36).) Hence Ib $\to 2(1+\delta)(2-\alpha)^{-1}\epsilon^{2-\alpha}$ as $k \to \infty$ by the dominated convergence theorem. We therefore conclude that

$$
\lim_{\epsilon\to 0} \limsup_{k\to\infty} \text{I} = 0.
$$

For II, note that

$$
\begin{aligned}
\text{II} &\leq 2d(k) \int_0^{\epsilon} x\mathbf{E}\left(\sigma_2^{(k)}\right)^{\eta} x^{-\eta} dx / b(d(k))^{\eta} \\
&\leq (\text{const})\frac{d(k)}{b(d(k))^{\eta}}\epsilon^{2-\eta}.
\end{aligned}
$$

This goes to 0 as $k \to \infty$ since $\eta/\alpha > 1$. This completes the verification that Assumption 2 holds.

The reason that Assumption 3 holds is clear, so we turn to verifying why Assumption 4 holds. Referring to the form of Assumption 4 in (8.55), we see that

$$
\begin{aligned}
&d(k)\mathbf{E}\left(\left|X_i^{(k)}\right| 1_{[|X_i^{(k)}|>M]}\right) \\
&\quad \leq d(k)\mathbf{E}\left(\left|\frac{\tau_1^{(k)}}{b(d(k))}\right| 1_{[\tau_1^{(k)}>b(d(k))M]}\right)
\end{aligned}
$$

$$
\begin{aligned}
&+ d(k)\mathbf{E}\left(\left|\frac{\sigma_2^{(k)}}{b(d(k))}\right| 1_{[\sigma_2^{(k)} > b(d(k))M]}\right) \\
&= d(k)\int_M^\infty \mathbb{P}\left[\frac{\tau_1^{(k)}}{b(d(k))} > x\right]dx + d(k)M\mathbb{P}\left[\frac{\tau_1^{(k)}}{b(d(k))} > M\right] \\
&+ d(k)\int_M^\infty \mathbb{P}\left[\frac{\sigma_2^{(k)}}{b(d(k))} > x\right]dx + d(k)M\mathbb{P}\left[\frac{\sigma_2^{(k)}}{b(d(k))} > M\right];
\end{aligned}
$$

again using the definition of $x_0$ and $\delta$ from (8.42) of Condition (A), we have the bound, for large $M$,

$$
\begin{aligned}
&\leq (1+\delta)\int_M^\infty d(k)\bar{F}(b(d(k))x)dx + (1+\delta)d(k)M\bar{F}(b(d(k))M) \\
&+ c^{\vee}\frac{d(k)}{b(d(k))^{\eta}}\int_M^\infty x^{-\eta}dx + o(1)
\end{aligned}
$$

using (8.58). As $k \to \infty$, this is asymptotic to

$$
\sim (1+\delta)\int_M^\infty x^{-\alpha}dx + (1+\delta)MM^{-\alpha} = O(M^{-\alpha+1}),
$$

which converges to 0 as $M \to \infty$. This verifies that Assumption 4 holds and completes the proof of Theorem 8.2. □

## 8.4 Problems

**8.1.** A propos of Corollary 8.2 (p. 271), verify that if $0 = \beta < \alpha < 1$, we get

$$
\frac{\bar{F}(s)}{\bar{F}_{\mathrm{on}}(s)}M(st) \Rightarrow X_\alpha^{\leftarrow}(t).
$$

Therefore, taking into account (8.28) (p. 266), conclude that as $s \to \infty$,

$$
\frac{M(s)}{N(s)} \sim \bar{F}_{\mathrm{on}}(s) \stackrel{\mathbb{P}}{\to} 0.
$$

**8.2.** Evaluate the function $d(k)$ given in (8.47) (p. 277) for the case in which $\bar{F}$ is a pure Pareto, so that

$$
\bar{F}(x) = x^{-\alpha}, \quad x > 1.
$$

Verify that

$$
\begin{aligned}
b(t) &= t^{1/\alpha}, && t > 1, \\
H(t) &= t^{1-1/\alpha}, \\
H^{\leftarrow}(t) &= t^{\alpha/(\alpha-1)} \\
d(k) &\sim \left(\frac{1}{|m(k)|}\right)^{\alpha/(\alpha-1)}, && k \to \infty.
\end{aligned}
$$

**8.3.** Suppose $\{\tau_i, i \geq 1\}$ are iid nonnegative random variables with common distribution $F$, where $\bar{F}$ satisfies the regular variation assumptions of Condition (A). Similarly, suppose $\{\sigma_i, i \geq 1\}$ are iid nonnegative random variables with common distribution $A(x)$. Both $\tau_1$ and $\sigma_2$ are assumed to have finite means with $E(\tau_1) < E(\sigma_2)$, and $\mathbf{E}(\sigma_2)^\eta < \infty$ for some $\eta > \alpha$. Define

$$
X_i^{(k)} = \frac{\tau_i - \theta_k \sigma_{i+1}}{b(d(k))},
$$

and assume that

$$
\theta_k = \frac{E(\tau_1)}{E(\sigma_2)}(1 + \epsilon_k),
$$

where $\epsilon_k > 0$ and $\epsilon_k \to 0$ as $k \to \infty$. Verify that Conditions (A)–(C) (p. 275) hold for this setup.

**8.4.** Show that the distribution of $N_\infty^*$ in (8.30) (p. 267) can be specified by its Laplace functional

$$
\mathbb{E}(e^{-N_\infty^*(f)}) = \mathbb{E}\left(\exp\left\{-\iint_E (1 - e^{-f(X_\alpha(s),y)}) ds\, \nu_\alpha(dy)\right\}\right), \quad f \in C_K^+(\mathbb{E}).
$$

# Part IV

# More Statistics

# 9

# Additional Statistics Topics

This chapter surveys some additional statistical topics and presents analysis of several data sets to illustrate the techniques. One focus is multivariate inference: We consider methods for estimating the limit measure $\nu$ and the angular measure $S$. These methods require statistical techniques for transforming the multivariate data to the *standard* case. We also consider the *coefficient of tail dependence* and an elaborating concept called *hidden regular variation*, which aid in considering models possessing asymptotic independence. Finally, we consider a standard time-series tool called the *sample correlation function* and discuss its properties in the case of a stationary time series with heavy-tailed marginal distribtions.

First, we consider in one dimension the asymptotic normality of estimators of the tail index of regular variation.

## 9.1 Asymptotic normality

A key inference issue is to estimate the index of a distribution $F$ satisfying $\bar{F} \in \mathrm{RV}_{-\alpha}$ based on a random sample from the distribution. To prove asymptotic normality of estimators, we follow the approach of Section 4.3 (p. 78) and first prove asymptotic normality of the tail empirical measure; then from this we extract asymptotic normality for estimators.

### 9.1.1 Asymptotic normality of the tail empirical measure

As in Section 4.3, suppose $\{X_j, j \geq 1\}$ are iid, nonnegative random variables with common distribution $F(x)$, where $\bar{F} \in \mathrm{RV}_{-\alpha}$ for $\alpha > 0$. Continue with the notation in (4.10)–(4.12) (p. 78). Define the tail empirical process,

$$W_n(y) = \sqrt{k}\left(\frac{1}{k}\sum_{i=1}^{n} \epsilon_{X_i/b(n/k)}(y^{-1/\alpha}, \infty] - \frac{n}{k}\bar{F}(b(n/k)y^{-1/\alpha})\right) \tag{9.1}$$
$$= \sqrt{k}(\nu_n(y^{-1/\alpha}, \infty] - \mathbf{E}(\nu_n(y^{-1/\alpha}, \infty])), \quad y \geq 0.$$

**Theorem 9.1.** *Suppose* (4.10)–(4.12) *(p. 78) hold. Then as* $n \to \infty$, $k = k(n) \to \infty$, $n/k \to \infty$,

$$W_n \Rightarrow W$$

*in* $D[0, \infty)$, *where* $W$ *is Brownian motion on* $[0, \infty)$.

*Remark* 9.1. Note that, because of regular variation, as $n \to \infty$, $k/n \to 0$,

$$\mathbf{E}\nu_n(y^{-1/\alpha}, \infty] = \frac{n}{k}\bar{F}(b(n/k)y^{-1/\alpha}) \to (y^{-1/\alpha})^{-\alpha} = y. \tag{9.2}$$

For applications to such things as the asymptotic normality of the Hill estimator and other estimators derived from the tail empirical measure, we would prefer the centering in (9.1) be $y$. However, to make this substitution in (9.1) requires knowing or assuming that

$$\lim_{n\to\infty} \sqrt{k}\left(\frac{n}{k}\bar{F}(b(n/k)y^{-1/\alpha}) - y\right) \tag{9.3}$$

exists and is finite. This is one of the origins of the need for *second-order regular variation.* See Problems 3.15–3.17 (p. 67ff) as well as [90, 101, 109, 111–115, 136, 149–152, 235].

*Remark* 9.2. The proof to follow is based on Donsker's theorem, Theorem 3.3 (p. 54), given in Section 3.4.2. Other proofs have been given in [252] and related material and proofs considered earlier in [58–61, 80, 81, 126, 211, 213].

*Proof.* The proof uses Donsker's theorem and then Vervaat's lemma (p. 59), especially (3.28). We proceed in a series of steps.

STEP 1: *Renewal theory.* Suppose $\{Y_n, n \geq 1\}$ are iid, nonnegative random variables with $\mathbf{E}(Y_j) = \mu$, and $\mathrm{Var}(Y_j) = \sigma^2$. Set $S_n = \sum_{i=1}^{n} Y_i$. Then from Donsker's theorem,

$$\frac{S_{[nt]} - [nt]\mu}{\sigma\sqrt{n}} \Rightarrow W(t)$$

in $D[0, \infty)$, where $W(\cdot)$ is a standard Brownian motion. Since for any $M > 0$,

$$\sup_{0\leq t\leq M} \frac{|nt\mu - [nt]\mu|}{\sqrt{n}} \to 0,$$

it is also true that in $D[0, \infty)$

$$\frac{S_{[nt]} - nt\mu}{\sigma\sqrt{n}} \Rightarrow W(t).$$

In preparation for applying Vervaat's lemma, divide the numerator and denominator by $n\mu$ to get

$$c_n(X_n(t) - t) := \frac{\left(\frac{S_{[nt]}}{n\mu} - t\right)}{\sigma n^{-1/2}/\mu} \Rightarrow W(t).$$

This implies the result for $X_n^{\leftarrow}(\cdot)$ and we need to evaluate this process:

$$\begin{aligned} X_n^{\leftarrow}(t) &= \inf\{s : X_n(s) \geq t\} \\ &= \inf\{s : S_{[ns]}/n\mu \geq t\} = \inf\{s : S_{[ns]} \geq tn\mu\} \\ &= \inf\left\{\frac{j}{n} : S_j \geq tn\mu\right\} = \frac{1}{n}N(tn\mu), \end{aligned}$$

some version of the renewal counting function. (*Truth in labeling*: This $N(t)$ could differ by 1 from the $N(t)$ defined in (7.57) and (7.58) (p. 245), but we ignore this.) The conclusion from Vervaat's lemma is

$$\sqrt{n}\frac{\mu}{\sigma}\left(\frac{1}{n}N(n\mu t) - t\right) \Rightarrow W(t) \quad (\text{in } D[0,\infty))$$

or, changing variables $s = \mu t$,

$$\sqrt{n}\frac{\mu}{\sigma}\left(\frac{1}{n}N(ns) - \frac{s}{\mu}\right) \Rightarrow W\left(\frac{s}{\mu}\right) \stackrel{d}{=} \frac{1}{\sqrt{\mu}}W(s).$$

Cleaning this up just a bit gives

$$\sqrt{n}\frac{\mu^{3/2}}{\sigma}\left(\frac{1}{n}N(ns) - \frac{s}{\mu}\right) \Rightarrow W(s) \quad \text{in } D[0,\infty). \tag{9.4}$$

As a special case, consider the homogeneous Poisson process on $[0,\infty)$. Let

$$\Gamma_n = E_1 + \cdots + E_n$$

be a sum of $n$ iid standard exponential random variables. In this case, $\mu = \sigma = 1$ and

$$\sqrt{k}\left(\frac{1}{k}N(ks) - s\right) \Rightarrow W(s) \quad (k \to \infty) \tag{9.5}$$

in $D[0,\infty)$.

STEP 2: *Approximation.* The definition of $N(t)$ in (7.58) (p. 245) is in terms of an infinite series. For the Poisson process special case, we want to truncate the infinite series to a finite sum. Toward this end, we prove that for any $T > 0$, as $n \to \infty$, $k \to \infty$, and $k/n \to 0$,

$$\sup_{0 \le s \le T} \sqrt{k} \left| \frac{1}{k} \sum_{i=1}^{\infty} 1_{[\Gamma_i \le ks]} - \frac{1}{k} \sum_{i=1}^{n} 1_{[\Gamma_i \le ks]} \right| \xrightarrow{P} 0. \tag{9.6}$$

The idea is that $\Gamma_i$ is localized about its mean, and any term with $i$ too far from $k$ is unlikely; also, $i > n$ gives a term that is negligible. More formally, the difference in (9.6) is

$$\begin{aligned} \sup_{0 \le s \le T} \frac{1}{\sqrt{k}} \sum_{i=n+1}^{\infty} 1_{[\Gamma_i \le ks]} &\le \frac{1}{\sqrt{k}} \sum_{i=n+1}^{\infty} 1_{[\Gamma_i \le kT]} \\ &= \frac{1}{\sqrt{k}} \sum_{i=1}^{\infty} 1_{[\Gamma_n + \Gamma_i' \le kT]}, \end{aligned}$$

where $\Gamma_i' = \sum_{l=1}^{i} E_{l+n}$. Now for any $\delta > 0$,

$$P\left[ \frac{1}{\sqrt{k}} \sum_{i=1}^{\infty} 1_{[\Gamma_n + \Gamma_i' \le kT]} > \delta \right] \le P[\Gamma_n \le kT] = P\left[ \frac{\Gamma_n}{n} \le \frac{k}{n} T \right]$$

and since $k/n \to 0$, for any $\eta > 0$, we ultimately have this last term bounded by

$$\le P\left[ \frac{\Gamma_n}{n} \le 1 - \eta \right] \to 0$$

by the weak law of large numbers.

*Conclusion*: Combining (9.6), the definition of $N$, and (9.5), we get

$$\sqrt{k} \left( \frac{1}{k} \sum_{i=1}^{n} 1_{[\Gamma_i \le ks]} - s \right) \Rightarrow W(s) \quad (k \to \infty) \quad (k/n \to 0) \tag{9.7}$$

in $D[0, \infty)$.

STEP 3: *Time change*. Define

$$\phi_n(s) = \frac{n}{k} \bar{F}(b(n/k) s^{-1/\alpha}) \frac{\Gamma_{n+1}}{n}, \quad s > 0,$$

so that from regular variation and the weak law of large numbers,

$$\sup_{0 \le s \le T} |\phi_n(s) - s| \xrightarrow{P} 0 \tag{9.8}$$

for any $T > 0$. Therefore, using Proposition 3.1 (p. 57), joint convergence holds in $D[0,\infty) \times D[0,\infty)$:

$$\left(\sqrt{k}\left(\frac{1}{k}\sum_{i=1}^{n} 1_{[\Gamma_i \le k\cdot]} - (\cdot)\right), \phi_n(\cdot)\right) \Rightarrow (W, e) \quad (e(t) = t).$$

Applying composition, we arrive at

$$\sqrt{k}\left(\frac{1}{k}\sum_{i=1}^{n} 1_{[\Gamma_i \le k\phi_n(s)]} - \phi_n(s)\right) \Rightarrow W(s) \tag{9.9}$$

in $D[0,\infty)$.

STEP 4: *Probability integral transform.* The $\Gamma$s have the property that

$$\left(\frac{\Gamma_1}{\Gamma_{n+1}}, \dots, \frac{\Gamma_n}{\Gamma_{n+1}}\right) \stackrel{d}{=} \left(1 - \frac{\Gamma_n}{\Gamma_{n+1}}, \dots, 1 - \frac{\Gamma_1}{\Gamma_{n+1}}\right) \stackrel{d}{=} (U_{1:n}, \dots, U_{n:n}),$$

where

$$U_{1:n} \le \cdots \le U_{n:n}$$

are the order statistics in increasing order of $n$ iid $U(0,1)$ random variables $U_1, \dots, U_n$. (A proof is in [135, 262].)

Consider the normalized sum from (9.9):

$$\begin{aligned}
\frac{1}{k}\sum_{i=1}^{n} 1_{[\Gamma_i \le k\phi_n(s)]} &= \frac{1}{k}\sum_{i=1}^{n} 1_{[\frac{\Gamma_i}{k} \le \frac{1}{k}\bar{F}(b(n/k)s^{-1/\alpha})\Gamma_{n+1}]} \\
&= \frac{1}{k}\sum_{i=1}^{n} 1_{[\frac{\Gamma_i}{\Gamma_{n+1}} \le \bar{F}(b(n/k)s^{-1/\alpha})]} = \frac{1}{k}\sum_{i=1}^{n} 1_{[F(b(n/k)s^{-1/\alpha}) \le 1 - \frac{\Gamma_i}{\Gamma_{n+1}}]} \\
&= \frac{1}{k}\sum_{i=1}^{n} 1_{[b(n/k)s^{-1/\alpha} \le F^{\leftarrow}(1 - \frac{\Gamma_i}{\Gamma_{n+1}})]} \stackrel{d}{=} \frac{1}{k}\sum_{i=1}^{n} 1_{[b(n/k)s^{-1/\alpha} \le F^{\leftarrow}(U_{i:n})]} \\
&= \frac{1}{k}\sum_{i=1}^{n} 1_{[b(n/k)s^{-1/\alpha} \le F^{\leftarrow}(U_i)]} \stackrel{d}{=} \frac{1}{k}\sum_{i=1}^{n} 1_{[b(n/k)s^{-1/\alpha} \le X_i]} \\
&= \frac{1}{k}\sum_{i=1}^{n} 1_{[\frac{X_i}{b(n/k)} \ge s^{-1/\alpha}]} = \nu_n[s^{-1/\alpha}, \infty],
\end{aligned}$$

where the equality in distribution is in $D[0,\infty)$.

Also,

$$\sqrt{k}\sup_{0\le s\le T}\left|\frac{n}{k}\bar{F}(b(n/k)s^{-1/\alpha})\frac{\Gamma_{n+1}}{n}-\frac{n}{k}\bar{F}(b(n/k)s^{-1/\alpha}\right|$$
$$=\sup_{0\le s\le T}\frac{n}{k}\bar{F}(b(n/k)s^{-1/\alpha})\sqrt{k}\left|\frac{\Gamma_{n+1}}{n}-1\right|$$
$$=O(1)\sqrt{\frac{k}{n}}\left|\frac{\Gamma_{n+1}-n}{\sqrt{n}}\right|=O(1)o(1)O_p(1),$$

from the central limit theorem, and this $\stackrel{P}{\to} 0$.

This proves the desired result by appeal to (9.9) since the last statement removes the difference between $\phi_n(s)$ and $\mathbf{E}(\nu_n[s^{-1/\alpha},\infty])$. □

From this result we can recover Theorem 4.1 (p. 79) and its consequences.

### 9.1.2 Asymptotic normality of the Hill estimator

Recall that the Hill estimator $H_{k,n}$ was given in (4.3) (p. 74), and its relation to the tail empirical measure as an integral of the measure in (4.16) (p. 80) was discussed in Theorem 4.2 (p. 81).

What implications can be drawn for the Hill estimator from the asymptotic normality of the tail empirical measure given in Theorem 9.1? We continue to suppose only that $\bar{F}\in RV_{-\alpha}$. Centering the tail empirical measure to zero expectation results in a centering for the Hill estimator which is not the desired centering $1/\alpha$. The following is adapted from [100]. Another formulation with a random center is in [58].

**Proposition 9.1.** *Suppose, as in Theorem* 9.1, *that* $\bar{F}\in RV_{-\alpha}$. *Then in* $\mathbb{R}$,

$$\sqrt{k}\left(H_{k,n}-\int_{X_{(k)}}^{\infty}\frac{n}{k}\bar{F}(s)\frac{ds}{s}\right)\Rightarrow\int_1^{\infty}W(x^{-\alpha})\frac{dx}{x}=\frac{1}{\alpha}\int_0^1 W(s)\frac{ds}{s}. \tag{9.10}$$

*Proof.* For typing ease, set $\gamma=1/\alpha$. Write (9.1) as

$$\sqrt{k}\left(\nu_n\left((x^{-\gamma},\infty]\right)-\frac{n}{k}\bar{F}\left(b\left(\frac{n}{k}\right)x^{-\gamma}\right)\right)\Rightarrow W(x) \tag{9.11}$$

in $D[0,\infty)$ and set

$$V_n(x):=\frac{n}{k}\bar{F}\left(b\left(\frac{n}{k}\right)x^{-\gamma}\right),$$

so that $V_n$ is nondecreasing and $V_n(x)\to x$ and $V_n^{\leftarrow}(x)\to x$ locally uniformly as $n\to\infty$. We have

$$\sqrt{k}(\nu_n((V_n^{\leftarrow}(y))^{-\gamma},\infty]-y)\Rightarrow W(y) \tag{9.12}$$

in $D[0,\infty)$. Applying Vervaat's lemma (p. 59), we get by inversion that

$$\sqrt{k}((\nu_n(V_n^{\leftarrow}(y))^{-\gamma},\infty])^{\leftarrow} - y) \Rightarrow -W(y) \tag{9.13}$$

in $D[0,\infty)$. Evaluating the left side of (9.13) yields

$$\sqrt{k}\left(\frac{n}{k}\bar{F}(X_{([kx])}) - x\right) \Rightarrow -W(x) \tag{9.14}$$

in $D[0,\infty)$ and, in fact (see (3.28) (p. 60)), the convergence in (9.1) and (9.14) are joint in $D(0,\infty] \times D[0,\infty)$. Observe that (9.14) implies that

$$\frac{n}{k}\bar{F}(X_{([kx])}) \Rightarrow x$$

in $D[0,\infty)$ and from this or (4.18), we get

$$\frac{X_{([kx])}}{b(n/k)} \Rightarrow x^{\gamma}. \tag{9.15}$$

Note that (9.1), written with a change of variable, and (9.15) with $x = 1$ hold jointly:

$$\left(\sqrt{k}\left(\frac{1}{k}\sum_{i=1}^{n}\epsilon_{X_i/b(n/k)}(x,\infty] - \frac{n}{k}\bar{F}(b(n/k)x)\right), \frac{X_{(k)}}{b(n/k)}\right)$$
$$\Rightarrow (W(x^{-\alpha}), 1).$$

Apply the composition map $(x(t), p) \mapsto x(tp)$ to get

$$\sqrt{k}\left(\frac{1}{k}\sum_{i=1}^{n}\epsilon_{X_i/X_{(k)}}(x,\infty] - \frac{n}{k}\bar{F}(X_{(k)}x)\right) \Rightarrow W(x^{-\alpha}). \tag{9.16}$$

The final step is to justify application of the map

$$x \mapsto \int_1^{\infty} x(s)\frac{ds}{s}. \tag{9.17}$$

If this application can be justified, we get

$$\sqrt{k}\left(H_{k,n} - \int_{X_{(k)}}^{\infty}\frac{n}{k}\bar{F}(s)\frac{ds}{s}\right) \Rightarrow \int_1^{\infty} W(x^{-\alpha})\frac{dx}{x}, \tag{9.18}$$

as desired. □

**Blood and guts**

We now justify use of the map (9.17) to get the conclusion (9.18). The argument is based on the second converging together theorem, Theorem 3.5 (p. 56) and is adapted from [252].

Pick $M$ large. In (9.16), apply the map

$$x \mapsto \int_1^M x(s)\frac{ds}{s}. \tag{9.19}$$

Since the integration is over a finite region, this is a continuous map, and applying it to (9.16) gives

$$\sqrt{k}\left(\int_1^M \frac{1}{k}\sum_{i=1}^n \epsilon_{X_i/X_{(k)}}(s,\infty]\frac{ds}{s} - \int_1^M \frac{n}{k}\bar{F}(X_{(k)}s)\frac{ds}{s}\right) \Rightarrow \int_1^M W(s^{-\alpha})\frac{ds}{s}.$$

Now, as $M \to \infty$,

$$\int_1^M W(s^{-\alpha})\frac{ds}{s} \Rightarrow \int_1^\infty W(s^{-\alpha})\frac{ds}{s}.$$

It remains to verify (3.20) (p. 56). This translates to showing, for any $\delta > 0$, that

$$\lim_{M\to\infty}\limsup_{n\to\infty}\mathbb{P}\left[\sqrt{k}\left|\int_M^\infty \frac{1}{k}\sum_{i=1}^n \epsilon_{X_i/X_{(k)}}(s,\infty]\frac{ds}{s} - \int_M^\infty \frac{n}{k}\bar{F}(X_{(k)}s)\frac{ds}{s}\right| > \delta\right] = 0.$$

Rewrite the probability as

$$\mathbb{P}\left[\sqrt{k}\left|\int_M^\infty \frac{1}{k}\sum_{i=1}^n (\epsilon_{X_i/X_{(k)}}(s,\infty] - \bar{F}(X_{(k)}s))\frac{ds}{s}\right| > \delta\right]$$
$$\le \mathbb{P}\left[\sqrt{k}\int_M^\infty \left|\frac{1}{k}\sum_{i=1}^n \epsilon_{X_i/X_{(k)}}(s,\infty] - \frac{n}{k}\bar{F}(X_{(k)}s)\right|\frac{ds}{s} > \delta\right].$$

Make the change of variable $u = sX_{(k)}/b(n/k)$ and we get

$$= \mathbb{P}\left[\sqrt{k}\int_{MX_{(k)}/b(n/k)}^\infty \left|\frac{1}{k}\sum_{i=1}^n \epsilon_{X_i/b(n/k)}(u,\infty] - \frac{n}{k}\bar{F}(b(n/k)u)\right|\frac{du}{u} > \delta\right]. \tag{9.20}$$

Pick some small $\eta > 0$ and decompose the probability according to whether $|X_{(k)}/b(n/k) - 1| > \eta$ occurs or not. The former has probability going to zero as $n \to \infty$ from (4.53). So an upper bound for the probability in (9.20) is

$$\leq \mathbb{P}\left[\sqrt{k}\int_{M(1-\eta)}^{\infty}\left|\frac{1}{k}\sum_{i=1}^{n}\epsilon_{X_i/b(n/k)}(u,\infty] - \frac{n}{k}\bar{F}(b(n/k)u)\right|\frac{du}{u} > \delta\right] + o(1).$$

Neglect the $o(1)$ term. Setting $M' = M(1-\eta)$, we get from Chebychev's inequality the upper bound

$$\leq \frac{k}{\delta^2}\mathbf{E}\left(\int_{M'}^{\infty}\left|\frac{1}{k}\sum_{i=1}^{n}[\epsilon_{X_i/b(n/k)}(u,\infty] - \bar{F}(b(n/k)u)]\right|\frac{du}{u}\right)^2.$$

Moving the square inside the integral, we get the further bound

$$\leq \frac{k}{\delta^2}\mathbf{E}\int_{M'}^{\infty}\left(\frac{1}{k}\sum_{i=1}^{n}[\epsilon_{X_i/b(n/k)}(u,\infty] - \bar{F}(b(n/k)u)]\right)^2\frac{du}{u},$$

and moving the expectation inside the integral yields

$$\begin{aligned}
&\leq \frac{k}{\delta^2}\int_{M'}^{\infty}\frac{1}{k^2}n\,\mathrm{Var}(\epsilon_{X_i/b(n/k)}(u,\infty])\frac{du}{u}\\
&\leq \frac{1}{\delta^2}\int_{M'}^{\infty}\frac{n}{k}\bar{F}(b(n/k)u)\frac{du}{u}\\
&\to \frac{1}{\delta^2}\int_{M'}^{\infty}u^{-\alpha-1}du = (\mathrm{const})(M')^{-\alpha} \quad (n\to\infty),
\end{aligned}$$

where we used Karamata's theorem, and as $M \to \infty$, this converges to 0. □

### Removing the random centering

The convergence in (9.10) is a consequence of Theorem 9.1 and requires only that $\bar{F}$ is regularly varying. How do we replace this random centering $\int_{X_{(k)}}^{\infty}\frac{n}{k}\bar{F}(s)\frac{ds}{s}$ with a deterministic centering? Since

$$X_{(k)}/b(n/k) \xrightarrow{P} 1,$$

we hope we can replace the random centering with $\int_{b(n/k)}^{\infty}\frac{n}{k}\bar{F}(s)\frac{ds}{s}$ in (9.10). In order to replace the random centering by this deterministic one, we thus need

$$\sqrt{k}\left(\int_{X_{(k)}}^{\infty}\frac{n}{k}\bar{F}(s)\frac{ds}{s} - \int_{b(n/k)}^{\infty}\frac{n}{k}\bar{F}(s)\frac{ds}{s}\right) \Rightarrow V \tag{9.21}$$

for some limit $V$. We can achieve (9.21) in a variety of ways [100], and we only cite the simplest method [72, 100, 213], which assumes a smoother and slightly stronger form of regular variation, namely the von Mises condition. See (2.26) (p. 31), as well as [90, 105, 260, 292, 293].

**Proposition 9.2.** *Suppose $\bar{F} \in \mathrm{RV}_{-\alpha}$, $\alpha > 0$, and, additionally, the von Mises condition*

$$\lim_{t\to\infty} \alpha(t) := \lim_{t\to\infty} \frac{tF'(t)}{1-F(t)} = \alpha \tag{9.22}$$

*holds, where $F'$ is the density of $F$. Then we have*

$$\sqrt{k}\left(\int_{X_{(k)}}^{\infty} \frac{n}{k}\bar{F}(s)\frac{ds}{s} - \int_{b(n/k)}^{\infty} \frac{n}{k}\bar{F}(s)\frac{ds}{s}\right) \Rightarrow -\frac{1}{\alpha}W(1), \tag{9.23}$$

*and thus*

$$\begin{aligned}\sqrt{k}&\left(H_{k,n} - \int_{b(n/k)}^{\infty} \frac{n}{k}\bar{F}(s)\frac{ds}{s}\right)\\ &\Rightarrow \int_1^{\infty} W(s^{-\alpha})\frac{ds}{s} - \frac{1}{\alpha}W(1) \stackrel{d}{=} \frac{1}{\alpha}W(1),\end{aligned} \tag{9.24}$$

*so that $H_{k,n}$ is asymptotically normal with asymptotic mean $\int_{b(n/k)}^{\infty} \frac{n}{k}\bar{F}(s)s^{-1}ds$ and variance $\gamma^2 = 1/\alpha^2$.*

*Proof.* If we show that the difference in (9.21) converges to $-W(1)/\alpha$, then the fact that (9.14) is jointly convergent with (9.18) will yield the conclusion expressed in terms of the Brownian motion. So we concentrate on showing that (9.22) implies that the difference in (9.21) converges to $-W(1)/\alpha$.

The idea is this: If the von Mises condition (9.22) holds, then we get for the difference

$$\begin{aligned}\sqrt{k}\int_{X_{(k)}}^{b(n/k)} \frac{n}{k}\bar{F}(s)\frac{ds}{s} &\sim \frac{\sqrt{k}}{\alpha}\int_{X_{(k)}}^{b(n/k)} \frac{n}{k}F'(s)ds\\ &= \frac{\sqrt{k}}{\alpha}\left(\frac{n}{k}\bar{F}(X_{(k)}) - \frac{n}{k}\bar{F}(b(n/k))\right)\\ &= \frac{\sqrt{k}}{\alpha}\left(\frac{n}{k}\bar{F}(X_{(k)}) - 1\right)\\ &\Rightarrow -\frac{1}{\alpha}W(1)\end{aligned}$$

from (9.14).

More precisely, proceed as follows: First, observe that

$$[X_{(k)} > b(n/k)] = \left[\frac{n}{k}\bar{F}(X_{(k)}) - 1 \leq 0\right],$$

and recall that $F' \geq 0$. Then we write

$$\begin{aligned}
\sqrt{k}\int_{X_{(k)}}^{b(n/k)} \frac{n}{k}\bar{F}(s)\frac{ds}{s} &= \sqrt{k}\int_{X_{(k)}}^{b(n/k)} \frac{n}{k}\bar{F}(s)\frac{ds}{s} 1_{[X_{(k)} \leq b(n/k)]} \\
&\quad + \sqrt{k}\int_{X_{(k)}}^{b(n/k)} \frac{n}{k}\bar{F}(s)\frac{ds}{s} 1_{[X_{(k)} > b(n/k)]} \\
&\leq \sqrt{k}\left(\frac{n}{k}\bar{F}(X_{(k)} - 1\right) \bigvee_{s \in [X_{(k)}, b(n/k)]} \frac{1}{\alpha(s)} 1_{[X_{(k)} \leq b(n/k)]} \\
&\quad + \sqrt{k}\left(\frac{n}{k}\bar{F}(X_{(k)} - 1\right) \bigwedge_{s \in [b(n/k), X_{(k)}]} \frac{1}{\alpha(s)} 1_{[X_{(k)} > b(n/k)]} \\
&\Rightarrow \frac{-W(1)}{\alpha} 1_{[-W(1) \geq 0]} + \frac{-W(1)}{\alpha} 1_{[-W(1) < 0]} \\
&= \frac{-W(1)}{\alpha}.
\end{aligned}$$

A lower bound can be constructed in the same way.

The equality in distribution assertion in (9.24) is covered by the next lemma. □

**Lemma 9.1.** *The random variable*

$$\int_0^1 W(s)\frac{ds}{s} - W(1)$$

*is* $N(0, 1)$.

*Proof.* The integral is a Gaussian random variable (it is a limit of linear combinations of Gaussian random variables), so we just calculate the variance: We use

$$\mathbf{E}(W(s)W(t)) = s \wedge t.$$

Then

$$\begin{aligned}
&\mathbf{E}\left(\int_0^1 W(s)\frac{ds}{s} - W(1)\right)^2 \\
&= \mathbf{E}\left(\int_0^1 W(s)\frac{ds}{s}\int_0^1 W(u)\frac{du}{u} - 2\int_0^1 W(s)W(1)\frac{ds}{s} + W(1)^2\right)
\end{aligned}$$

$$= 2 \iint_{0 \le s < u \le 1} (s \wedge u) \frac{ds}{s} \frac{du}{u} - 2 \int_0^1 s \frac{ds}{s} + 1$$
$$= 2 \int_{u=0}^1 \left( \int_{s=0}^u s \frac{ds}{s} \right) \frac{du}{u} - 2 + 1 = 1. \qquad \square$$

### Centering by $1/\alpha$

Equation (9.24), giving asymptotic normality for the Hill estimator is difficult to apply. For one thing, the limit depends on the unknown parameter $\alpha$, but even more seriously, the centering depends on $n$ and does not suggest any sort of confidence interval for $\alpha$. To remedy this, the concept of second-order regular variation is typically used. See Problems 3.15–3.17 (p. 67) and [90, 101, 136, 150, 151, 235]. We only present a result that is readily proven.

**Proposition 9.3.** *Suppose that* $\bar{F} \in \mathrm{RV}_{-\alpha}$ *and, additionally, assume that*

$$\lim_{n\to\infty} \sqrt{k} \left( \frac{n}{k} \bar{F}(b(n/k)y) - y^{-\alpha} \right) = 0 \tag{9.25}$$

*locally uniformly in* $(0, \infty]$ *and*

$$\lim_{n\to\infty} \sqrt{k} \int_1^\infty \left( \frac{n}{k} \bar{F}(b(n/k)s) - s^{-\alpha} \right) \frac{ds}{s} = 0. \tag{9.26}$$

*Then*

$$\sqrt{k} \left( H_{k,n} - \frac{1}{\alpha} \right) \Rightarrow \frac{1}{\alpha} \left[ \int_0^1 W(s) \frac{ds}{s} - W(1) \right] \stackrel{d}{=} \frac{1}{\alpha} W(1). \tag{9.27}$$

*Remark* 9.3. Conditions (9.25) and (9.26) can be more elegantly subsumed under a single second-order regular variation condition. However, no matter how phrased, the conditions involve assumptions about detailed tail information in excess of what is likely to be known in practice.

*Proof.* First, condition (9.25) allows one to rephrase the result of Theorem 9.1 as

$$\sqrt{k}(\nu_n(y^{-\gamma}, \infty] - y) \Rightarrow W(y^{-\gamma})$$

in $D[0, \infty)$. Follow the steps using Vervaat's lemma, which led to (4.53), but this time, due to the different centering, we obtain

$$\sqrt{k} \left( \left( \frac{X_{(k)}}{b(n/k)} \right)^{-\alpha} - 1 \right) \Rightarrow -W(1), \tag{9.28}$$

and after an application of the delta method, this becomes

$$\sqrt{k}\left(\frac{X_{(k)}}{b(n/k)} - 1\right) \Rightarrow \frac{1}{\alpha}W(1). \tag{9.29}$$

Now (9.26) implies that

$$\sqrt{k}\left(\int_{b(n/k)}^{\infty} \frac{n}{k}\bar{F}(s)\frac{ds}{s} - \frac{1}{\alpha}\right) \to 0,$$

so if we can prove (9.21) with the right limit, then we will be done since we can then replace the centering in (9.23) with $1/\alpha$, as desired.

The difference in (9.21) is

$$\begin{aligned}\sqrt{k}\int_{X_{(k)}}^{b(n/k)} \frac{n}{k}\bar{F}(s)\frac{ds}{s} &= \sqrt{k}\int_{X_{(k)}/b(n/k)}^{1} \frac{n}{k}\bar{F}(b(n/k)s)\frac{ds}{s} \\ &= \sqrt{k}\frac{n}{k}\bar{F}(b(n/k)s(n))\left(-\log\frac{X_{(k)}}{b(n/k)}\right),\end{aligned}$$

where $s(n)$ is between $X_{(k)}/b(n/k)$ and 1, so that $s(n) \xrightarrow{P} 1$. This implies that $\frac{n}{k}\bar{F}(b(n/k)s(n) \xrightarrow{P} 1$. Furthermore,

$$\begin{aligned}\sqrt{k}\left(-\log\frac{X_{(k)}}{b(n/k)}\right) &= \sqrt{k}\left(-\log\left(1-\left(1-\frac{X_{(k)}}{b(n/k)}\right)\right)\right) \\ &= \sqrt{k}\left(1-\frac{X_{(k)}}{b(n/k)}\right) + o_p(1).\end{aligned}$$

Applying (9.29) gives convergence in distribution to $-\frac{1}{\alpha}W(1)$. □

## Conclusions

Here is a brief summary:

- With just the assumption of regular variation, the Hill estimator requires a random centering to become asymptotically normal.

- With just a bit more than the assumption of regular variation, such as the von Mises condition, the Hill estimator centered by a deterministic function of the sample size becomes asymptotically normal.

- With regular variation and conditions akin to second-order regular variation (here expressed via (9.25) and (9.26)) to control departure of the tail empirical mean measure from a Pareto function,

$$\sqrt{k}\left(H_{k,n} - \frac{1}{\alpha}\right) \Rightarrow N\left(0, \frac{1}{\alpha^2}\right).$$

  Using the delta method, this last result implies that

$$\sqrt{k}(H_{k,n}^{-1} - \alpha) \Rightarrow N(0, \alpha^2).$$

- *Addendum*: When phrased correctly, consistency of the Hill estimator is equivalent to regular variation of the distribution tail, as astutely noted by Mason [212]. Properly phrased, asymptotic normality of Hill's estimator is equivalent to second-order regular variation [146].

## 9.2 Estimation for multivariate heavy-tailed variables

We now consider some aspects of inference in the multivariate heavy-tailed case. Typically, one-dimensional marginal distributions will not be tail equivalent. One can estimate tail indices of the one-dimensional marginal distributions relatively easily, but it is much more difficult to obtain information about the dependence structure.

### 9.2.1 Dependence among extreme events

Given multivariate heavy-tailed data, how do we assess independence? How do we decide if the data come from a model with asymptotic independence or asymptotic dependence? If neither extreme case holds, can we estimate the angular measure and generate useful estimates of probabilities of extreme events or remote failure regions?

One of the ways to assess dependence is with sample (cross-)correlations. In heavy-tailed modeling, there is no guarantee that theoretical moments such as correlations exist, but sample versions will always exist. However, correlation is a somewhat crude summary of dependence that is most informative only between jointly normal variables. It is simple but not subtle. It is a meat cleaver that does not distinguish between the dependence between large values and the dependence between small values. We will seek alternative methods while later inquiring in Section 9.5 (p. 340) if the sample correlation function has useful time-series implications for heavy-tailed time series.

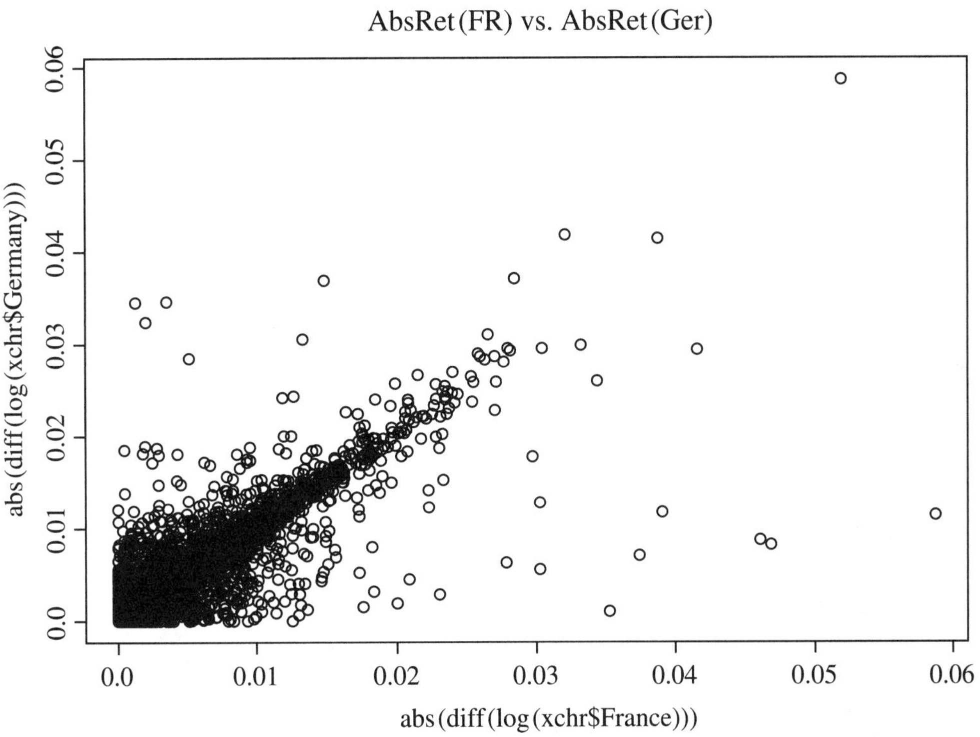


**Fig. 9.1.** Scatter plot of daily absolute log-returns of the French franc against the German deutschmark.

## Example: Modeling of exchange rates

The file fm-exch1.dat included with the program *Xtremes* [238], gives daily spot exchange rates of the currencies of France, Germany, Japan, Switzerland, and the UK against the US dollar over a period of 6041 days from January 1971 to February 1994. For what follows, we emphasize that the reference currency is the US dollar.

Figure 9.1 gives a scatter plot of the daily absolute log-returns for the French franc against the daily absolute log-returns for the German mark. Observe that small absolute log-returns for one currency are matched by a wide range of values for the other currency. However, visually, dependence increases as the size of the absolute log-returns for the pair increases. Even more pronounced effects of this sort are visible for three-hour returns.

The pattern varies, however, between different exchange rate processes. For example, the dependence among large daily absolute log-returns between Japan and Germany is much less pronounced than between France and Germany. Similar patterns hold if daily absolute log-returns are replaced by squared log-returns.

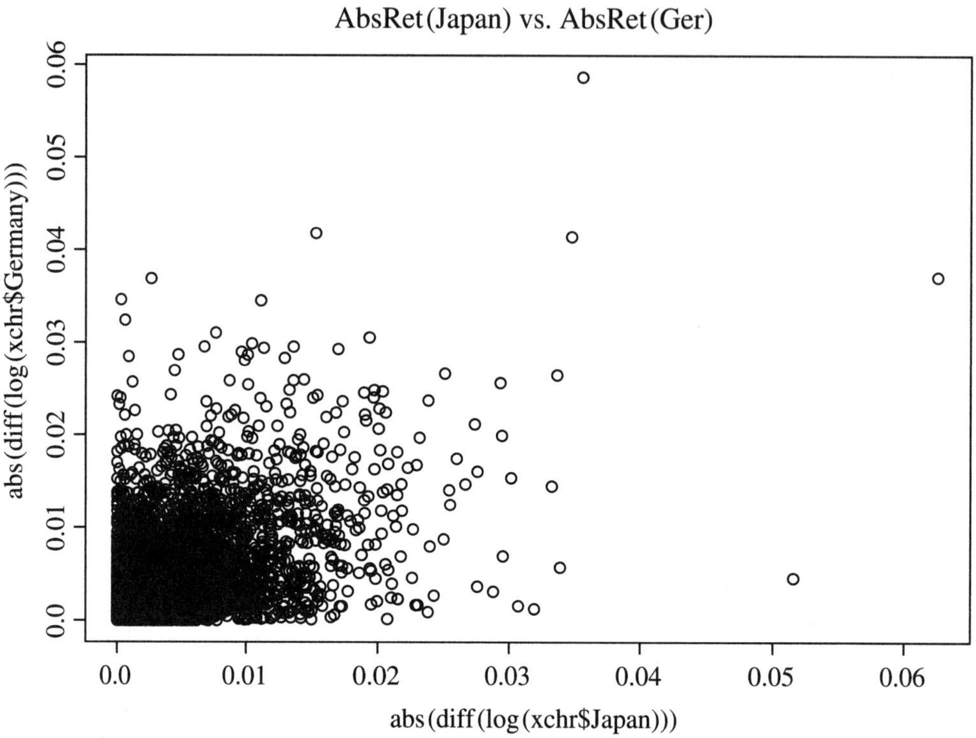


**Fig. 9.2.** Scatter plot of daily absolute log-returns of the Japanese yen against the German mark.

These plots offer more information than the crude numerical summaries, such as correlations. The cross-correlations are summarized in Table 9.1.

| | absresidFr | absresidGer | absresidJap |
|---|---|---|---|
| absresidFr | 1.0000000 | 0.8323291 | 0.4411682 |
| absresidGer | 0.8323291 | 1.0000000 | 0.4553256 |
| absresidJap | 0.4411682 | 0.4553256 | 1.0000000 |

**Table 9.1.** Cross-correlations between the daily absolute log-returns of the French franc, German mark, and Japanese yen.

The crude nature of the correlation summaries is emphasized by the fact that if we compute correlation of only those (franc, mark) return pairs corresponding to an absolute log franc return $< .005$, we get a value of 0.480, as opposed to the correlation of all the pairs of 0.832.

We look at the tails of the squared log-returns for France and Germany individually. The reason for looking at squared log-returns is that log-returns are frequently modeled

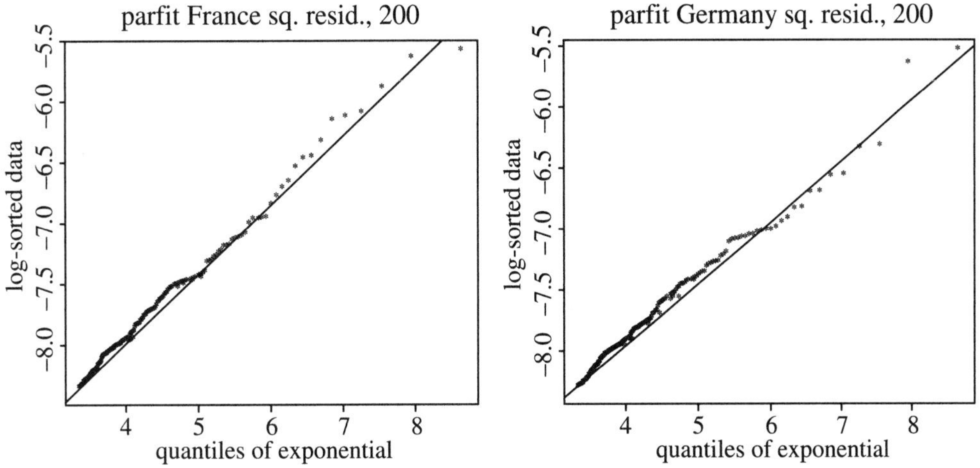


**Fig. 9.3.** QQ plot fitting of $\alpha$ to squared log-returns for French exchange rates (left) and German exchange rates.

as ARCH or GARCH processes of the vector form

$$\boldsymbol{R}_t = \sqrt{\boldsymbol{V}_t}\boldsymbol{\epsilon}_t, \quad t \geq 1,$$

where in the $d$-dimensional case, $\boldsymbol{R}_t = (R_{t1}, \dots, R_{td})$ are the returns and $\boldsymbol{\epsilon}_t = (\epsilon_{t1}, \dots, \epsilon_{td})$ are iid vectors of iid $N(0, 1)$ random variables. Also, $\boldsymbol{V}_t$ is a conditional covariance matrix that is modeled in various ways. Assuming that ARCH or GARCH modeling is justified leads to the theoretical conclusion that tails are asymptotically pure Pareto. See [14, 70, 94, 148, 184].

Based on a combination of QQ plots and Hill plots, we conclude that squared log-returns of France and Germany are each heavy tailed with

$$\alpha_{\text{Germany}^2} = 1.98, \qquad \alpha_{\text{France}^2} = 1.75.$$

The QQ plots are given in Figure 9.3. These values are consistent with usual finance estimates of $\alpha$, which range in the parameter region $(2, 4)$ since if $(\alpha_{\text{Germany}^2}, \alpha_{\text{France}^2}) = (1.98, 1.75)$, then $(\alpha_{\text{Germany}}, \alpha_{\text{France}}) = 2(1.98, 1.75) = (3.96, 3.50)$.

### 9.2.2 Estimation in the standard case

Let $\{\boldsymbol{Z}_j, 1 \leq j \leq n\}$ be a random sample of nonnegative random vectors whose common distribution $F$ is multivariate regularly varying. We assume that each component can be scaled with the same function. From Theorem 6.1 (p. 173), multivariate regular variation is equivalent to the existence of $b(t) \to \infty$ and a Radon measure $\nu$ on $\mathbb{E} = [\boldsymbol{0}, \boldsymbol{\infty}] \setminus \{\boldsymbol{0}\}$

such that $t\mathbb{P}[\frac{\mathbf{Z}_1}{b(t)} \in \cdot] \stackrel{v}{\to} \nu$. From (6.17) of Theorem 6.2 (p. 179), this implies that as $n \to \infty, k \to \infty, k/n \to 0$,

$$\frac{1}{k}\sum_{i=1}^{n} \epsilon_{\mathbf{Z}_i/b(n/k)} \Rightarrow \nu. \tag{9.30}$$

In polar coordinate form, this is (see (6.18) (p. 180))

$$\frac{1}{k}\sum_{i=1}^{n} \epsilon_{(R_i/b(\frac{n}{k}),\boldsymbol{\Theta}_i)} \Rightarrow c\nu_\alpha \times S. \tag{9.31}$$

If we ignore the fact that $b(\cdot)$ is unknown, then the left side of (9.30) could be regarded as a consistent estimator of the limit measure $\nu$. Likewise, (9.31) provides a way to get a consistent estimator of the angular probability measure $S$, since

$$\frac{\sum_{i=1}^{n} \epsilon_{(R_i/b(n/k),\Theta_i)}([1,\infty] \times \cdot)}{\sum_{i=1}^{n} \epsilon_{R_i/b(n/k)}[1,\infty]} \Rightarrow S(\cdot). \tag{9.32}$$

What can we do about the fact that $b(\cdot)$ is unknown? Evaluating (9.31) along the marginal of $R$ by inserting $\aleph_+$, we get

$$\frac{1}{k}\sum_{i=1}^{n} \epsilon_{R_i/b(\frac{n}{k})} \Rightarrow c\nu_\alpha,$$

and the method of Theorem 4.2 (p. 81) gives

$$R_{(k)}/b(n/k) \stackrel{P}{\to} 1,$$

where $R_{(k)}$ is the $k$th largest among $R_1, \dots, R_n$. So a suitable choice of $\hat{b}(n/k)$ would be $R_{(k)}$. This allows us to replace $b(n/k)$ with $\hat{b}(n/k)$ in (9.31) and hence in (9.32), which rids us of the unknown $b(n/k)$. A scaling argument justifies the substitution as, for instance, in (4.22) (p. 83). See [99] for other details.

If we truly knew we were in the standard case, then Theorem 6.5 (p. 204) permits $b(t) = t$, and the problem of estimating $b(\cdot)$ disappears. In practice, where the standard case assumption might be approximately but not exactly true, it usually works better to scale with $R_{(k)}$ rather than $n/k$.

Recall that the ability to scale each component by the same function $b(t)$ in (9.30) means that for $j = 1, \dots, d$,

$$\mathbb{P}[Z_1^{(j)} > x] \sim c^{(j)} x^{-\alpha} L(x),$$

and for those $j$ such that $c^{(j)} > 0$, we have comparable tails. This is a rather special situation. In practice, we suspect that jointly multivariate regularly varying tails will never have the same $\alpha$s, and so we need a strategy for this more realistic case.

### 9.2.3 Estimation in the nonstandard case

In practice, it is unusual to conclude for multivariate heavy-tailed data that the $\alpha$s of each component are the same; recall for the exchange-rate return data that

$$\alpha_{\text{Germany}^2} = 1.98, \qquad \alpha_{\text{France}^2} = 1.75.$$

So suppose the nonstandard regular variation conditions (6.41) and (6.42) of Theorem 6.5 (p. 204) hold. For convenience, here are the conditions again: The global condition (6.42), with $\boldsymbol{b}(t) = (b^{(1)}(t), \dots, b^{(d)}(t))$, is

$$t\mathbb{P}\left[\frac{\boldsymbol{Z}_1}{\boldsymbol{b}(t)} \in \cdot\right] \xrightarrow{v} \nu, \tag{9.33}$$

where we recall that our convention that division of one vector by another means the vector of ratios of corresponding components. The marginal condition (6.41) is stated, assuming we have chosen $b^{(j)}(t)$ so that

$$t\mathbb{P}[Z_1^{(j)} > b^{(j)}(t)x] = x^{-\alpha_j}, \quad x > 0; \quad j = 1, \dots, d. \tag{9.34}$$

How do we deal with this realistic case? There are several possible strategies.

#### Live with diversity

Define

$$\hat{b}^{(j)}(n/k) = Z_{(k)}^{(j)}, \quad j = 1, \dots, d$$

the $k$th largest of the $j$th components in the sample of size $n$. Using (4.17) of Theorem 4.2 (p. 81), we have

$$\frac{\hat{b}^{(j)}(n/k)}{b^{(j)}(n/k)} \xrightarrow{P} 1, \quad j = 1, \dots, d. \tag{9.35}$$

The nonstandard regular variation conditions (9.33) and (9.34), along with (5.16) of Theorem 5.3 (p. 138), give

$$\frac{1}{k}\sum_{i=1}^{n} \epsilon_{\boldsymbol{Z}_i/\boldsymbol{b}(n/k)} \Rightarrow \nu. \tag{9.36}$$

Combining (9.35) and (9.36) and the scaling argument coupled with Lemma 6.1, then gives

$$\frac{1}{k}\sum_{i=1}^{n} \epsilon_{\boldsymbol{Z}_i/\hat{\boldsymbol{b}}(n/k)} \Rightarrow \nu. \tag{9.37}$$

This removes the unknown function $\boldsymbol{b}(\cdot)$ and allows estimation with the left side of (9.37) as the surrogate of $\nu$. However, it does not permit estimation of the angular measure $S$, which first requires standardization.

## Be crude!

Are you ever tempted to give up a night at the opera in favor of reruns of *The Simpsons*? Then this method may be for you, and it works pretty well. A way to make the tails of the marginal distributions roughly the same is to *assume* for $j = 1, \dots, d$ that the $j$th marginal tail is asymptotically equivalent to a Pareto tail with index $\alpha_j$ (so assume there is no annoying slowly varying perturbation) and then take appropriate powers of the component random variables so that the result of this operation is that each transformed component has regular variation index with $\alpha = 1$. The idea is that if $X$ is a random variable whose tail is asymptotically Pareto, that is,

$$\mathbb{P}[X > x] \sim x^{-\alpha}, \quad x \to \infty,$$

then

$$\mathbb{P}[X^{\alpha} > x] = \mathbb{P}[X > x^{1/\alpha}] \sim x^{-1}, \quad x \to \infty.$$

So the assumption that slowly varying functions are absent from the marginal tail expressions allows us to hope that

$$\frac{1}{k}\sum_{i=1}^{n} \epsilon_{((Z_i^{(j)}/\hat{b}^{(j)}(n/k))^{\alpha_i};\, j=1,\dots,d)} \tag{9.38}$$

is estimating a standard limit measure $\nu_*$ (see (6.44) (p. 205)), and this would give us a method to estimate the angular measure $S$ associated with $\nu_*$. In practice, $\boldsymbol{\alpha} = (\alpha_1, \dots, \alpha_d)$ is replaced by estimates $\hat{\boldsymbol{\alpha}}$, obtained from the one-dimensional marginal data. This has the potential to introduce significant errors and is a drawback.

The form of (9.38) is not the obvious one, which would use the points $\{((Z_i^{(j)})^{\alpha_i}/(n/k);\, j = 1, \dots, d);\, 1 \le i \le n\}$. However, the form given in (9.38) seems more robust to departures from the assumptions of approximate Pareto tails and works better in practice where marginal $\alpha$s need to be estimated. Typically, the division by the order statistics $\hat{b}^{(j)}(n/k)$ does a better job of scaling the sample.

A strong defense of this method is that it is not really different from the multivariate peaks-over-threshold philosophy (Section 6.3 (p. 183)), which assumes that all large observations are distributed by the limit distribution. In the heavy-tailed case, this philosophy forces us to believe that marginally, each component has a Pareto tail from some point thereafter, which is exactly what the method of this section assumes. So maybe this method, while mathematically simple, is not so crude after all!

## The ranks method

A method based on ranks (see [38, 89, 161, 167, 268]) overcomes some of the drawbacks of the previous multivariate methods. The ranks method does not require estimation of the marginal $\alpha$s, yet it achieves transformation to the standard case, allowing estimation of the angular measure. It is simple to program. The drawback is that the transformation destroys the iid property of the sample, and asymptotic analysis is sophisticated.

Continue to suppose the global and marginal conditions (9.33), (9.34), so that

$$\nu_n(\cdot) = \frac{1}{k}\sum_{i=1}^{n} \epsilon_{\left(\frac{Z_i^{(1)}}{b^{(1)}(n/k)},\ldots,\frac{Z_i^{(d)}}{b^{(d)}(n/k)}\right)}(\cdot) \Rightarrow \nu(\cdot) \tag{9.39}$$

in $M_+(\mathbb{E})$, where $\mathbb{E} = [\mathbf{0}, \infty] \setminus \{\mathbf{0}\}$. From marginal regular variation (9.34), we know from (4.18) (p. 82) that for each $j$,

$$\frac{Z^{(j)}_{(\lceil kt^{(j)}\rceil)}}{b^{(j)}(n/k)} \xrightarrow{P} (t^{(j)})^{-1/\alpha_j}$$

in $D(0, \infty]$. Because convergence is to a constant limit, we may append this to (9.39) to get

$$\left(\nu_n, \left(\frac{Z^{(j)}_{(\lceil kt^{(j)}\rceil)}}{b^{(j)}(n/k)}; j = 1, \ldots, d\right)\right) \Rightarrow (\nu, ((t^{(j)})^{-1/\alpha_j}; j = 1, \ldots, d)) \tag{9.40}$$

in $M_+(\mathbb{E}) \times D(0, \infty] \times \cdots \times D(0, \infty]$.

Recall from (6.46) (p. 205) that

$$\nu([\mathbf{0}, \boldsymbol{x}^{1/\boldsymbol{\alpha}}]^c) =: \nu_*([\mathbf{0}, \boldsymbol{x}]^c) \tag{9.41}$$

is standard with $\nu_*(t\cdot) = t^{-1}\nu_*(\cdot)$. Assume that $\boldsymbol{t}$ is a continuity point of $\nu([\mathbf{0}, \cdot]^c)$, and apply the continuous map

$$(\nu, \boldsymbol{t}) \mapsto \nu([\mathbf{0}, \boldsymbol{t}]^c)$$

to (9.40) to get

$$\nu_n\left(\left[\mathbf{0}, \left(\frac{Z^{(j)}_{(\lceil kt^{(j)}\rceil)}}{b^{(j)}(n/k)}; j = 1, \ldots, d\right)\right]^c\right) \Rightarrow \nu([\mathbf{0}, \boldsymbol{t}^{-1/\boldsymbol{\alpha}}]^c). \tag{9.42}$$

Unpack the left side of (9.42). We have

$$\nu_n\left(\left[\mathbf{0}, \left(\frac{Z^{(j)}_{(\lceil kt^{(j)}\rceil)}}{b^{(j)}(n/k)}; j = 1, \ldots, d\right)\right]^c\right)$$

$$= \frac{1}{k}\sum_{i=1}^{n} 1_{\left[\frac{Z_i^{(j)}}{b^{(j)}(n/k)} \le \frac{Z^{(j)}_{(\lceil kt^{(j)}\rceil)}}{b^{(j)}(n/k)};\, j=1,\dots,d\right]^c} = \frac{1}{k}\sum_{i=1}^{n} 1_{\left[Z_i^{(j)} \le Z^{(j)}_{(\lceil kt^{(j)}\rceil)};\, j=1,\dots,d\right]^c}. \tag{9.43}$$

The indicated set on the right side of (9.43) says that for each $j$, the number of $Z_l^{(j)}$s, for $1 \le l \le n$, that are at least $Z_i^{(j)}$ is at least $kt^{(j)}$. For each fixed $j$, define the rank (some traditions would vote, with justification, for the name antirank) of $Z_i^{(j)}$,

$$r_i^{(j)} = \sum_{l=1}^{n} 1_{\left[Z_l^{(j)} \ge Z_i^{(j)}\right]}, \quad j = 1, \dots, d, \tag{9.44}$$

as the number of $j$th components bigger than $Z_i^{(j)}$. Rephrase (9.43) as

$$\frac{1}{k}\sum_{i=1}^{n} 1_{\left[r_i^{(j)} \ge kt^{(j)};\, j=1,\dots,d\right]^c}.$$

Change variables

$$\mathbf{s} \mapsto \mathbf{t}^{-1}$$

to get from (9.43)

$$\frac{1}{k}\sum_{i=1}^{n} 1_{\left[r_i^{(j)} \ge k(s^{(j)})^{-1};\, j=1,\dots,d\right]^c} \Rightarrow \nu([\mathbf{0}, \mathbf{s}^{1/\alpha}]^c)$$

or

$$\frac{1}{k}\sum_{i=1}^{n} 1_{\left[\frac{k}{r_i^{(j)}} \le (s^{(j)});\, j=1,\dots,d\right]^c} \Rightarrow \nu([\mathbf{0}, \mathbf{s}^{1/\alpha}]^c)$$

or, applying Lemma 6.1 (p. 174),

$$\frac{1}{k}\sum_{i=1}^{n} \epsilon_{\left(\frac{k}{r_i^{(j)}};\, j=1,\dots,d\right)} \Rightarrow \nu_* \tag{9.45}$$

in $M_+(\mathbb{E})$.

We summarize.

**Proposition 9.4.** *Suppose* $\mathbf{Z}_1, \dots, \mathbf{Z}_n$ *are an iid sample whose common distribution satisfies the marginal and global regular variation conditions* (9.33) *and* (9.34). *Then* (9.45) *holds with* $\nu_*$ *being standard and satisfying* (9.41).

The vectors of ranks are not independent, and this makes obtaining asymptotic distributions more difficult. See [127].

### Estimation of the angular measure

Once we have transformed to the standard case, we can use the methods of Section 9.2.2. For estimating the angular measure $S$ using the ranks method, we proceed as follows:

- Transform the data using the rank transform (9.44):

$$(\mathbf{Z}_1, \dots, \mathbf{Z}_n) \mapsto \{(r_i^{(j)}; j = 1, \dots, d); 1 \le i \le n\} \\ = \{\boldsymbol{r}_i, 1 \le i \le n\}. \tag{9.46}$$

- Apply the polar coordinate transformation

$$\text{POLAR}\left(\frac{k}{r_i^{(j)}}; j = 1, \dots, d\right) = (R_{i,k}, \boldsymbol{\Theta}_{i,k}),$$

  and then from (9.45), we get

$$\frac{1}{k}\sum_{i=1}^{n} \epsilon_{(R_{i,k}, \boldsymbol{\Theta}_{i,k})} \Rightarrow c\nu_1 \times S.$$

- Consequently,

$$\frac{1}{k}\sum_{i=1}^{n} \epsilon_{(R_{i,k}, \boldsymbol{\Theta}_{i,k})}((1, \infty] \times \cdot) \Rightarrow cS,$$

  and we estimate $S$ with

$$\hat{S}_n(\cdot) = \frac{\sum_{i=1}^{n} \epsilon_{(R_{i,k}, \boldsymbol{\Theta}_{i,k})}((1, \infty] \times \cdot)}{\sum_{i=1}^{n} \epsilon_{R_{i,k}}((1, \infty])} \Rightarrow S. \tag{9.47}$$

The interpretation of (9.32) and (9.47) is that the empirical probability measure of those $\boldsymbol{\Theta}$s whose radius vector is greater than 1 approximates $S$. Apart from normalization of the plot, if we consider the points

$$\{\boldsymbol{\Theta}_{i,k} : R_{i,k} > 1\}$$

and make a density plot, we should get an estimate of the density of $S(\cdot)$. A notable mode in the density at $\pi/4$ reveals a tendency toward dependence. Modes at 0 and $\pi/2$ show a tendency toward independence, or at least asymptotic independence. Of course, since we do not know that $S$ has a density, we could proceed by making plots for the distribution function $S[0, \theta]$, $0 \le \theta \le \pi/2$. However, often density estimate plots are striking and show qualitative behavior effectively.

### 9.2.4 How to choose $k$; the Stărică plot

A famous Yiddish haiku poet once wrote

> Oy Vay!
> How do we choose $k$?

So far in this chapter, we have ignored the problem of threshold selection or, equivalently, the criteria for choosing the $k$th largest radius vector of the polar transformed data, which serves as the threshold. The Achilles heel of many heavy-tailed methods is the choice of $k$. A scaling method suggested by C. Stărică [286, 287] can help with the selection of $k$. The method does not have a strong theoretical base, but it seems useful. It is readily programmed, and using either visual inspection or automation, it can produce a $k$ value as input to another program.

Imagine that we have transformed to the standard case by one of the methods outlined in Section 9.2.3. Suppose $\nu_*$ is a standard limit measure with the scaling property (6.44) (p. 205). Suppose $\hat{\nu}_{*,n}$ is the estimator of $\nu_*$. This estimator is dependent on $k$. We pick $k$ so that $\nu_{*,n}$ mimics the scaling. We try to use the set

$$\aleph^{>} = \{\boldsymbol{x} \in \mathbb{E} : \|\boldsymbol{x}\| \geq 1\},$$

thinking that such a set encompasses information from all directions. So for a fixed $k$, we graph

$$\left\{ \frac{u\hat{\nu}_{*,n}(u\aleph^{>})}{\hat{\nu}_{*,n}(\aleph^{>})}, u \geq 0.1 \right\}, \tag{9.48}$$

which we call the Stărică plot. The idea [286] is the that ratio should be roughly constant and equal to 1 for $u$ in a neighborhood of 1 if $k$ is chosen wisely so that $\hat{\nu}_{*,n}$ is close to $\nu_*$. If $u$ is too small, we are using too many small observations from the center of the distribution, which are not likely to carry accurate information about the tail.

In practice, we make the Stărică plot (9.48) for various values of $k$ and choose the $k$ that seems to have the plot most closely hugging the horizontal line at height 1. Alternatively, one could do a search procedure through various values of $k$ to find the optimal choice in the sense of minimizing distance of the ratio from 1 or maximizing the occupation time of the ratio curve in a neighborhood of 1.

To illustrate this method, suppose we estimate $\nu_*$ with

$$\hat{\nu}_{*,n} := \frac{1}{k} \sum_{i=1}^{n} \epsilon_{\boldsymbol{Z}_i/\hat{b}(n/k)}.$$

Make the Stărică plot as follows. Evaluate (9.48):

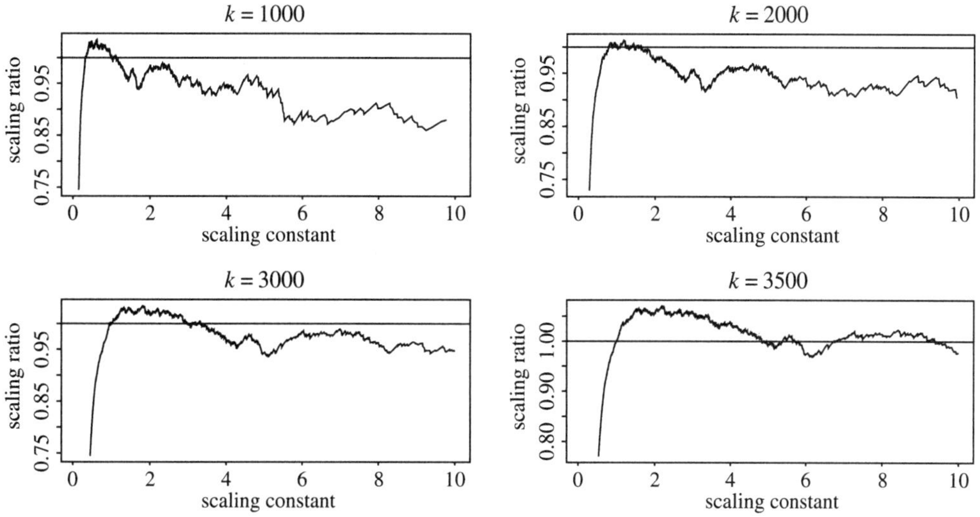


**Fig. 9.4.** Four views of the Stărică plot for 5000 simulated Pareto independent pairs, each with $\alpha = 1$.

$$\frac{u\hat{\nu}_{*,n}(u\aleph^{>})}{\hat{\nu}_{*,n}(\aleph^{>})} = \frac{u\frac{1}{k}\sum_{i=1}^{n}\epsilon_{\mathbf{Z}_i/\hat{b}(n/k)}(u\aleph^{>})}{\frac{1}{k}\sum_{i=1}^{n}\epsilon_{\mathbf{Z}_i/\hat{b}(n/k)}(\aleph^{>})},$$

where $\hat{b}(n/k) = R_{(k)}$, which is the $k$th largest value in the one-dimensional set $\{\|\mathbf{Z}_i\|, 1 \leq i \leq n\}$. Since the sum in the denominator of the ratio is $k$, if we set $u = R_{(j)}/\hat{b}(n/k)$, we get for the ratio $\frac{R_{(j)}}{R_{(k)}} \cdot \frac{j}{k}$. So, to get the Stărică plot, we graph

$$\left\{\left(\frac{R_{(j)}}{R_{(k)}}, \frac{R_{(j)}}{R_{(k)}} \cdot \frac{j}{k}\right), 1 \leq j \leq n\right\}. \tag{9.49}$$

But look closely at values of the abscissa in a neighborhood of 1. The plots will look different for various values of $k$, and either some experimentation can be done to visually choose a good value of $k$ or the procedure can be automated by searching for $k$ which optimizes a distance measure.

Figure 9.4 gives four views of a Stărică plot for 5000 independent pairs of iid observations simulated from a Pareto distribution with $\alpha = 1$ and using $k = 1000, 2000, 3000, 3500$. The choices $k = 1000, 2000$ are inferior, and $k = 3000$ seems to be a good choice. Note that the scaling plot is for one-dimensional data as the method reduces the problem to one dimension by focusing on radius vectors.

If we use the ranks method and approximate $\nu_*$ by the left side of (9.45) (p. 312), then we choose a $k$, compute $R_{i,k}$ with order statistics

$$R_{(1),k} \geq R_{(2),k} \geq \cdots \geq R_{(n),k},$$

and then plot

$$\left\{\left(R_{(j),k}, \frac{R_{(j),k}j}{\sum_{i=1}^{n} 1_{[R_{i,k}\geq 1]}}\right), j = 1, \ldots, n\right\}.$$

## 9.3 Examples

Here we offer some examples illustrating strong qualitative differences in the properties of our data. Recall that when viewing the density plots of the angular measure, a notable mode in the density at $\pi/4$ reveals a tendency toward dependence, while modes at 0 and $\pi/2$ show a tendency toward independence, or at least asymptotic independence. (Review Section 6.5.1 (p. 192).) Additional analyses of the type presented here are in [64].

### 9.3.1 Internet data

This section continues the discussion begun in Sections 1.3.1 (p. 3) (especially Example 1.1) and 7.3.2 (p. 238).

Internet file transfers are subject to delays, and although one expects larger file transfers to encounter more delays, this is overly simplistic [38]. Large file transfers, while comparatively rare, comprise a significant fraction of all the bytes transferred on the Internet and hence are important for understanding the impact of diverse networking technologies such as routing, congestion control, and server design on end-user performance measures. For HTTP (web browsing) responses, the joint behavior of large values of three variables—size of response (abbreviated $F$), time duration of response ($L$), and throughput (syonym: rate = size/time and abbreviated $R$)—can be considered. All three quantities are typically heavy tailed, but for some data sets, for example, the BU data, rate and duration tend to be asymptotically independent, and for others size and rate tend to be asymptotically independent. See also [215, 267]. For applied probability models of network inputs, the assumptions made about the dependence structure of $(F, L, R)$ will dramatically affect model properties [68].

#### Boston University data

In Section 7.3.2 (p. 238), we studied trivariate data $(F, L, R)$ of download file sizes, download durations, and download rates obtained from a measurement study at Boston University [51, 53, 54, 63]. The evidence presented in Section 7.3.2 supported some form of asymptotic independence for the pair $(L, R)$. We see this supported by the estimate of the angular measure after the rank transform takes the two heavy-tailed variables to the standard case. Figure 9.5 shows an estimate for the density of $S$ with

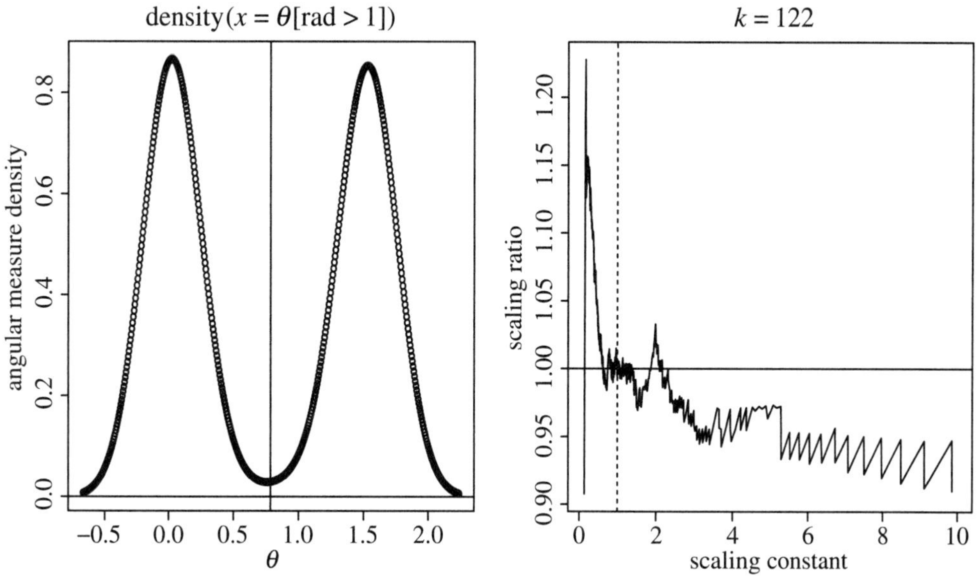


**Fig. 9.5.** BU: Estimate of the angular density (left) and Stărică plot (right) for the $k$ chosen by optimization.

most mass concentrated near 0 and $\pi/2 \approx 1.57$. The right plot shows the Stărică plot for the value of $k = 122$ chosen by minimizing a distance measure for the scaling plot as a function of $k$.

### Internet HTTP response data

This data set consists of HTTP response data describing Internet transmissions observed during a four-hour period from 1:00pm–5:00pm on April 26, 2001 at the University of North Carolina. The data set consists of responses (bytes) in the stated time period whose size is in excess of 100,000 bytes coupled with the time required for transmission (seconds). There were 21,829 such transmissions. The data sets were obtained from the University of North Carolina Computer Science Distributed and Real-Time Systems Group, which was then under the direction of Don Smith and Kevin Jeffay. As mentioned in Section 7.3.2 (p. 238), this data set offers contrasting behavior to what was observed in the previous discussion of the Boston University data.

To confirm the heavy-tailed nature of the marginal distributions, we estimated the marginal $\alpha$s for the transmission rate and size variables. A combination of Hill, altHill, and QQ plots were used and we chose values of $k = 150, 250$ for the size and rate variables, respectively. Estimates of $\alpha$ were relatively stable around the values of

$$\alpha_R = 1.8 \quad \text{and} \quad \alpha_F = 2.1$$

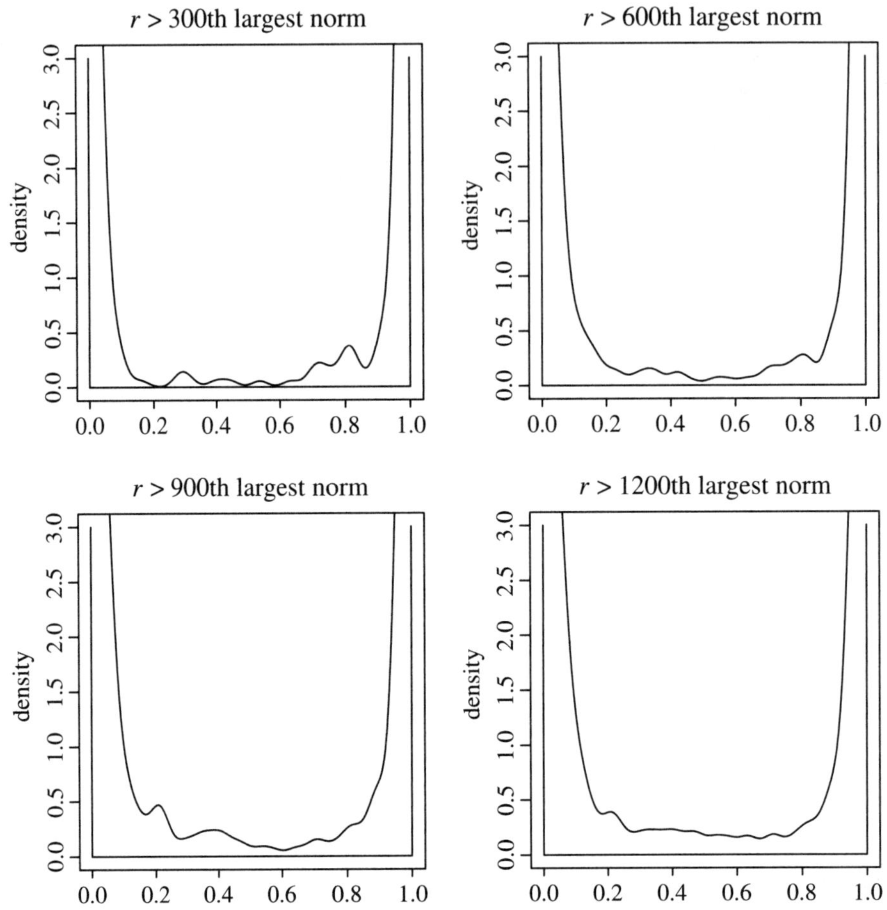


**Fig. 9.6.** UNC: Estimates corresponding to different thresholds of the angular measure density for the Internet transmission size and rate data. The horizontal axis represents angle normalized to [0, 1].

in the ranges of $k$ in $[50, 3000]$ and $[50, 400]$, respectively. Figure 9.6 presents the estimates of the density of the angular measure $S$ using the rank-transform method for various choices of $k$. The plots are rather stable against the choice of $k$, and each indicates a piling of mass at the extreme points.

### 9.3.2 Exchange rates

Figure 9.7 shows the angular density estimate for the German and French exchange rate returns relative to the US dollar using ranks. Based on the Stărică plot applied to the points in (9.45), we used $k = 4500$. Note the mode at approximately $\pi/4$, which

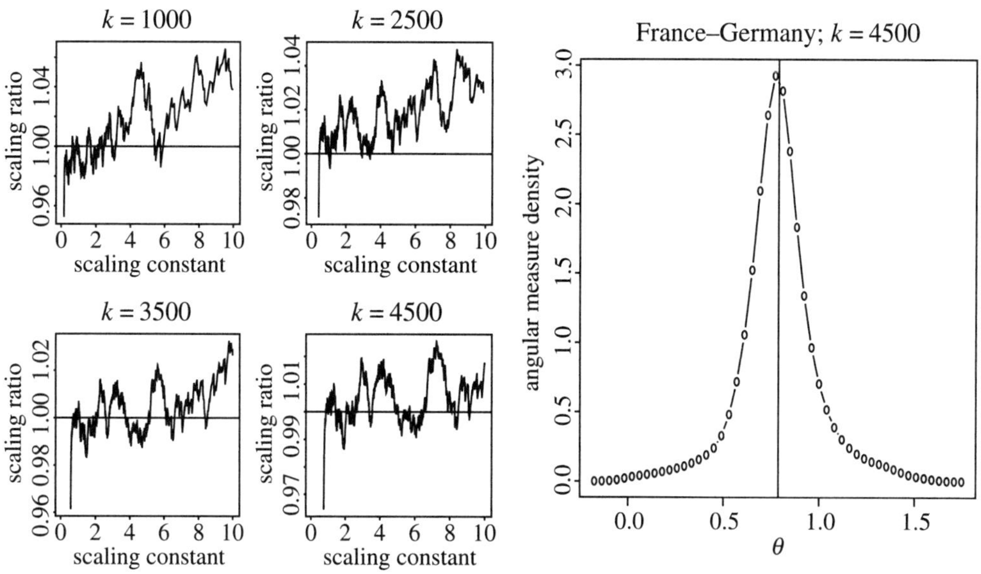


**Fig. 9.7.** Returns from the German mark vs. French franc relative to the US dollar: Stărică plot (left) and angular density estimation.

is indicated by the vertical line. This bears out the promise of the plot in Figure 9.1, where apparently large values seemed to be highly dependent.

Contrast this with the returns from the German mark jointly with those of the Japanese yen. Large values are much less dependent and, in fact, appear to exhibit asymptotic independence. Figure 9.8 shows the Stărică plot that helps in choosing $k = 1000$, and on the right there is the density estimate that shows a clear tendency towards having two modes at 0 and $\pi/2$.

### 9.3.3 Insurance

An insurance company keeps records of auto and fire claims per windstorm over the period January 1, 1990–December 31, 2000; the data set consists of 736 records. The left side of Figure 9.9 shows a scatter plot of the (auto, fire) claim data. The range of the data is quite broad, as shown in Table 9.2, so it is not surprising that the scatter plot is relatively uninformative. The right side of Figure 9.9 shows the scatter plot after a logarithmic transform of each component.

*Marginal analysis*: The marginal tails are rather heavy for these data, and a combination of QQ and Hill plotting gives the estimates

$$\alpha_{\text{AUTO}} = .92 \quad (k = 200), \qquad \alpha_{\text{FIRE}} = .70 \quad (k = 100),$$

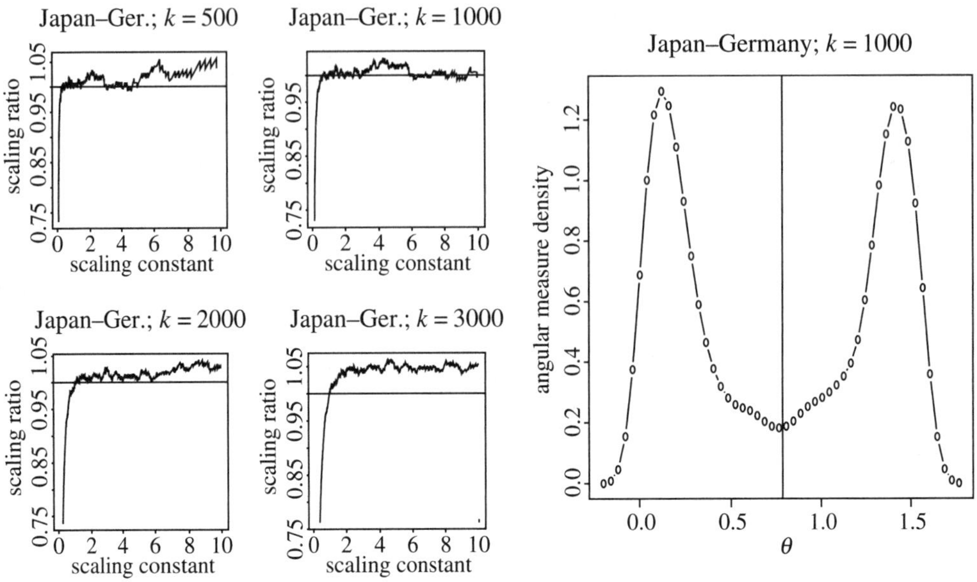


**Fig. 9.8.** Returns from the German mark vs. Japanese yen: Stărică plot (left) and angular density estimation.

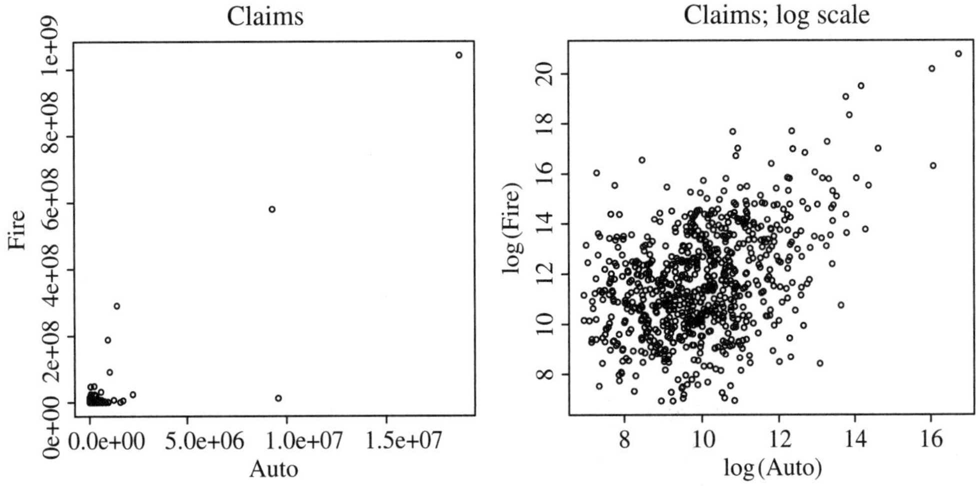


**Fig. 9.9.** Insurance data: Scatter plot of the auto-fire claim data in (left) natural scale and (right) logarithmic scale.

with the accompanying Hill plots in Figure 9.10.

The estimate of the angular density given in Figure 9.11 indicates a bimodal density with modes in the interior of $(0, \pi/2)$. The asymptotic dependence presumably reflects the fact that both auto and fire claims are affected by the severity of the windstorm.

| | Min. | First quarter | Median | Mean | Third quarter | Max. |
|---|---|---|---|---|---|---|
| Auto | 1052 | 7685 | 19910 | 121500 | 51310 | 18740000 |
| Fire | 1.022e+03 | 2.549e+04 | 9.780e+04 | 3.869e+06 | 4.976e+05 | 1.041e+09 |

**Table 9.2.** Summary statistics for each marginal data set.

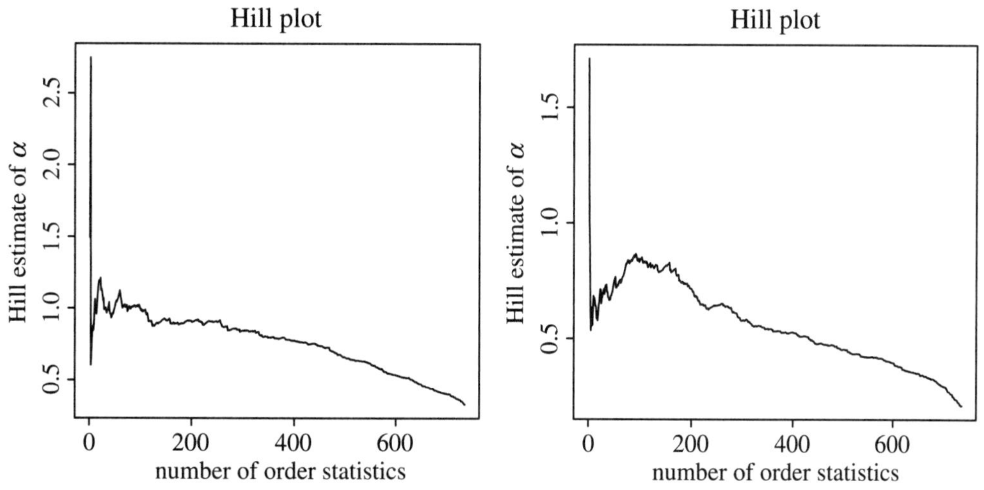


**Fig. 9.10.** Hillplots of the auto (left) and fire (right) claims data.

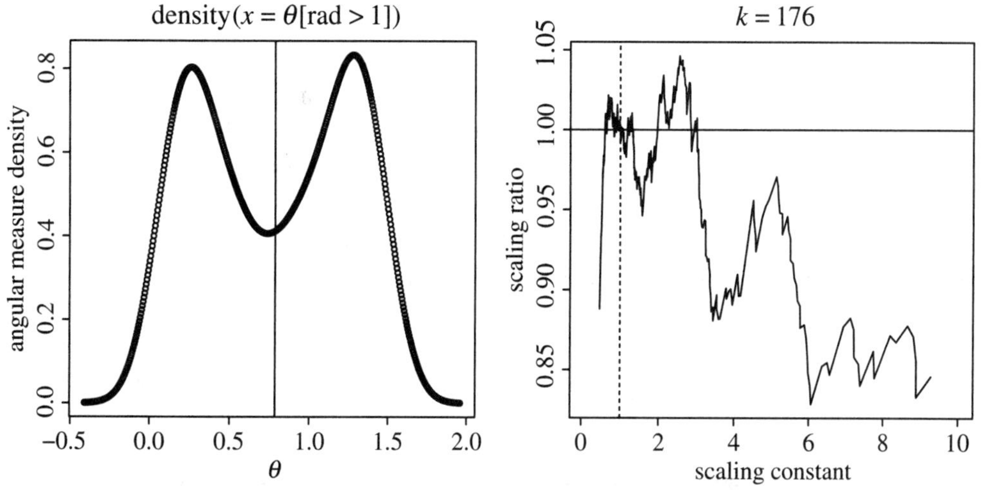


**Fig. 9.11.** Insurance data: Estimate of the angular density (left) after the rank transform and Stărică plot (right) corresponding to $k = 176$.

## 9.4 The coefficient of tail dependence and hidden regular variation

This section builds on information about asymptotic independence contained in Section 6.5.1 (p. 192) and Problems 6.9–6.12 (p. 209).

By now, it should be clear that semiparametric statistical analysis of the tails of multivariate heavy-tailed probability distributions relies on asymptotic regular variation arguments, which force us into the distributional tails. Regular variation methods provide a large class of models that in standard form is indexed by the set of probability measures on the unit sphere. However, the methods have difficulty in distinguishing between asymptotic independence and exact independence and, in the presence of either, may fail to provide satisfactory estimates of probabilities of remote critical sets, such as failure regions (reliability), overflow regions (networks or hydrology), out-of-compliance regions (environmental protection).

To illustrate the problem, consider the following scenario, which requires estimation of the probability of noncompliance.

*Example* 9.1 (*estimate the probability of noncompliance*). Supppose that the vector $\boldsymbol{Z} = (Z^{(1)}, \dots, Z^{(d)})$ represents concentrations of a specific pollutant at $d$ locations, or $\boldsymbol{Z}$ could represent concentations of different pollutants at a single site. Environmental agencies set standards by requiring that critical levels $\boldsymbol{t}_0 = (t_0^{(1)}, \dots, t_0^{(d)})$ not be exceeded at each of the $d$ sites, so that noncompliance is represented by the event

$$[\text{noncompliance}] = [\boldsymbol{Z} \le \boldsymbol{t}_0]^c = \bigcup_{j=1}^{d} [Z^{(j)} > t_0^{(j)}].$$

Noncompliance results in a fine or withdrawal of government support; it has various economic and political implications, none of which is desirable. How do we estimate the probability of noncompliance? Suppose, for simplicity, that $d = 2$.

Assume only that the distribution of $\boldsymbol{Z}$ satisfies (perhaps nonstandard) regular variation with limit measure $\nu$ and that asymptotic independence is present. For estimation, we would proceed as follows: The probability of noncompliance is

$$\mathbb{P}\left\{\bigcup_{j=1}^{2} [Z^{(j)} > t_0^{(j)}]\right\} = \sum_{j=1}^{2} P[Z^{(j)} > t_0^{(j)}] - \mathbb{P}[Z^{(1)} > t_0^{(1)}, Z^{(2)} > t_0^{(2)}]. \quad (9.50)$$

Assuming only asymptotic independence is present, one would be inclined to neglect the joint probability on the right since it is negligible compared with the univariate probabilities. However, just neglecting the joint probability seems rather crude, and one would prefer a refinement that allows estimation of what to subtract. □

In response to this weakness, Ledford and Tawn [198–201] developed theory and models offering a richer description of asymptotic independence behavior, which led to the concept of the *coefficient of tail dependence*. See also [49]. These ideas were repackaged into a construct consistent with regular variation in [161, 217, 266].

### 9.4.1 Hidden regular variation

Recall that in Section 6.1.1, we emphasized that the definition of regular variation for functions was most naturally formulated on cones. Although most applications use the cone $[\mathbf{0}, \infty] \setminus \{\mathbf{0}\}$, which was the setting of Section 6.1.4, we have had occasion to use the cone $[-\infty, \infty] \setminus \{\mathbf{0}\}$. Now we make use of other cones.

Suppose, as usual, that $\boldsymbol{Z}$ is a $d$-dimensional random vector satisfying the regular variation condition given in one of its equivalent forms in Theorem 6.1 on $\mathbb{E} = [\mathbf{0}, \infty] \setminus \{\mathbf{0}\}$. So $\boldsymbol{Z}$ satisfies

$$t\mathbb{P}\left[\frac{\boldsymbol{Z}}{b(t)} \in \cdot\right] \xrightarrow{v} \nu, \tag{9.51}$$

where the scaling function $b(t) \uparrow \infty$, with $b(\cdot) \in RV_{1/\alpha}$, and the nonnegative, nonzero limit measure $\nu$ has the scaling property

$$\nu(c\cdot) = c^{-\alpha}\nu(\cdot), \quad c > 0, \tag{9.52}$$

on $\mathbb{E}$.

Hidden regular variation allows for another regular variation property on a subcone. Although not the only choice, the most useful subcone is $\mathbb{E}^0$ defined as

$$\mathbb{E}^0 := \{\boldsymbol{s} \in \mathbb{E} : \text{for some } 1 \le i < j \le d, s^{(i)} \wedge s^{(j)} > 0\}, \tag{9.53}$$

that is, points of $\mathbb{E}$ such that at most $d - 2$ coordinates are 0. For $i = 1, \dots, d$, define the basis vectors

$$\boldsymbol{e}_i = (0, \dots, 0, 1, 0, \dots, 0),$$

so that the axes originating at $\mathbf{0}$ are $\mathbb{L}_i := \{t\boldsymbol{e}_i, t > 0\}$, $i = 1, \dots, d$. Then

$$\mathbb{E}^0 = \mathbb{E} \setminus \bigcup_{i=1}^{d} \mathbb{L}_i.$$

If $d = 2$, we have $\mathbb{E}^0 = (0, \infty]^2$. Since $\mathbb{E}^0 \subset \mathbb{E}$, the compact subsets of $\mathbb{E}^0$ are specified by Proposition 6.1 (p. 171).

## Definition of hidden regular variation

The random vector $\boldsymbol{Z}$ has a distribution possessing *hidden regular variation* on $\mathbb{E}^0$ if, in addition to (9.51), there exists a nondecreasing function $b^0(t) \uparrow \infty$ such that $b(t)/b^0(t) \to \infty$, and there exists a measure $\nu^0 \neq 0$ that is Radon on $\mathbb{E}^0$ and such that

$$t\mathbb{P}\left[\frac{\boldsymbol{Z}}{b^0(t)} \in \cdot\right] \xrightarrow{v} \nu^0 \tag{9.54}$$

on the cone $\mathbb{E}^0$. Then there exists $\alpha^0 \geq \alpha$ such that $b^0 \in RV_{1/\alpha^0}$ and $\nu^0$ and $\alpha^0$ satisfy the analogue of (9.52) on $\mathbb{E}^0$.

The definition can be reformulated in terms of polar coordinates. The reformulation of (9.51) is (6.8) (p. 173), where $S$, the angular measure, is a probability measure on $\aleph_+ = \mathbb{E} \cap [\boldsymbol{0}, \infty)$. In polar form, (9.54) becomes

$$t\mathbb{P}\left[\left(\frac{R}{b^0(t)}, \boldsymbol{\Theta}\right) \in \cdot\right] \xrightarrow{v} \nu_{\alpha^0} \times S^0 \tag{9.55}$$

in $M_+((0, \infty] \times \aleph^0)$ and

$$\aleph^0 := \aleph \cap \mathbb{E}^0.$$

An important point is that $\aleph^0$ is not compact and therefore $S^0$, the angular measure, is not necessarily finite. We can only guarantee that it is Radon.

## Topology is destiny

The reason for the name *hidden* regular variation is that the relatively crude normalization necessary for convergence on the axes is too large to get things correct in the interior of the cone and obliterates or hides the fine structure that may be present in the interior. A normalization of smaller order, namely $b^0(\cdot)$, is necessary on the cone $\mathbb{E}^0$. The definition of compactness on $\mathbb{E}^0$ makes $[Z^{(i)} > b^0(t), Z^{(j)} > b^0(t)]$ a typical relatively compact set, and because at least two conditions are required, probabilities tend to be relatively small; hence $b^0(t)$ should be relatively small. By contrast, the topology on $\mathbb{E}$ makes $[Z^{(1)} > b(t)x]$ relatively compact without a second condition. So $b(t)$ must be relatively big to accomodate the relatively big marginal probability of the event $[Z^{(1)} > b(t)x]$.

Note that if both (9.51) and (9.54) hold, then for any $1 \leq i < j \leq d$ and $\delta > 0$, we have

$$t\mathbb{P}[Z^{(i)} > b(t)\delta, Z^{(j)} > b(t)\delta] = t\mathbb{P}\left[\frac{Z^{(i)} \wedge Z^{(j)}}{b^0(t)} > \frac{b(t)}{b^0(t)}\delta\right] \to 0$$

as $t \to \infty$, since $b(t)/b^0(t) \to \infty$. This means that $\nu$ concentrates on the axes $\bigcup_{i=1}^d \mathbb{L}_i$ originating at $\mathbf{0}$ and $\nu(\mathbb{E}^0) = 0$, which is exactly the condition for *asymptotic independence*.

*Property* 9.1. Thus if the distribution of $\boldsymbol{Z}$ possesses hidden regular variation, then asymptotic independence is present as well.

*Remark* 9.4. Here are some comments on the definition:

- The normalization in (9.54) of each component of the random vector is by the same function. One could allow different normalizations $b_i^0(t)$, $i = 1, \dots, d$, provided $b(t)/b_i^0(t) \to \infty$ for $i = 1, \dots, d$. The case of different normalizing functions can be reduced to (9.54) by monotone transformations. For the time being, we continue with the standard case. This issue is considered again in Section 9.4.4 (p. 332).
- Another choice of subcone of $\mathbb{E}$ is $[0, \infty]^{d-1} \times (0, \infty]$. This is used in [162].
- The definition of hidden regular variation uses two cones. In principle, one could have more cones $\mathbb{E} \supset \mathbb{E}^0 \supset \mathbb{E}^1 \supset \cdots \supset \mathbb{E}^k$ with regular variation of progressively smaller order present in each. A simple example in which $d = 3 = k$ is given in Problem 9.3.

### 9.4.2 A simple characterization

Define

$$\boldsymbol{e}_i^{-1} = (\infty, \dots, \infty, 1, \infty, \dots, \infty)$$

for the vector whose $i$th component is 1 and whose other components are $\infty$, and set

$$(\mathbb{E}^0)^{-1} = (\mathbf{0}, \boldsymbol{\infty}] \setminus \bigcup_{i=1}^{d} \{t\boldsymbol{e}_i^{-1}, 0 < t \le \infty\}.$$

So $(\mathbb{E}^0)^{-1}$ consists of points all of whose components are positive and such that at least two components are finite.

We may characterize hidden regular variation by using max- and min-linear combinations of the form

$$\bigvee_{i=1}^{d} s_i Z^{(i)} \quad \text{and} \quad \bigwedge_{i=1}^{d} a_i Z^{(i)}$$

for $\boldsymbol{s} \in [\mathbf{0}, \boldsymbol{\infty}) \setminus \{\mathbf{0}\}$ and $\boldsymbol{a} \in (\mathbb{E}^0)^{-1}$. This will provide a diagnostic for statistically identifying when hidden regular variation is present.

**Proposition 9.5.** *Suppose $b(\cdot) \in \mathrm{RV}_{1/\alpha}$ and $b^0(\cdot) \in \mathrm{RV}_{1/\alpha^0}$, $0 < \alpha \le \alpha^0$, are functions such that $b(t)/b^0(t) \to \infty$ as $t \to \infty$. Suppose, in addition, that the marginal distributions of the random vector $\boldsymbol{Z} \ge \boldsymbol{0}$ satisfy*

$$\lim_{t\to\infty} t\mathbb{P}\left[\frac{Z^{(i)}}{b(t)} > x\right] = x^{-\alpha}, \quad x > 0.$$

*Then $\boldsymbol{Z}$ has a distribution possessing hidden regular variation iff we have the following:*

(i) *For all $\boldsymbol{s} \in [\boldsymbol{0}, \infty) \setminus \{\boldsymbol{0}\}$, we have*

$$\lim_{t\to\infty} t\mathbb{P}\left[\frac{\bigvee_{i=1}^d s^{(i)} Z^{(i)}}{b(t)} > x\right] = c(\boldsymbol{s})x^{-\alpha}, \quad x > 0, \tag{9.56}$$

*for some function $c(\boldsymbol{s}) > 0$.*

(ii) *For all $\boldsymbol{a} \in (\mathbb{E}^0)^{-1}$, we have*

$$\lim_{t\to\infty} t\mathbb{P}\left[\frac{\bigwedge_{i=1}^d a^{(i)} Z^{(i)}}{b^0(t)} > x\right] = d(\boldsymbol{a})x^{-\alpha^0}, \quad x > 0. \tag{9.57}$$

*The hidden regular variation of the distribution of $\boldsymbol{Z}$ takes the following form:*

(a) *As $t \to \infty$,*

$$t\mathbb{P}\left[\frac{\boldsymbol{Z}}{b(t)} \in \cdot\right] \to \nu_{(0)}(\cdot) \tag{9.58}$$

*in $M_+(\mathbb{E})$, where $\nu_{(0)} \ne 0$ and $\nu_{(0)}$ concentrates on the axes through $\boldsymbol{0}$ in the sense that*

$$\nu_{(0)}(\mathbb{E}^0) = 0.$$

*(Note that, apart from constants, $\nu_{(0)}$ is the limit measure identified in Section 6.5.1 (p. 192) for asymptotic independence.)*

(b) *As $t \to \infty$,*

$$t\mathbb{P}\left[\frac{\boldsymbol{Z}}{b^0(t)} \in \cdot\right] \to \nu^0(\cdot) \ne 0 \tag{9.59}$$

*in $M_+(\mathbb{E}^0)$.*

*Remark* 9.5.

(1) The proof shows that $\nu_{(0)}$ in (9.58) is

$$\nu_{(0)}([\boldsymbol{0}, \boldsymbol{x}]^c) = \sum_{i=1}^d c(\boldsymbol{e}_i)(x^{(i)})^{-\alpha}.$$

Also, we obtain for $\nu^0$ in (9.59),

$$\nu^0(\boldsymbol{x}, \infty] = d(\boldsymbol{x}^{-1}).$$

(2) Hidden regular variation requires that for every $\boldsymbol{s} \geq 0$, $\boldsymbol{s} \neq 0$, $\bigvee_{i=1}^d s^{(i)} Z^{(i)}$ has a distribution with a regularly varying tail of index $\alpha$, and for every $\boldsymbol{a} \in (\mathbb{E}^0)^{-1}$, $\bigwedge_{i=1}^d a^{(i)} Z^{(i)}$ has a regularly varying distribution tail of index $\alpha^0$. In particular, when $d = 2$, hidden regular variation means that both $Z^{(1)} \vee Z^{(2)}$ and $Z^{(1)} \wedge Z^{(2)}$ have regularly varying tail probabilities.

(3) Ledford and Tawn [198–201] call $\eta = 1/\alpha^0$ the coefficient of tail dependence and consider it as a measure of dependence in asymptotic independence.

*Proof.* Assume (i) and (ii). Let $\boldsymbol{s} = \boldsymbol{e}_i = (0, \dots, 1, \dots, 0)$, and apply (9.56) to get

$$t\mathbb{P}\left[\frac{Z^{(i)}}{b(t)} > x\right] \to c(0, \dots, 1, \dots, 0)x^{-\alpha}, \quad x > 0.$$

Then for $\boldsymbol{x} > \boldsymbol{0}$,

$$t\sum_{i=1}^{d} \mathbb{P}\left[\frac{Z^{(i)}}{b(t)} > x^{(i)}\right] \geq tP\left\{\left[\frac{\boldsymbol{Z}}{b(t)} \leq \boldsymbol{x}\right]^c\right\} = tP\left\{\bigcup_{i=1}^{d}\left[\frac{Z^{(i)}}{b(t)} > x^{(i)}\right]\right\}$$
$$\geq t\sum_{i=1}^{d} P\left[\frac{Z^{(i)}}{b(t)} > x^{(i)}\right] - tP\left\{\bigcup_{1\leq i<j\leq d}\left[\frac{Z^{(i)}}{b(t)} > x^{(i)}, \frac{Z^{(j)}}{b(t)} > x^{(j)}\right]\right\}; \quad (9.60)$$

both extremes in the inequalities converge to

$$\to \sum_{i=1}^{d} c(\boldsymbol{e}_i)(x^{(i)})^{-\alpha} + 0.$$

The reason for convergence to 0 in the second term of the right side of (9.60) is that

$$tP\left\{\bigcup_{1\leq i<j\leq d}\left[\frac{Z^{(i)}}{b(t)} > x^{(i)}, \frac{Z^{(j)}}{b(t)} > x^{(j)}\right]\right\}$$
$$\leq t\sum_{1\leq i<j\leq d} P\left[\frac{Z^{(i)}}{b(t)} > x^{(i)}, \frac{Z^{(j)}}{b(t)} > x^{(j)}\right]$$
$$= t\sum_{1\leq i<j\leq d} P\left[\frac{Z^{(i)}}{b^0(t)} > \frac{b(t)}{b^0(t)}x^{(i)}, \frac{Z^{(j)}}{b^0(t)} > \frac{b(t)}{b^0(t)}x^{(j)}\right]$$

$$= t \sum_{1 \le i < j \le d} P\left[\frac{(x^{(i)})^{-1}Z^{(i)} \wedge (x^{(j)})^{-1}Z^{(j)}}{b^0(t)} > \frac{b(t)}{b^0(t)}\right]$$
$$\to 0$$

using (9.57) and the fact that $b(t)/b^0(t) \to \infty$.

From (9.60), we identify the limit measure

$$\nu_{(0)}([\mathbf{0}, \boldsymbol{x}]^c) = \sum_{i=1}^{d} c(\boldsymbol{e}_i)(x^{(i)})^{-\alpha},$$

and $\nu_{(0)}$ concentrates on $\mathbb{E} \setminus \mathbb{E}^0$. To identify $\nu^0$, note that for $\mathbf{0} < \boldsymbol{x} < \infty$,

$$t\mathbb{P}\left[\frac{\boldsymbol{Z}}{b^0(t)} > \boldsymbol{x}\right] = t\mathbb{P}\left[\frac{\wedge_{i=1}^d (x^{(i)})^{-1}Z^{(i)}}{b^0(t)} > 1\right];$$

using (9.57), this converges to $d(\boldsymbol{x}^{-1})$. This identifies

$$\nu^0(\boldsymbol{x}, \infty] = d(\boldsymbol{x}^{-1}).$$

Conversely, if (a) and (b) hold, then for $\boldsymbol{s} \in [\mathbf{0}, \infty) \setminus \{\mathbf{0}\}$ and $x > 0$, we have

$$t\mathbb{P}\left[\frac{\bigvee_{i=1}^d s^{(i)}Z^{(i)}}{b(t)} > x\right] = tP\left\{\bigcup_{i=1}^{d}\left[\frac{Z^{(i)}}{b(t)} > (s^{(i)})^{-1}x\right]\right\}$$
$$= tP\left\{\left[\frac{\boldsymbol{Z}}{b(t)} \le x\boldsymbol{s}^{-1}\right]^c\right\}$$
$$\to \nu_{(0)}([\mathbf{0}, x\boldsymbol{s}^{-1}) = x^{-\alpha}\nu_{(0)}([\mathbf{0}, \boldsymbol{s}^{-1})$$
$$= c(\boldsymbol{s})x^{-\alpha}.$$

Also for $\boldsymbol{a} \in (\mathbb{E}^0)^{-1}$ and $x > 0$, we get

$$t\mathbb{P}\left[\frac{\bigwedge_{i=1}^d a^{(i)}Z^{(i)}}{b^0(t)} > x\right] = t\mathbb{P}\left[\frac{\boldsymbol{Z}}{b^0(t)} > x\boldsymbol{a}^{-1}\right]$$
$$\to \nu^0(x\boldsymbol{a}^{-1}, \infty] = x^{-\alpha^0}\nu^0(\boldsymbol{a}^{-1}, \infty]$$
$$= d(\boldsymbol{a}^{-1})x^{-\alpha^0}. \qquad \square$$

When the one-dimensional components of $\boldsymbol{Z}$ are tail equivalent with each component having regularly varying tail probabilities, multivariate regular variation on $\mathbb{E}^0$ implies regular variation on $\mathbb{E}$ and hence hidden regular variation.

**Corollary 9.1.** *Suppose for* $\alpha^0 > 0$ *that* (9.54) *holds in* $M_+(\mathbb{E}^0)$ *and* $b^0 \in \mathrm{RV}_{1/\alpha^0}$. *Suppose further that for each* $i = 1, \dots, d$,

$$\lim_{t\to\infty} t\mathbb{P}\left[\frac{Z^{(i)}}{b(t)} > x\right] = x^{-\alpha}, \quad x > 0,$$

*for* $0 < \alpha \le \alpha^0$ *and some* $b \in \mathrm{RV}_{1/\alpha}$ *with* $b(t)/b^0(t) \to \infty$. *Then hidden regular variation holds; that is,* (9.51) *holds with* $\nu = \nu_{(0)}$ *on* $\mathbb{E}$.

*Proof.* For $\boldsymbol{a} \in (\mathbb{E}^0)^{-1}$ and $x > 0$,

$$\begin{aligned} t\mathbb{P}\left[\frac{\bigwedge_{i=1}^d a^{(i)}Z^{(i)}}{b^0(t)} > y\right] &= t\mathbb{P}\left[\frac{\bigwedge_{i:a^{(i)}<\infty} a^{(i)}Z^{(i)}}{b^0(t)} > y\right] \\ &\to \nu^0\left\{\boldsymbol{x} \in \mathbb{E}^0 : \bigwedge_{i:a^{(i)}<\infty} a^{(i)}x^{(i)} > y\right\} \\ &= y^{-\alpha^0}\nu^0\left\{\boldsymbol{x} \in \mathbb{E}^0 : \bigwedge_{i=1}^d a^{(i)}x^{(i)} > 1\right\} = d(\boldsymbol{a})y^{-\alpha^0}. \end{aligned}$$

So (9.57) holds. For (9.56), we have for $\boldsymbol{s} \in [\boldsymbol{0}, \boldsymbol{\infty}) \setminus \{\boldsymbol{0}\}$ and $x > 0$,

$$\begin{aligned} t\mathbb{P}\left[\bigvee_{i=1}^d s^{(i)}\frac{Z^{(i)}}{b(t)} > x\right] &= t\mathbb{P}\left[\bigvee_{i:s^{(i)}>0} s^{(i)}\frac{Z^{(i)}}{b(t)} > x\right] \\ &\le t\sum_{i:s^{(i)}>0} \mathbb{P}\left[\frac{Z^{(i)}}{b(t)} > x(s^{(i)})^{-1}\right] \\ &\to t\sum_{i:s^{(i)}>0} (x(s^{(i)})^{-1})^{-\alpha} = x^{-\alpha}\sum_{i=1}^d (s^{(i)})^{\alpha}. \end{aligned}$$

To verify that this upper bound is indeed the limit, observe that for $s(i) > 0$, $s(j) > 0$,

$$\begin{aligned} &t\mathbb{P}\left[\frac{Z^{(i)}}{b(t)} > x(s^{(i)})^{-1}, \frac{Z^{(j)}}{b(t)} > x(s^{(j)})^{-1}\right] \\ &\quad = t\mathbb{P}\left[\frac{Z^{(i)}}{b^0(t)} > \frac{b(t)}{b^0(t)}x(s^{(i)})^{-1}, \frac{Z^{(j)}}{b^0(t)} > \frac{b(t)}{b^0(t)}x(s^{(j)})^{-1}\right] \to 0, \end{aligned}$$

since $b(t)/b^0(t) \to \infty$.

The result follows from Proposition 9.5. □

### 9.4.3 Two examples

The following examples illustrate methods for obtaining hidden regular variation by means of independence and mixtures and also clarify that $\nu^0$ may be finite or infinite on $\{\boldsymbol{x} \in \mathbb{E}^0 : \|\boldsymbol{x}\| > 1\}$. This means that $S^0$ may be finite or infinite on $\aleph^0$.

*Example* 9.2. Let $d = 2$. Define on $[0, \infty)^2$,

$$\mathbb{LEB}((\mathbf{0}, \boldsymbol{x}]) = x^{(1)} x^{(2)};$$

that is, $\mathbb{LEB}$ is Lebesgue measure. Apply the transform $T_1 : \boldsymbol{x} \mapsto \boldsymbol{x}^{-1}$ to get the measure $\nu^0$ on $(0, \infty]^2$ given by

$$\nu^0[\boldsymbol{x}, \infty] = \mathbb{LEB} \circ T_1^{-1}([\boldsymbol{x}, \infty]) = (x^{(1)} x^{(2)})^{-1}, \quad \boldsymbol{x} > \mathbf{0},$$

and $\nu^0$ has density

$$f^0(\boldsymbol{x}) = (x^{(1)} x^{(2)})^{-2}, \quad \boldsymbol{x} > \mathbf{0}.$$

Define $\mathbf{Z} = (Z^{(1)}, Z^{(2)})$ iid and Pareto distributed with

$$P[Z^{(1)} > x] = x^{-1}, \quad x > 1, \quad i = 1, 2.$$

Set

$$b(t) = t, \quad b^0(t) = \sqrt{t},$$

so that $b(t)/b^0(t) \to \infty$. Then on $\mathbb{E}$,

$$tP\left[\frac{\mathbf{Z}}{b(t)} \in \cdot\right] \xrightarrow{v} \nu_{(0)},$$

the measure giving zero mass to the interior of $\mathbb{E}$. On $\mathbb{E}^0$,

$$tP\left[\frac{\mathbf{Z}}{b^0(t)} \in \cdot\right] \xrightarrow{v} \nu^0.$$

To check this last statement, note for $\boldsymbol{x} > \mathbf{0}$ and large $t$ that

$$\begin{aligned} tP\left[\frac{\mathbf{Z}}{b^0(t)} > \boldsymbol{x}\right] &= tP\left[Z^{(1)} > \sqrt{t}x^{(1)}, Z^{(2)} > \sqrt{t}x^{(2)}\right] \\ &= \sqrt{t}\left(\sqrt{t}x^{(1)}\right)^{-1} \sqrt{t}\left(\sqrt{t}x^{(2)}\right)^{-1} \\ &= (x^{(1)} x^{(2)})^{-1} = \nu^0[\boldsymbol{x}, \infty]. \end{aligned}$$

In this case, $\nu^0$ is infinite on $\{\boldsymbol{x} \in \mathbb{E}^0 : \|\boldsymbol{x}\| > 1\}$ since for any $\delta > 0$,

$$\nu^0(\{\boldsymbol{x} \in \mathbb{E}^0 : \|\boldsymbol{x}\| > 1\}) \geq \nu^0([2, \infty] \times [\delta, \infty]) = 2^{-1}\delta^{-1} \to \infty$$

as $\delta \to 0$. □

*Example* 9.3. Assume $d = 2$ and consider three independent random quantities $B$, $\boldsymbol{Y}$, $\boldsymbol{U}$. Suppose $B$ is a Bernoulli random variable with

$$P[B = 0] = P[B = 1] = 1/2$$

and $\boldsymbol{Y} = (Y^{(1)}, Y^{(1)})$ is iid with common standard Pareto distribution

$$P[Y^{(1)} > x] = x^{-1}, \quad x > 1.$$

Set $b(t) = t$. Suppose $\boldsymbol{U}$ has a multivariate regularly varying distribution on $\mathbb{E}$ and that there exist $\alpha^0 > 1$ and $b^0(t) \in \mathrm{RV}_{1/\alpha^0}$ and a nonzero measure $\nu^0$ that is Radon on $\mathbb{E}$ such that

$$tP\left[\frac{\boldsymbol{U}}{b^0(t)} \in \cdot\right] \to \nu^0 \neq 0.$$

Define

$$\boldsymbol{Z} = B\boldsymbol{Y} + (1 - B)\boldsymbol{U}. \tag{9.61}$$

For $\boldsymbol{x} > \boldsymbol{0}$,

$$\begin{aligned} tP[\boldsymbol{Z} > b^0(t)\boldsymbol{x}] &= \frac{t}{2}P[\boldsymbol{Y} > b^0(t)\boldsymbol{x}] + \frac{t}{2}P[\boldsymbol{U} > b^0(t)\boldsymbol{x}] \\ &= \mathrm{I} + \mathrm{II}. \end{aligned}$$

Now for II we have

$$\mathrm{II} \to \frac{1}{2}\nu^0((\boldsymbol{x}, \infty]).$$

For I, from the fact that $\boldsymbol{Y}$ is iid Pareto, we get that

$$\mathrm{I} = \frac{t}{2}(b^0(t))^{-2}\frac{1}{x^{(1)}x^{(2)}}$$

converges to 0 iff $t/(b^0(t))^2 \to 0$. This will be the case if $1 < \alpha^0 < 2$.

Each one-dimensional marginal distribution tail is regularly varying with index $-1$, which follows from

$$P[Z^{(1)} > x] = \frac{1}{2}P[Y^{(1)} > x] + \frac{1}{2}P[U^{(1)} > x].$$

For $x > 0$, the first term on the right of the equality is in $\mathrm{RV}_{-1}$ and the second term is in $\mathrm{RV}_{-\alpha^0}$. The second term decays more quickly, and therefore for $i = 1, 2$,

$$P[Z^{(i)} > x] \in \mathrm{RV}_{-1}.$$

From Corollary 9.1 (p. 329), $\boldsymbol{Z}$ possesses hidden regular variation.

So on the cone $\mathbb{E}$, $\boldsymbol{Z}$ is regularly varying with limit measure $\nu_{(0)}$ and on $\mathbb{E}^0$, the hidden measure is $\nu^0$.

Note for this example that $\nu^0$ is finite on $\{\boldsymbol{x} \in \mathbb{E}^0 : \|\boldsymbol{x}\| > 1\}$. □

*Remark* 9.6. With the proper notion of a multivariate version of *tail equivalence* [256], any distribution possessing hidden regular variation is tail equivalent to a mixture as in Example 9.3. See [217].

### 9.4.4 Detection of hidden regular variation

Can one statistically detect the phenomenon of hidden regular variation? One point to this inquiry is that a technique that detects hidden regular variation also confirms asymptotic independence.

#### A first step

Proposition 9.5 suggests a data diagnostic, which increases our confidence that hidden regular variation is present. Follow these steps: Assuming the data are from the iid model, do the following:

- Perform the rank transform (9.46) given on p. 313. This converts the problem to standard form and $\alpha = 1$.
- As suggested by Proposition 9.5, for the resulting data vectors, take the minimum component of each vector.
- If these are data from a heavy-tailed distribution with index $\alpha^0 > 1 = \alpha$, there is no evidence against the hypothesis of hidden regular variation and hence against the hypothesis of asymptotic independence.

How does this work in practice? See Figure 9.12 for two examples in which hidden regular variation seems to be present. The left plot of Figure 9.12 is a Hill plot for the minimum component of the rank-transformed (buL, buR) data. The right plot is the Hill plot for the minimum component for the rank-transformed HTTP response (size, rate) data.

In contrast, we see in Figure 9.13 for the insurance data that a Hill plot of the component minima after the rank transform does not indicate an $\alpha^0$ distinguishable from 1. This is expected since we did not believe asymptotic independence was present for these data.

#### But wait! Why does the rank transform preserve hidden regular variation?

The outline in the previous section is simple, but it requires an explanation of why the rank transform preserves hidden regular variation. The definition of hidden regular

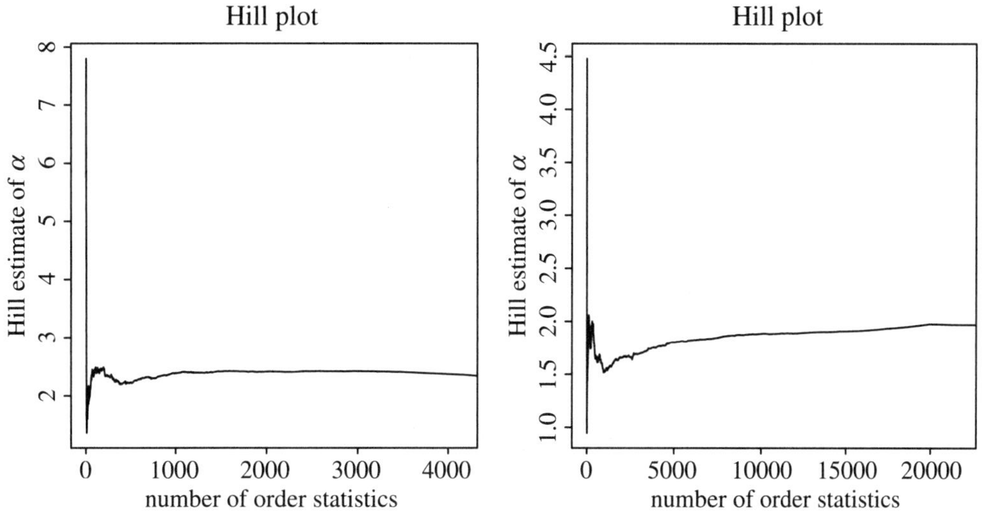


**Fig. 9.12.** Left: Estimate of $\alpha^0$ for buL vs. buR indicating an estimate of $\alpha^0 = 2.3$. Right: Estimate of $\alpha^0$ for HTTP response size vs. response rate indicating an $\alpha^0$ of about 1.8.

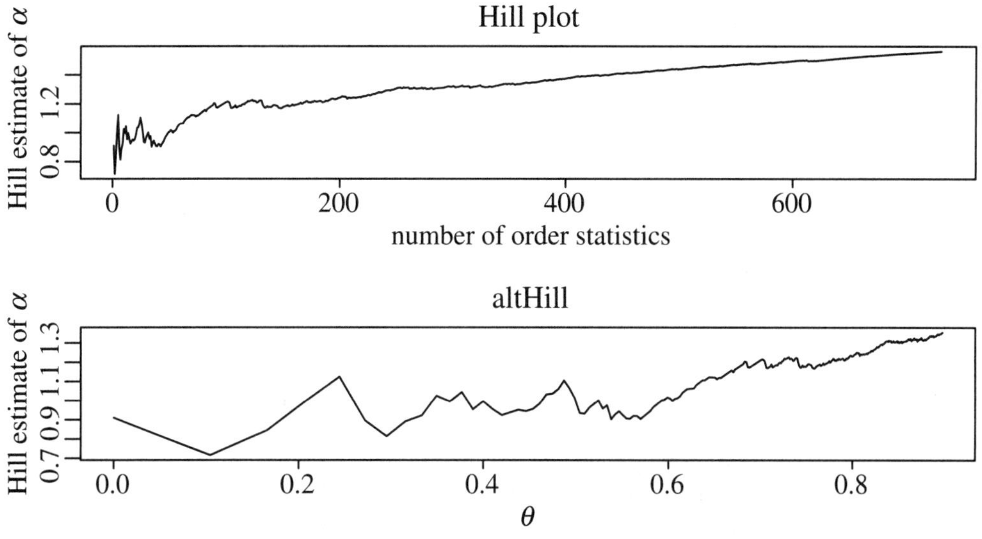


**Fig. 9.13.** Hill and altHill plots for the rank-transformed auto and fire claim data confirms the earlier impression of lack of asymptotic independence.

variation, given in Section 9.4.1 (p. 324), assumes that regular variation scaling is identical for each component. This is the correct assumption for theoretical developments but not for applications.

So start from the broader definition of regular variation, allowing unequal scaling in each component, which is discussed in Section 6.5.6 (p. 203). Theorem 6.5 (p. 204) discusses the simple theory of how to convert the unequal scaling to the standard case. We assume that there exist functions $b^{(j)}(t) \uparrow \infty$ as $t \to \infty$, such that for a Radon measure $\nu$ on $\mathbb{E}$,

$$tP\left[\left(\frac{Z^{(j)}}{b^{(j)}(t)}, j = 1, \dots, d\right) \in \cdot\right] = tP\left[\frac{\boldsymbol{Z}}{\boldsymbol{b}(t)} \in \cdot\right] \xrightarrow{v} \nu \tag{9.62}$$

on $\mathbb{E}$. Also suppose that the marginal convergences satisfy

$$tP\left[\frac{Z^{(j)}}{b^{(j)}(t)} > x\right] \to \nu_{\alpha^{(j)}}(x, \infty] = x^{-\alpha^{(j)}}, \tag{9.63}$$

where $\alpha^{(j)} > 0$, $j = 1, \dots, d$. Then $b^{(j)}(t) \in RV_{1/\alpha^{(j)}}$, and we can and do assume that each $b^{(j)}(t)$ is continuous and strictly increasing. (See Proposition 2.6(vii) (p. 32).)

We say that the distribution of $\boldsymbol{Z}$ has *hidden regular variation* if in addition to (9.62) or (6.43) (p. 205), we have the following property after transforming to the standard case: There must exist a function $b^0(t) \in RV_{1/\alpha^0}$ with $b^0(t) \uparrow \infty$, $\alpha^0 \geq 1$ and

$$\lim_{t\to\infty} \frac{t}{b^0(t)} = \infty, \tag{9.64}$$

such that on $\mathbb{E}^0$,

$$tP\left[\left(\frac{b^{(j)\leftarrow}(Z^{(j)})}{b^0(t)}, j = 1, \dots, d\right) \in \cdot\right] \xrightarrow{v} \nu^0(\cdot) \tag{9.65}$$

for some Radon measure $\nu^0$ on $\mathbb{E}^0$. Note that (9.65) is equivalent to

$$tP\left[\left(\frac{Z^{(j)}}{b^{(j)}(b^0(t))}, j = 1, \dots, d\right) \in \cdot\right] \xrightarrow{v} \tilde{\nu}^0(\cdot) \tag{9.66}$$

on $\mathbb{E}^0$, where $\nu^0$ and $\tilde{\nu}^0$ are related by

$$\tilde{\nu}^0(\boldsymbol{x}, \infty] = \nu^0(\boldsymbol{x}^{\boldsymbol{\alpha}}, \infty], \quad \boldsymbol{x} \in \mathbb{E}^0. \tag{9.67}$$

The measure $\nu^0$ is also homogeneous on $\mathbb{E}^0$,

$$\nu^0(t\cdot) = t^{-\alpha^0}\nu^0(\cdot),$$

but $\nu^0$ can be either finite or infinite on $E^0$ (see Section 9.4.3 (p. 330)), and when we transform $\nu^0$ to polar coordinates, we get

$$\nu^0\left\{\boldsymbol{x} \in \mathbb{E} : \|\boldsymbol{x}\| > r, \frac{\boldsymbol{x}}{\|\boldsymbol{x}\|} \in \cdot\right\} = r^{-\alpha^0} S^0(\cdot), \tag{9.68}$$

where $S^0$ is a Radon measure on Borel subsets of

$$\aleph^0 = \aleph \cap \mathbb{E}^0.$$

The hidden angular measure $S^0$ is not necessarily finite. Since the region

$$\aleph_{\text{INV}} := \left\{\boldsymbol{x} \in \mathbb{E}^0 : \bigwedge_{j=1}^{d} x^{(j)} \geq 1\right\}$$

is a compact subset of $\mathbb{E}^0$ and hence will always have finite hidden measure, we can (and do) always choose $b^0(t)$ so that

$$\nu^0(\aleph_{\text{INV}}) = 1. \tag{9.69}$$

How can we use the rank transform to estimate the hidden measure $S^0$? To find the hidden angular measure, we expect we have to use points

$$\left\{\left(\frac{k}{r_i^{(j)}}; j = 1, \ldots, d\right); i = 1, \ldots, n\right\}$$

thresholded at a lower level than in (9.47) (p. 313). Since $b^0(t)/t \to 0$, it seems plausible to use the points

$$\left\{\left(\frac{\frac{k}{r_i^{(j)}}}{b^0(n/k)/(n/k)}; j = 1, \ldots, d\right); i = 1, \ldots, n\right\}$$
$$= \left\{\left(\frac{n/r_i^{(j)}}{b^0(n/k)}; j = 1, \ldots, d\right); i = 1, \ldots, n\right\}.$$

This scheme yields the hidden measure [161].

**Proposition 9.6.** *Assume that $\mathbf{Z}_1, \ldots, \mathbf{Z}_n$ is an iid sample from a distribution on $[0, \infty)^d$ that possesses both regular and hidden regular variation, so that* (9.62) *and* (9.65) *hold. Then we have*

$$\frac{1}{k}\sum_{i=1}^{n} \epsilon_{\left(\frac{n/r_i^{(j)}}{b^0(n/k)}; j = 1, \ldots, d\right)} \Rightarrow \nu^0 \tag{9.70}$$

*in $M_+(\mathbb{E}^0)$, where we recall that $\nu^0$ is given in* (9.65).

*Proof.* The argument mimics the proof of Proposition 9.4 (p. 312) for estimating $\nu$ or $S$. Observe for $\boldsymbol{x} \in \mathbb{E}^0$ that

$$\frac{1}{k}\sum_{i=1}^{n} \epsilon_{\left(\frac{n/r_i^{(j)}}{b^0(n/k)};\, j=1,\dots,d\right)}[\boldsymbol{x},\infty] = \frac{1}{k}\sum_{i=1}^{n} 1_{\left[n(x^{(j)})^{-1}b^0(n/k)^{-1} \geq r_i^{(j)};\, j=1,\dots,d\right]}$$
$$= \frac{1}{k}\sum_{i=1}^{n} 1_{\left[\frac{Z_i^{(j)}}{b^{(j)}(b^0(n/k))} \geq \frac{Z^{(j)}_{(\lceil n(x^{(j)}b^0(n/k))^{-1}\rceil)}}{b^{(j)}(b^0(n/k))};\, j=1,\dots,d\right]}.$$

We claim (see p. 336 for the proof) that for each $j = 1, \dots, d$,

$$\frac{Z^{(j)}_{(\lceil n(x^{(j)}b^0(n/k))^{-1}\rceil)}}{b^{(j)}(b^0(n/k))} \xrightarrow{P} (x^{(j)})^{1/\alpha^{(j)}}. \tag{9.71}$$

Using this to scale the convergence in (9.66), we get from the Theorem 5.3 (p. 138) that

$$\frac{1}{k}\sum_{i=1}^{n} \epsilon_{\left(\frac{n/r_i^{(j)}}{b^0(n/k)};\, j=1,\dots,d\right)}[\boldsymbol{x},\infty] \Rightarrow \tilde{\nu}^0[\boldsymbol{x}^{1/\boldsymbol{\alpha}},\infty] = \nu^0[\boldsymbol{x},\infty],$$

where we used (9.67) for the last equality. This suffices to prove the result modulo the claim. □

*Claim.* Assume that $\boldsymbol{Z}_1, \dots, \boldsymbol{Z}_n$ is an iid sample from a distribution on $[0,\infty)^d$ that possesses both regular and hidden regular variation, so that (9.62) and (9.65) hold. Then (9.71) holds in $D[0,\infty)$ for each $j = 1, \dots, d$.

*Proof.* We have, for each $j = 1, \dots, d$,

$$\frac{b^0(n/k)}{n}\sum_{i=1}^{n} \epsilon_{Z_i^{(j)}/b^{(j)}(b^0(n/k))} \Rightarrow \nu_{\alpha^{(j)}}$$

in $M_+(0,\infty]$ using Theorem 5.3 (p. 138) (see (5.16)). In particular,

$$\frac{b^0(n/k)}{n}\sum_{i=1}^{n} \epsilon_{Z_i^{(j)}/b^{(j)}(b^0(n/k))}(x^{-1},\infty] \Rightarrow x^{\alpha^{(j)}}$$

in $D[0,\infty)$. This is a sequence of nondecreasing functions converging to a continuous limit, and so the inverse functions converge as well. This yields the claim statement. □

## Estimating the hidden angular measure

If one converts (9.70) to polar coordinates in order to estimate $S^0$, one gets the analogue of (9.47) (p. 313). Transform this way: Apply the polar coordinate transformation

$$\text{POLAR}\left(\frac{1}{r_i^{(j)}};\, j=1,\dots,d\right) = (R_i, \boldsymbol{\Theta}_i).$$

Then, assuming $S^0$ is finite (otherwise one has to restrict $\boldsymbol{\Theta}_i$ to a compact subset of $\aleph^0$),

$$\frac{\sum_{i=1}^n 1_{[R_i \geq n^{-1} b^0(n/k)]}\epsilon_{\boldsymbol{\Theta}_i}}{\sum_{i=1}^n 1_{[R_i \geq n^{-1} b^0(n/k)]}} \Rightarrow S^0 \tag{9.72}$$

in $M_+(\mathbb{E}^0)$. Since $b^0(n/k)$ is unknown for statistical purposes, it must be estimated before we can regard (9.72) as a suitable estimate of $S^0$.

Recall that

$$\aleph_{\text{INV}} := \left\{\boldsymbol{x} \in \mathbb{E}^0 : \bigwedge_{j=1}^d x^{(j)} \geq 1\right\}$$

is the set of vectors all of whose components are at least 1. Define

$$m_i = \bigwedge_{j=1}^d \frac{1}{r_i^{(j)}}, \qquad i = 1,\dots,n,$$

and further suppose that

$$m_{(1)} \geq m_{(2)} \geq \cdots \geq m_{(n)}$$

is the ordering of $m_1,\dots,m_n$ with the biggest first. The next result removes the unknown $b^0(n/k)$ and replaces it by a statistic.

**Proposition 9.7.** *Assume $\mathbf{Z}_1,\dots,\mathbf{Z}_n$ is an iid sample from a distribution on $[0,\infty)^d$ that possesses both regular and hidden regular variation, so that* (9.62) *and* (9.65) *hold and continue to assume that $\nu^0(\aleph_{\text{INV}}) = 1$. Then we have in $M_+(\mathbb{E}^0)$,*

$$\widehat{\nu^0} := \frac{1}{k}\sum_{i=1}^n \epsilon_{\left(\frac{1/r_i^{(j)}}{m_{(k)}},\, 1 \leq j \leq d\right)} \Rightarrow \nu^0. \tag{9.73}$$

*Proof.* On $D[0,\infty)$, we have from Proposition 9.6 and continuous mapping that

$$\eta_n(t) := \frac{1}{k}\sum_{i=1}^k \epsilon_{\frac{n}{b^0(n/k)}\wedge_{j=1}^d \frac{1}{r_i^{(j)}}}(t^{-1},\infty] \Rightarrow \nu^0\left\{\boldsymbol{x} : \wedge_{j=1}^d x^{(j)} \geq t^{-1}\right\}$$

$$= t^{\alpha^0} \nu^0(\aleph_{\mathrm{INV}}) = t^{\alpha^0} =: \eta_\infty(t).$$

Therefore, we also have in $D[0, \infty)$ that the inverse processes converge:

$$\eta_n^{\leftarrow}(s) \Rightarrow \eta_\infty^{\leftarrow}(s) = s^{1/\alpha^0}.$$

Unpack the left-hand side. We have

$$\begin{aligned}
\eta_n^{\leftarrow}(s) &= \inf\{u : \eta_n(u) \geq s\} \\
&= \inf\left\{u : \sum_{i=1}^{n} \epsilon_{\frac{n}{b^0(n/k)} m_i}(u^{-1}, \infty] \geq ks\right\} \\
&= \left(\sup\left\{v : \sum_{i=1}^{n} \epsilon_{\frac{n}{b^0(n/k)} m_i}(v, \infty] \geq ks\right\}\right)^{-1} \\
&= \frac{b^0(n/k)}{n}\left(\sup\left\{w : \sum_{i=1}^{n} \epsilon_{m_i}(w, \infty] \geq ks\right\}\right)^{-1} \\
&= \frac{b^0(n/k)}{n} m_{([ks])}^{-1}.
\end{aligned}$$

Therefore, we see that

$$\frac{n}{b^0(n/k)} m_{([ks])} \Rightarrow s^{-1/\alpha^0} \tag{9.74}$$

in $D(0, \infty]$.

The rest is a scaling argument. Couple (9.74) with $s = 1$ with (9.70) and compose to get in $D(\mathbb{E}^0)$,

$$\begin{aligned}
\frac{1}{k}\sum_{i=1}^{n} &\epsilon_{\left(\frac{n/r_i^{(j)}}{b^0(n/k)},\, j=1,\ldots,d\right)}\left[\frac{n}{b^0(n/k)} m_{(k)} \boldsymbol{x}, \infty\right] \\
&= \frac{1}{k}\sum_{i=1}^{n} \epsilon_{\left(\frac{1/r_i^{(j)}}{m_{(k)}},\, j=1,\ldots,d\right)}[\boldsymbol{x}, \infty] \Rightarrow \nu^0(\boldsymbol{x}, \infty],
\end{aligned}$$

as required. □

This suggests a way forward around the problem of the unknown function $b^0(n/k)$ in (9.72): We replace $n^{-1}b^0(n/k)$ with $m_{(k)}$. We can then write the analogue of (9.72). If $\nu^0$ is infinite, let $\aleph^0(K)$ be a convenient compact subset of $\aleph^0$. For $d = 2$, where $\aleph$ can be parameterized as $\aleph = [0, \pi/2]$ and $\aleph^0 = (0, \pi/2)$, we can set $\aleph^0(K) = [\delta, \pi/2 - \delta]$ for some small $\delta > 0$. Then we have from Proposition 9.7 that

$$\frac{\sum_{i=1}^{n} 1_{[R_i \geq m_{(k)}, \boldsymbol{\Theta}_i \in \aleph^0(K)]} \epsilon_{\boldsymbol{\Theta}_i}}{\sum_{i=1}^{n} 1_{[R_i \geq m_{(k)}, \boldsymbol{\Theta}_i \in \aleph^0(K)]}} \Rightarrow S^0\left(\cdot \bigcap \aleph^0(K)\right). \tag{9.75}$$

If $\nu^0$ is finite, we can replace $\aleph^0(K)$ with $\aleph^0$, as was done in (9.72).

Thus, to summarize, we proceed as follows when estimating $S^0$:

1. Replace the heavy-tailed multivariate sample $\mathbf{Z}_1, \ldots, \mathbf{Z}_n$ with the $n$ vectors of antiranks $\mathbf{r}_1, \ldots, \mathbf{r}_n$, where

$$r_i^{(j)} = \sum_{l=1}^{n} 1_{[Z_l^{(j)} \geq Z_i^{(j)}]}; \quad j = 1, \ldots, d; \quad i = 1, \ldots, n.$$

2. Compute the normalizing factors

$$m_i = \bigwedge_{j=1}^{d} \frac{1}{r_i^{(j)}}; \quad i = 1, \ldots, n,$$

and their order statistics

$$m_{(1)} \geq \cdots \geq m_{(n)}.$$

3. Compute $\{(R_i, \boldsymbol{\Theta}_i); i = 1, \ldots, n\}$, which are the the polar coordinates of $\{(1/r_i^{(j)}; j = 1, \ldots, d); i = 1, \ldots, n\}$.

4. Estimate $S^0$ using the $\boldsymbol{\Theta}_i$ corresponding to $R_i \geq m_{(k)}$.

This is rather ambitious for a statistical procedure and results may be no more than suggestive. One obvious problem is choice of $k$. If a particular choice of $k$ is good for estimating $S$, is it also a good choice for estimating $S^0$?

To see how this might work in practice, consider again the HTTP response data discussed in Section 9.3.1 (p. 317). Following the outline above, we plot the estimated hidden measure for $k = 1000$, 1150, 1200, and 1250 in Figure 9.14. These plots show stability of the estimated measure for these values of $k$. The $L_1$-norm has been used in the polar coordinate transformation and the range of $\boldsymbol{\Theta}_i$ is $[0, 1]$. Again, since the measure may be infinite, we have bounded the interval on which we estimate the measure away from 0 and 1 and show the kernel density estimate on the interval $[0.1, 0.9]$. An edge correction has been applied so that the density integrates to 1 on this interval. All plots show the hidden measure to be bimodal with peaks around 0.2 and 0.85.

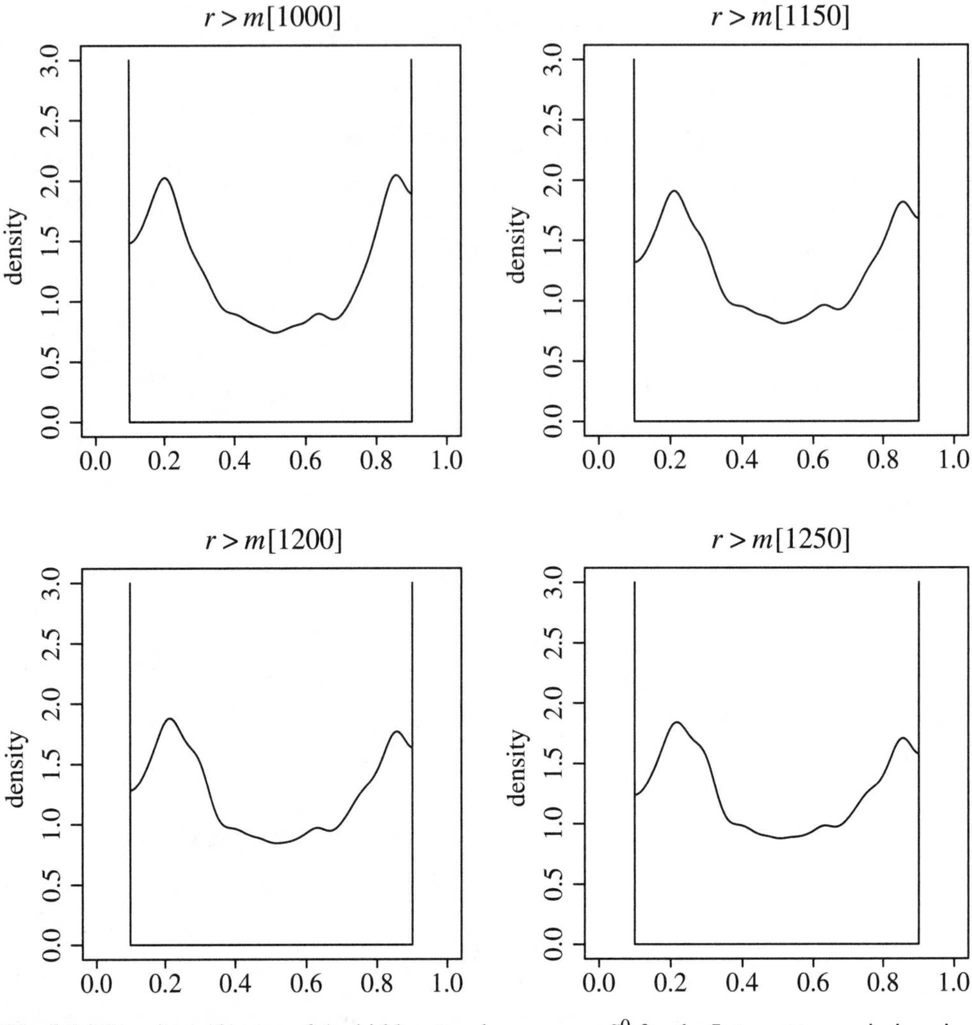


**Fig. 9.14.** Density estimates of the hidden angular measure $S^0$ for the Internet transmission size and rate data.

## 9.5 The sample correlation function

### 9.5.1 Overview

There are numerous data sets from the disciplines of data networks, finance, and economics, subsets of which appear to be stationary and compatible with the assumption of heavy-tailed marginal distributions. A key question, of course, is how to fit time-series models to data that require heavy-tailed marginal distributions. In the traditional setting of a stationary time series with finite variance, every purely nondeterministic process can be expressed as a linear process driven by an uncorrelated input sequence. Brockwell and Davis [31] ably summarize the conventional wisdom. For such time series, the autocorrelation function (acf) can be well approximated by that of a finite-order ARMA$(p, q)$ model. In particular, one can choose an autoregressive model of order $p$ (AR$(p)$) such that the acf of the two models agree for lags $1, \ldots, p$ [31, p. 240]. So when finite variance models are considered from a second-order point of view, linear models are sufficient for data analysis. The sample acf has traditionally been an important tool for analyzing linear time series.

In the heavy-tailed case, we have no such confidence that linear models are sufficiently flexible and rich enough for modeling purposes. Many theoretical attempts to study heavy-tailed time-series models have concentrated effort on ARMA models or infinite-order moving averages, despite little evidence that such models would actually fit heavy-tailed data. Understandably, these attempts are motivated by the desire to see how well classical ARMA models and methods perform in the new context of heavy-tailed modeling.

So what place does the sample autocorrelation have in heavy-tailed modeling? Its use is pervasive and almost reflexive. It is programmed into all respectable statistics packages. The acf is often plotted with 95% confidence windows computed on the possibly erroneous assumption that the sample acf is asymptotically normally distributed.

In the non-Gaussian modeling world, the case for the sample autocorrelation as a model selection tool is weak, and, in fact, its use for heavy-tailed model selection can be quite deceptive and can lead to erroneous conclusions. This has been richly documented by many authors. See [12, 48, 70, 71, 77, 134, 224, 227, 248, 249, 253, 265, 267]. Although there are many pitfalls associated with using the sample autocorrelation function for non-Gaussian data, it is still a most common method for checking for independence. In this section we will outline how the acf can be used for this purpose.

The sample acf of the stationary sequence $X_1, X_2, \ldots$ based on observing $X_1, \ldots, X_n$ is defined as

$$\hat{\rho}(h) = \frac{\sum_{i=1}^{n-h}(X_i - \bar{X})(X_{i+h} - \bar{X})}{\sum_{i=1}^{n}(X_i - \bar{X})^2}, \quad h = 0, 1, 2, \ldots, n-1, \tag{9.76}$$

where $\bar{X}$ is the sample mean. The independence test consists of plotting $\hat{\rho}(h)$ for various lags $h$ and checking if the values are all close to 0. Typically, we plot at lags $h = 1, \ldots, 25$, which should be adequate to uncover evidence against independence. Of course, it is essential to give precision to the phrase *close to* 0. Asymptotic distributions for $(\hat{\rho}(n), 1 \leq n \leq h)$ help provide some needed precision.

Consider the following two cases:

(a) *The variances are finite.* Then standard $L_2$ theory applies, and Bartlett's formula from classical time-series analysis [31] provides asymptotic normality for $\hat{\rho}(h)$;

under the null hypothesis of independence, one constructs for each $\hat{\rho}(h)$ a 95% confidence interval. This leads to the window that many mature statistical packages, such as Splus or R, automatically plot. If the acf spikes at various lags protrude from the window for less than 5% of the lags, we shrug comfortably and think so far, so good. No evidence against independence has turned up using this technique.

(b) *The variances are infinite*. If the data are heavy tailed with infinite variance, the mathematical correlations do not exist. However, under the null hypothesis of independence, formulas of Davis and Resnick [31, 73–75] provide for asymptotic distributions of $\hat{\rho}(h)$ given by the distribution of the ratio of stable random variables. The distribution cannot be calculated explicitly, but percentiles of the distribution can be simulated and incorporated into a routine. Then a window is plotted and the procedure outlined in (a) can be carried out with this new window.

When heavy tails are present, and especially when the data are positive, as is frequently the case, it makes little sense to center at $\bar{X}$, and the following heavy-tailed version is used:

$$\hat{\rho}_H(h) = \frac{\sum_{t=1}^{n-h} X_t X_{t+h}}{\sum_{t=1}^{n} X_t^2}. \tag{9.77}$$

We now outline how to use $\hat{\rho}(h)$ and $\hat{\rho}_H(h)$ as a diagnostic for detecting departures from the iid assumption.

We assume the stationary sequence $\{X_t\}$ is $\{Z_t\}$, an iid sequence of random variables satisfying the global regular variation assumptions as discussed in (6.38)–(6.40) (p. 202). This means that

$$P[|Z_1| > x] = x^{-\alpha} L(x) \quad (x \to \infty), \tag{9.78}$$

$$\frac{P[Z_1 > x]}{P[|Z_1| > x]} \to p \qquad (x \to \infty), \tag{9.79}$$

and $\alpha > 0$, $L$ is slowly varying, and $0 \leq p \leq 1$. Note that if $Z_1 \geq 0$, then $p = 1$. If $\alpha < 2$, there is no finite second moment, and hence the mathematical correlations of the $Z$s do not exist. We assume $0 < \alpha < 2$ as our standing assumption for this section.

### 9.5.2 Limit theory

#### Preliminaries

We add an additional assumption that

$$\mathbf{E}(|Z_0|^{\alpha}) = \infty. \tag{9.80}$$

Refer to Problem 7.8 (p. 250). It follows from (9.80) and global regular variation that $Z_0 Z_1$ also satisfies global regular variation with index $-\alpha$:

$$\mathbb{P}[|Z_0 Z_1| > x] = x^{-\alpha}\tilde{L}(x) \qquad (x \to \infty), \tag{9.81}$$

$$\frac{\mathbb{P}[Z_0 Z_1 > x]}{\mathbb{P}[|Z_0 Z_1| > x]} \to \tilde{p} := p^2 + (1-p)^2 \quad (x \to \infty) \tag{9.82}$$

with $p$ given by (9.79). Define the quantile functions

$$b_n = b(n) = \left(\frac{1}{\mathbb{P}[|Z_0| > \cdot]}\right)^{\leftarrow}(n), \qquad \tilde{b}_n = \tilde{b}(n) = \left(\frac{1}{\mathbb{P}[|Z_0 Z_1| > \cdot]}\right)^{\leftarrow}(n).$$

Because of (9.80), $Z_0 Z_1$ has a heavier tail than $Z_0$:

$$\frac{\mathbb{P}[|Z_0 Z_1| > t]}{\mathbb{P}[|Z_0| > t]} \to \infty \quad (t \to \infty), \tag{9.83}$$

since

$$\begin{aligned}\frac{\mathbb{P}[|Z_0 Z_1| > t]}{\mathbb{P}[|Z_0| > t]} &\geq \frac{\mathbb{P}[|Z_0| > t/|Z_1|, 0 < |Z_1| \leq M]}{\mathbb{P}[|Z_0| > t]} \\ &= \int_0^M \frac{\mathbb{P}[|Z_0| > t/y]}{\mathbb{P}[|Z_0| > t]} \mathbb{P}[|Z_1| \in dy].\end{aligned}$$

From Fatou's lemma and the previous line,

$$\liminf_{t\to\infty} \frac{\mathbb{P}[|Z_0 Z_1| > t]}{\mathbb{P}[|Z_0| > t]} \geq \int_0^M y^{\alpha} \mathbb{P}[|Z_1| \in dy]$$

for any $M$. Letting $M \to \infty$ and using the assumption $\mathbf{E}(|Z_0|^{\alpha}) = \infty$ gives (9.83). Taking inverses, we get

$$\lim_{n\to\infty} \frac{\tilde{b}_n}{b_n} = \infty. \tag{9.84}$$

**Point process limits**

For $d \geq 1$, define the vector

$$\boldsymbol{X}_{n,i} = (b_n^{-1} Z_i, \tilde{b}_n^{-1} Z_i Z_{i+1}, \ldots, \tilde{b}_n^{-1} Z_i Z_{i+d}),$$

and based on these vectors, define the random point measures

$$N_n = \sum_{i=1}^{n} \epsilon_{\boldsymbol{X}_{n,i}}, \quad n \geq 1,$$

in $M_p(\mathbb{E})$, where $\mathbb{E} = [-\infty, \infty]^{d+1} \setminus \{\mathbf{0}\}$. If $\{N_n\}$ has a point process limit $N_\infty$, it is clear the points of this limit must lie on the axes, since for any $M$, $x > 0$, $y > 0$, and all large $n$,

$$
\begin{aligned}
n\mathbb{P}[|Z_1| > b_n x, |Z_1 Z_2| > \tilde{b}_n y] \\
&\le n\mathbb{P}[|Z_1| > b_n M] + n\mathbb{P}[|Z_1| \le b_n M, |Z_1| > b_n x, |Z_1 Z_2| > \tilde{b}_n y] \\
&\le M^{-\alpha} + o(1) + n\mathbb{P}[|Z_1| > b_n x]\mathbb{P}\left[|Z_2| > \frac{\tilde{b}_n y}{b_n M}\right] \\
&\to M^{-\alpha} + x^{-\alpha} \cdot 0 \quad (n \to \infty) \\
&\to 0 \quad (M \to \infty),
\end{aligned} \tag{9.85}
$$

where we applied (9.84). On the other hand, for any $M > 0$ and $x > 0$, $y > 0$,

$$
\begin{aligned}
n\mathbb{P}[|Z_1 Z_2| > \tilde{b}_n x, |Z_2 Z_3| > \tilde{b}_n y] \\
&\le n\mathbb{P}[|Z_2| > b_n M] + n\mathbb{P}[|Z_2| \le b_n M, |Z_1 Z_2| > \tilde{b}_n x, |Z_2 Z_3| > \tilde{b}_n y] \\
&\le M^{-\alpha} + o(1) + n\mathbb{P}[|Z_1 Z_2| > \tilde{b}_n x] P\left[|Z_3| > \frac{\tilde{b}_n y}{b_n M}\right] \\
&\to M^{-\alpha} \quad (n \to \infty) \\
&\to 0 \quad (M \to \infty).
\end{aligned} \tag{9.86}
$$

This shows that any limit point measure must have an expected number of points in regions away from the axes equal to 0. Note where the assumption $\mathbf{E}(|Z_1|^\alpha) = \infty$ was used.

We now show that $N_n$ has a weak limit, which we call $N_\infty$. The description of $N_\infty$ is as follows: Let

$$N_\infty^{(i)} = \sum_k \epsilon_{j_k^{(i)}}; \quad i = 0, \dots, d,$$

be $d+1$ independent Poisson processes on $[-\infty, \infty] \setminus \{0\}$. The mean measure of $N_\infty^{(0)}$ has density

$$\alpha p x^{-\alpha-1} 1_{(0,\infty)}(x) + \alpha(1-p)|x|^{-\alpha-1} 1_{(-\infty,0)}(x),$$

corresponding to (9.78), (9.79). The mean measures of the iid Poisson processes $N_\infty^{(i)}$, $i = 1, \dots, d$, are all equal to the measure with density

$$\alpha \tilde{p} x^{-\alpha-1} 1_{(0,\infty)}(x) + \alpha(1-\tilde{p})|x|^{-\alpha-1} 1_{(-\infty,0)}(x),$$

corresponding to (9.81), (9.82). Let $\boldsymbol{e}_i$, $i = 0, \dots, d$, be the $(d+1)$-dimensional basis vectors with 0s everywhere except one slot in which there is a 1. Define

$$N_\infty = \sum_k \sum_{i=0}^{d} \epsilon_{j_k^{(i)} \boldsymbol{e}_i},$$

which results from taking the points of $N_\infty^{(i)}$ and laying them on the $i$th axis. $N_\infty$ is a Poisson process with all points on the axes.

**Proposition 9.8.** *Under the described conditions,*

$$N_n \Rightarrow N_\infty \tag{9.87}$$

*in* $M_p(\mathbb{E})$.

*Proof.* Our experience with the proof of Lemma 7.2 (p. 228) will be helpful here, as will Problem 6.16 (p. 210), since for each $n$, the family $\{\boldsymbol{X}_{n,i}, i \geq 1\}$ is $(d+1)$-dependent.

According to Problem 6.16, there are two conditions to verify. The first is (6.48) (p. 210); that is, as $n \to \infty$,

$$n\mathbb{P}[\boldsymbol{X}_{n,1} \in \cdot] \stackrel{v}{\to} \mathbf{E}(N_\infty(\cdot)). \tag{9.88}$$

Let $f \in C_K^+(\mathbb{E})$. As with Lemma 7.2, we use (9.85) and (9.86) to show that for any $f \in C_K^+(\mathbb{E})$,

$$\begin{aligned} n\mathbf{E}f(\boldsymbol{X}_{n,1}) &= n\mathbf{E}f(b_n^{-1}Z_1, \tilde{b}_n^{-1}Z_1Z_2, \ldots, \tilde{b}_n^{-1}Z_1Z_{1+d}) \\ &= n\mathbf{E}f(b_n^{-1}Z_1\boldsymbol{e}_0) + \sum_{l=1}^{d} n\mathbf{E}f(\tilde{b}_n^{-1}Z_1Z_{1+l}\boldsymbol{e}_l) + o(1). \end{aligned}$$

Now apply the global regular variation conditions (9.78) and (9.79) for $Z_1$ and the comparable conditions (9.81) and (9.82) for $Z_1Z_2$ to get (9.88).

The second condition from Problem 6.16 is

$$\lim_{k\to\infty} \limsup_{n\to\infty} n \sum_{i=2}^{[n/k]} \mathbf{E}(g(X_{n,1})g(X_{n,i})) = 0 \tag{9.89}$$

for $g \in C_K^+(\mathbb{E})$, $g \leq 1$. We have

$$\begin{aligned} & n \sum_{i=2}^{[n/k]} \mathbf{E}(g(X_{n,1})g(X_{n,i})) \\ &\quad = n \sum_{i=2}^{d+1} \mathbf{E}(g(X_{n,1})g(X_{n,i})) + n \sum_{i=d+2}^{[n/k]} \mathbf{E}(g(X_{n,1})g(X_{n,i})) = A + B, \end{aligned}$$

and because of $(d+1)$-dependence,

$$\begin{aligned} B &\leq n[n/k](\mathbf{E}(g(\boldsymbol{X}_{n,1})))^2 \\ &\sim \frac{1}{k}(n\mathbf{E}(g(\boldsymbol{X}_{n,1})))^2 \to \frac{1}{k}(\mathbf{E}(N_\infty(g)))^2 \quad (n \to \infty) \\ &\to 0 \quad (k \to \infty). \end{aligned}$$

The remaining term $A \to 0$ since each term in the finite sum is dominated by either a term governed by (9.85) or by (9.86). □

**Summing the points**

Modify (9.87) to get

$$\sum_{i=1}^{n} \epsilon_{(b_n^{-2}Z_i^2, \tilde{b}_n^{-1}Z_iZ_{i+1},\ldots,\tilde{b}_n^{-1}Z_iZ_{i+d})} \Rightarrow \sum_k \epsilon_{(j_k^{(0)})^2 \boldsymbol{e}_0} + \sum_k \sum_{i=1}^{d} \epsilon_{j_k^{(i)} \boldsymbol{e}_i}. \tag{9.90}$$

Apply the summation functional and the converging together technique explained in Section 7.2. Keep in mind that

$$P[Z_1^2 > x] \in \mathrm{RV}_{-\alpha/2}, \quad 0 < \frac{\alpha}{2} < 1,$$

so that by Problem 7.3, no centering is needed for $\sum_{i=1}^n Z_i^2$. The result of this technique applied to (9.90) is

$$\left( \frac{\sum_{i=1}^n Z_i^2}{b_n^2}, \frac{\sum_{i=1}^n (Z_i Z_{i+1} - \mu_n)}{\tilde{b}_n}, \ldots, \frac{\sum_{i=1}^n (Z_i Z_{i+d} - \mu_n)}{\tilde{b}_n} \right) \Rightarrow (S_0, S_1, \ldots, S_d), \tag{9.91}$$

where

$$\mu_n = \mathbf{E}\left( Z_1 Z_2 1_{[|Z_1 Z_2| \leq \tilde{b}_n]} \right)$$

and $S_1, \ldots, S_d$ are independent stable random variables constructed in Section 5.5.1 (p. 149). Because $\alpha/2 < 1$, $S_0$ is stable with index $\alpha/2$, obtained without needing centering, and has representation $\sum_k (j_k^{(0)})^2$. (See Section 5.5.2 (p. 153).) The presence of the centering $\mu_n$ in (9.91) is a nuisance we have to deal with. For now, divide the first component into the others to get

$$\frac{b_n^2}{\tilde{b}_n} \left( \hat{\rho}_H(j) - \frac{\mu_n}{\sum_{i=1}^n Z_i^2}; j = 1, \ldots, d \right) \Rightarrow \left( \frac{S_j}{S_0}; j = 1, \ldots, d \right), \tag{9.92}$$

where $\hat{\rho}_H(j)$ was defined in (9.77).

How do we cope with the centering and make something useful out of this limit theory?

### 9.5.3 The heavy-tailed sample acf; $\alpha < 1$

If we assume that $\alpha < 1$, then the centering by $\mu_n$ is not necessary, and (9.91) becomes

$$\left( \frac{\sum_{i=1}^n Z_i^2}{b_n^2}, \frac{\sum_{i=1}^n Z_i Z_{i+1}}{\tilde{b}_n}, \ldots, \frac{\sum_{i=1}^n Z_i Z_{i+d}}{\tilde{b}_n} \right) \Rightarrow (S_0, \tilde{S}_1, \ldots, \tilde{S}_d), \tag{9.93}$$

where $\tilde{S}_1, \ldots, \tilde{S}_d$ are independent stable subordinator random variables constructed in Section 5.5.2 (p. 153). Divide the first component into the others to get

$$\frac{b_n^2}{\tilde{b}_n}(\hat{\rho}_H(j); j = 1, \ldots, d) \Rightarrow \left(\frac{\tilde{S}_j}{S_0}; j = 1, \ldots, d\right). \tag{9.94}$$

So in the very heavy-tail case ($\mathbf{E}(|Z_1|) = \infty$), we use the heavy-tailed acf.

### 9.5.4 The classical sample acf: $1 < \alpha < 2$

For the case $1 < \alpha < 2$, the limit theory is more amenable to using the classical sample acf defined in (9.76) (p. 341). We proceed in a series of steps to redo the results leading to (9.91).

STEP 1. Since the mean $\mu := \mathbf{E}(Z_1)$ exists, we may recenter (9.91) so that the last $d$-components are centered to zero mean. This yields

$$\left(\frac{\sum_{i=1}^n Z_i^2}{b_n^2}, \frac{\sum_{i=1}^n (Z_i Z_{i+1} - \mu^2)}{\tilde{b}_n}, \ldots, \frac{\sum_{i=1}^n (Z_i Z_{i+d} - \mu^2)}{\tilde{b}_n}\right) \Rightarrow (S_0, S_1^0, \ldots, S_d^0), \tag{9.95}$$

where $(S_1^0, \ldots, S_d^0)$ are iid, zero mean, stable random variables.

STEP 2. We claim, as $n \to \infty$,

$$b_n^{-2} \sum_{i=1}^n Z_i^2 - b_n^{-2} \sum_{i=1}^n (Z_i - \bar{Z})^2 \xrightarrow{P} 0. \tag{9.96}$$

To verify this, use the identity

$$\sum_{i=1}^n (Z_i - \bar{Z})^2 = \sum_{i=1}^n Z_i^2 - n\bar{Z}^2,$$

so that the difference in (9.96) is $n\bar{Z}^2/b_n^2$. Because $\alpha > 1$, $\mathbf{E}(|Z_1|) < \infty$ and $\bar{Z}^2 \xrightarrow{P} (\mathbf{E}(Z_1))^2 < \infty$. So $\bar{Z}^2$ is stochastically bounded. However, $n/b_n^2 \in RV_{1-2/\alpha}$, and since $2/\alpha > 1$, we get $n/b_n^2 \to 0$, proving the assertion (9.96).

By Slutsky's theorem, Theorem 3.4 (p. 55), (9.96) allows us to modify (9.91) to get

$$\left(\frac{\sum_{i=1}^n (Z_i - \bar{Z})^2}{b_n^2}, \frac{\sum_{i=1}^n (Z_i Z_{i+1} - \mu^2)}{\tilde{b}_n}, \ldots, \frac{\sum_{i=1}^n (Z_i Z_{i+d} - \mu^2)}{\tilde{b}_n}\right)$$
$$\Rightarrow (S_0, S_1^0, \ldots, S_d^0), \tag{9.97}$$

STEP 3. We claim, for each $l = 1, \ldots, d$, that as $n \to \infty$,

$$\frac{1}{\tilde{b}_n}\sum_{i=1}^{n}(Z_i Z_{i+l} - \mu^2) - \frac{1}{\tilde{b}_n}\sum_{i=1}^{n}(Z_i - \bar{Z})(Z_{i+l} - \bar{Z}) \xrightarrow{P} 0. \tag{9.98}$$

To see this, write

$$\begin{aligned}\frac{1}{\tilde{b}_n}\sum_{i=1}^{n}(Z_i - \bar{Z})(Z_{i+l} - \bar{Z}) &= \frac{1}{\tilde{b}_n}\left(\sum_{i=1}^{n} Z_i Z_{i+l} - \bar{Z}\sum_{i=1}^{n} Z_{i+l}\right)\\ &= \frac{1}{\tilde{b}_n}\sum_{i=1}^{n}(Z_i Z_{i+l} - \mu^2) - \frac{1}{\tilde{b}_n}\left(\bar{Z}\sum_{i=1}^{n} Z_{i+l} - n\mu^2\right).\end{aligned}$$

Now

$$\begin{aligned}\frac{1}{\tilde{b}_n}\left(\bar{Z}\sum_{i=1}^{n} Z_{i+l} - n\mu^2\right) &= \sum_{i=1}^{n}\frac{(Z_{i+l} - \mu)}{b_n}\bar{Z}\frac{b_n}{\tilde{b}_n} + \frac{n\mu\bar{Z} - n\mu^2}{\tilde{b}_n}\\ &= A + B.\end{aligned}$$

For $A$, observe that $\bar{Z}$ converges to the mean and is stochastically bounded, $b_n^{-1}\sum_{i=1}^{n}(Z_{i+l} - \mu)$ converges to a stable law and is hence stochastically bounded, and $b_n/\tilde{b}_n \to 0$. This drives $A$ to zero. Write $B$ as

$$\mu\frac{b_n}{\tilde{b}_n}\left(\frac{\sum_{i=1}^{n}(Z_i - \mu)}{b_n}\right) \xrightarrow{P} 0$$

as $n \to \infty$. This yields (9.98) and allows us to rewrite (9.97) as

$$\left(\frac{\sum_{i=1}^{n}(Z_i - \bar{Z})^2}{b_n^2}, \frac{\sum_{i=1}^{n}(Z_i - \bar{Z})(Z_{i+1} - \bar{Z})}{\tilde{b}_n}, \ldots, \frac{\sum_{i=1}^{n}(Z_i - \bar{Z})(Z_{i+d} - \bar{Z})}{\tilde{b}_n}\right) \Rightarrow (S_0, S_1^0, \ldots, S_d^0). \tag{9.99}$$

Dividing the first component into the others, yields the following result.

**Proposition 9.9.** *If* $1 < \alpha < 2$,

$$\frac{b_n^2}{\tilde{b}_n}(\hat{\rho}(1), \ldots, \hat{\rho}(d)) \Rightarrow \left(\frac{S_1^0}{S_0}, \ldots, \frac{S_d^0}{S_0}\right)$$

*in* $\mathbb{R}^d$, *where* $(S_1^0, \ldots, S_d^0)$ *are iid, zero mean, stable random variables with index* $\alpha$ *and* $S_0$ *is independent of the rest and has index* $\alpha/2$.

### 9.5.5 Suggestions to use

Here are suggestions for how to use this theory in practice. With a null hypothesis that the data comes from an iid heavy-tailed model with $0 < \alpha < 2$, do the following:

1. Estimate $\alpha$. If $\alpha < 1$, use $\hat{\rho}_H(l)$, $l = 1, \ldots, d$. If $1 < \alpha < 2$, use the classical $\hat{\rho}(l)$, $l = 1, \ldots, d$. With a reasonably sized data set (say, $> 3000$), $d = 25$ usually provides enough lags.

2. If $1 < \alpha < 2$, find a quantile $q$ satisfying

$$\mathbb{P}\left[\left|\frac{b_n^2}{\tilde{b}_n}\hat{\rho}(1)\right| \leq q\right] \approx \mathbb{P}\left[\left|\frac{S_1^0}{S_0}\right| \leq q\right] = 0.95.$$

   If $\alpha < 1$, find a quantile $q$ satisfying

$$\mathbb{P}\left[\frac{b_n^2}{\tilde{b}_n}\hat{\rho}_H(1) \leq q\right] \approx \mathbb{P}\left[\frac{\tilde{S}_1}{S_0} \leq q\right] = 0.95.$$

3. Plot either $\hat{\rho}_H(l)$ or $\hat{\rho}(l)$, depending on the case, for $l = 1, \ldots, d$. Depending on the case, if

$$\hat{\rho}_H(l) > \frac{\tilde{b}_n}{b_n^2} q$$

   or

$$|\hat{\rho}(l)| > \frac{\tilde{b}_n}{b_n^2} q$$

   for more than 5% of the values of $l$, be suspicious of the iid assumption.

In practice, one will not know $b_n$ or $\tilde{b}_n$. If tails are close to pure Pareto, as suggested in Problem 9.13 (p. 356), then $b_n^2/\tilde{b}_n \sim (n/\log n)^{1/\alpha}$. Replace $\alpha$ with its estimate. Also, one can only get $q$ by simulation. This is easier said than done, because one must simulate the random variables given in (9.94) or Proposition 9.9. In principle, this is straightforward, but there are a gazillion parameterizations of stable laws; and each package has its own simulation method. So some patience is needed to get the correct values.

*Example* 9.4 (*call holding*). This data set consists of 4045 telephone call-holding times indexed according to the time of initiation of the call. The data set is about 15 years old, and it presumably reflects a time when telephone modem calls were common; the data are surprisingly heavy tailed. Figure 9.15 gives a time-series plot and a Hill plot for these data. The Hill plot is gratifyingly stable.

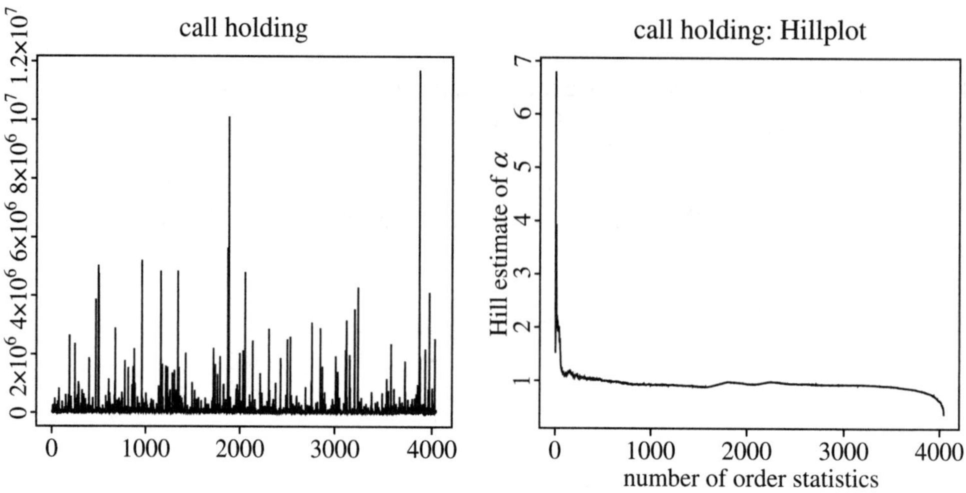


**Fig. 9.15.** Time-series and Hill plots for call-holding data.

The estimate of $\alpha$ in the call-holding example is between .9 and 1, and with a combination of QQ plotting and more careful examination, we settled on an estimated value of $\hat{\alpha} = 0.97$.

Figure 9.16 shows the classical acf for the call-holding data side by side with the heavy-tailed modification. Because of the many exceedances of the quantile line on the right of Figure 9.16, one should have reservations about modeling these data as iid.

## 9.6 Problems

**9.1 (Asymptotic normality of the Pickands estimator).** State and derive asymptotic normality for the Pickands estimator (review Section 4.5.2 (p. 93)) for positive $\gamma$. Compare asymptotic variances between the Hill and the Pickands estimators. (The asymptotic variance is given in (4.34) (p. 94).)

**9.2 (Asymptotic normality of the smooHill estimator).** Assuming iid nonnegative observations from a distribution with a regularly varying tail, state and derive asymptotic normality of the smooHill estimator discussed in Section 4.4.3 (p. 89).

**9.3 ([217]).** Suppose that $d = 3$ and $Z^{(1)}$, $Z^{(2)}$, $Z^{(3)}$ are iid Pareto random variables with parameter 1. Set $\mathbb{E}^1 = (\mathbf{0}, \boldsymbol{\infty}]$. Verify that hidden regular variation takes place on $\mathbb{E} \supset \mathbb{E}^0 \supset \mathbb{E}^1$ with $\alpha = 1$, $\alpha^0 = 2$, and $\alpha^1 = 3$.

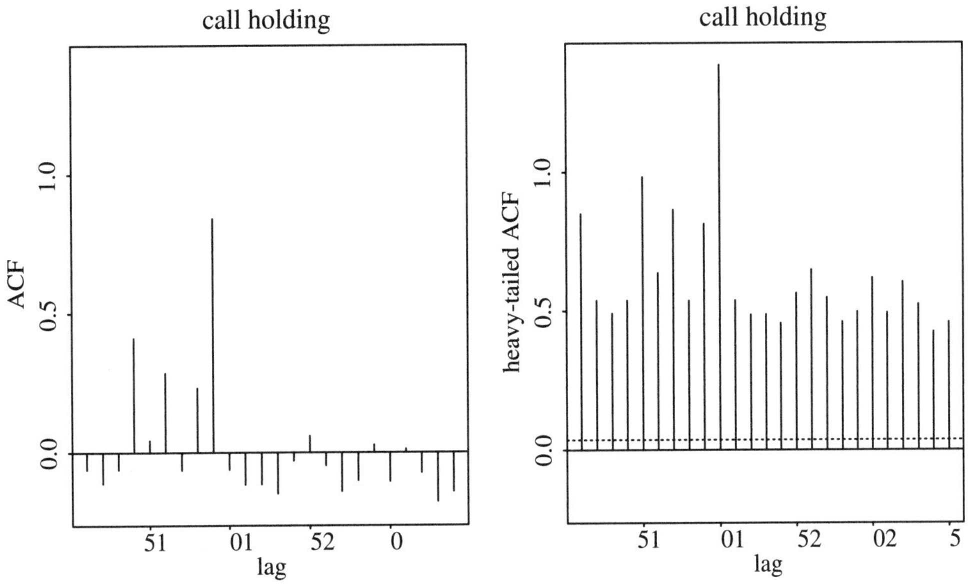


**Fig. 9.16.** Classical (left) and heavy-tailed (right) acf plots for call holding.

**9.4 ([49, 163, 217]).** Suppose that $d = 2$ and hidden regular variation holds; that is, (9.51) and (9.54) hold. Show that

$$\bar{\chi} := \lim_{t\to\infty} \frac{2\log P[Z^{(1)} > t]}{\log P[Z^{(1)} > t,\, Z^{(2)} > t]} - 1 = \frac{2\alpha - \alpha^0}{\alpha^0}.$$

**9.5 (Tail equivalence [217]).** Suppose $\boldsymbol{Y}$ and $\boldsymbol{Z}$ are $[\boldsymbol{0}, \infty)$-valued random vectors with distributions $F$ and $G$, respectively. $F$ and $G$ (or, by abuse of language, $\boldsymbol{Y}$ and $\boldsymbol{Z}$) *are tail equivalent on the cone* $\mathfrak{C} \subset \mathbb{E}$ if there exists a scaling function $b(t) \uparrow \infty$ such that

$$t\mathbb{P}\left[\frac{\boldsymbol{Y}}{b(t)} \in \cdot\right] = tF(b(t)\cdot) \xrightarrow{v} \nu \quad \text{and} \quad t\mathbb{P}\left[\frac{\boldsymbol{Z}}{b(t)} \in \cdot\right] = tG(b(t)\cdot) \xrightarrow{v} c\nu \tag{9.100}$$

in $M_+(\mathfrak{C})$ for some constant $c \in (0, \infty)$ and nonzero Radon measure $\nu$ on $\mathfrak{C}$. We shall write

$$\boldsymbol{Y} \overset{te(\mathfrak{C})}{\sim} \boldsymbol{Z}.$$

*Prove*: If $\{\boldsymbol{Y}_n, n \geq 1\}$ is an iid sample from $F$ and $\{\boldsymbol{Z}_n, n \geq 1\}$ is an iid sample from $G$, then

$$\sum_{i=1}^{n} \epsilon_{\boldsymbol{Y}_i/b(n)} \Rightarrow \text{PRM}(\nu) \quad \text{iff} \sum_{i=1}^{n} \epsilon_{\boldsymbol{Z}_i/b(n)} \Rightarrow \text{PRM}(c\nu) \tag{9.101}$$

in $M_p(\mathfrak{C})$, the space of Radon point measures on $\mathfrak{C}$.

**9.6 (Finite hidden measure [217]).** Suppose that $\boldsymbol{V}$ is regularly varying on $\mathbb{E}^0$ with index $\alpha^0$, scaling function $b^0(t)$, limit measure $\nu^0$, angular measure $S^0$ on $\aleph^0$. The following are equivalent:

(i) $S^0$ is finite on $\aleph^0$.

(ii) There exists a random vector $\boldsymbol{V}_*$ defined on $\mathbb{E}$ such that

$$\boldsymbol{V}_* \overset{te(\mathbb{E}^0)}{\sim} \boldsymbol{V},$$

and for $i = 1, \ldots, d$,

$$t\mathbb{P}[V_*^{(i)} > b^0(t)x] \to cx^{-\alpha^0}, \quad t \to \infty,$$

for some $c > 0$, so that each component $V_*^{(i)}$ has regularly varying tail probabilities with index $\alpha^0$.

(iii) There exists a random vector $\boldsymbol{V}_*$ defined on $\mathbb{E}$ such that

$$\boldsymbol{V}_* \overset{te(\mathbb{E}^0)}{\sim} \boldsymbol{V},$$

and $\boldsymbol{V}_*$ is regularly varying on the full cone $\mathbb{E}$ with scaling function $b^0(t)$, limit measure $\nu$, and

$$\nu|_{\mathbb{E}^0} = \nu^0;$$

that is, the restriction of $\nu$ to $\mathbb{E}^0$ is $\nu^0$.

(iv) There exists a random vector $\boldsymbol{V}_*$ defined on $\mathbb{E}$ such that

$$\boldsymbol{V}_* \overset{te(\mathbb{E}^0)}{\sim} \boldsymbol{V},$$

such that for any $\boldsymbol{s} \in [\boldsymbol{0}, \infty) \setminus \{\boldsymbol{0}\}$ and any $\boldsymbol{a} \in (\boldsymbol{0}, \infty] \setminus \bigcup_{i=1}^d \{t\boldsymbol{e}_i^{-1} : 0 < t < \infty\}$,

$$\bigvee_{i=1}^d s_i V_*^{(i)} \quad \text{and} \quad \bigwedge_{i=1}^d a_i V_*^{(i)}$$

are tail equivalent on $[0, \infty)$ and have regularly varying tail probabilities of index $\alpha^0$. (Recall that $\boldsymbol{e}_i^{-1} = (\infty, \ldots, \infty, 1, \infty, \ldots, \infty)$ is the vector whose $i$th component is 1 and whose other components are $\infty$.)

**9.7 ([217]).** Suppose that $\boldsymbol{Z}$ is a random vector of dimension $d > 2$, which is multivariate regularly varying on $\mathbb{E}$ with hidden regular variation on $\mathbb{E}^0$, having limit measures $\nu$, $\nu^0$. Define $\tilde{\mathbb{E}}$ and $\tilde{\mathbb{E}}^0$ to be the spaces corresponding to $\mathbb{E}$ and $\mathbb{E}^0$ in two dimensions; that is,

$$\tilde{\mathbb{E}} = [0,\infty]^2 \setminus \{(0,0)\}; \quad \tilde{\mathbb{E}}^0 = (0,\infty]^2.$$

Then for some pair $1 \le i < j \le d$, necessarily

$$\nu^0(\{\boldsymbol{x} \in \mathbb{E}^0 : x^{(i)} > 0, x^{(j)} > 0\}) \neq 0, \tag{9.102}$$

and for all such pairs, we must have $(Z^{(i)}, Z^{(j)})$ to be multivariate regularly varying on $\tilde{\mathbb{E}}$ with hidden regular variation on $\tilde{\mathbb{E}}^0$. Also, the hidden angular measure of $\boldsymbol{Z}$ is finite iff for all pairs $1 \le i < j \le d$, satisfying the condition (9.102), the hidden angular measure of $(Z^{(i)}, Z^{(j)})$ is finite.

**9.8 ([217]).** Suppose $U_i$, $i = 1, 2, 3$, are iid Pareto random variables on $[1, \infty)$ with parameter 1. Let $B_i$, $i = 1, 2$, be iid Bernoulli random variables with

$$P[B_i = 0] = P[B_i = 1] = \frac{1}{2},$$

and $\{B_i\}$ and $\{U_i\}$ are independent. Define

$$W = (1 - B_1)U_3, \qquad \boldsymbol{Z} = B_2(U_1, 0, W) + (1 - B_2)(0, U_2, W).$$

Then

$$Z^{(1)} = B_2 U_1, \qquad Z^{(2)} = (1 - B_2)U_2, \qquad Z^{(3)} = W$$

are identically distributed, and each random variable has a distribution that has an atom of size $\frac{1}{2}$ at 0 and a Pareto density with parameter 1 and total mass $\frac{1}{2}$ on $[1, \infty)$. The distribution of $\boldsymbol{Z}$ is supported on the planes where either of the first two coordinates vanish. Prove that $\boldsymbol{Z}$ is multivariate regularly varying on $\mathbb{E}$ with $\alpha = 1$ and $b(t) = t$ and that has hidden regular variation on $\mathbb{E}^0$ with $\alpha^0 = 2$ and $b^0(t) = \sqrt{t}$. However, $\nu^0(\{\boldsymbol{x} : x^{(1)} > 0, x^{(2)} > 0\}) = 0$. So hidden regular variation is not possible for the marginal distribution of $(Z^{(1)}, Z^{(2)})$. We have hidden regular variation on $\mathbb{E}^0$ but *not* $\mathbb{E}^{00} = (\boldsymbol{0}, \infty]$. So regular variation on $\mathbb{E}^0$ and $\mathbb{E}^{00}$ can be quite different.

**9.9 ([217]).** Suppose $\boldsymbol{Z}$ is a $d$-dimensional random vector defined on $[\boldsymbol{0}, \infty)$ that is multivariate regularly varying on $\mathbb{E}$ and has hidden regular variation on $\mathbb{E}^0$ with limit measure $\nu^0$. Then $\nu^0$ has finite angular measure iff for all pairs $(i, j)$, $i \neq j$, the function

$$\bar{G}_{i,j}(s) := \nu^0\left(\left\{\boldsymbol{x} \in \mathbb{E}^0 : \boldsymbol{x} \ge \boldsymbol{1}, \frac{x^{(i)}}{x^{(j)}} > s\right\}\right)$$

satisfies

$$\int_1^\infty s^{\alpha^0 - 1} \bar{G}_{i,j}(s) ds < \infty.$$

**9.10 (Examples where the hidden angular measure is infinite [217]).** Let $d = 2$. Suppose that $R$ is a Pareto random variable on $[1, \infty)$ with parameter 1 and $\boldsymbol{\Theta}$ is a random vector living on $\{\boldsymbol{x} \in \mathbb{E} : x^{(1)} \wedge x^{(2)} = 1\}$, and for some probability distribution $G$ concentrating on $[1, \infty)$, we have

$$\mathbb{P}[\boldsymbol{\Theta} \in \{1\} \times (s, \infty]] = \mathbb{P}[\boldsymbol{\Theta} \in (s, \infty] \times \{1\}] = \frac{1}{2}\bar{G}(s), \quad s \geq 1. \tag{9.103}$$

Suppose $R$ and $\boldsymbol{\Theta}$ are independent. Define

$$V = R \cdot \boldsymbol{\Theta}. \tag{9.104}$$

Verify that by (9.103), we have for $s > 1$,

$$\mathbb{P}[\Theta^{(1)} > s] = \mathbb{P}[\boldsymbol{\Theta} \in (s, \infty] \times \{1\}] = \frac{1}{2}\bar{G}(s) = \mathbb{P}[\boldsymbol{\Theta} \in \{1\} \times (s, \infty]] = \mathbb{P}[\Theta^{(2)} > s],$$

so that the marginals of $\boldsymbol{\Theta}$ are the same.

With $\boldsymbol{V}$ as specified, $\boldsymbol{V}$ is regularly varying on $\mathbb{E}^0$ with index $\alpha^0 = 1$. Verify that the hidden limit measure $\nu^0$ has infinite angular measure iff

$$\int_1^\infty \bar{G}(s)ds = \infty. \tag{9.105}$$

Furthermore, if (9.105) holds, we have the following:

(i) $\boldsymbol{V}$ is regularly varying on $\mathbb{E}$ with index $\alpha < 1$ (and hence possesses hidden regular variation with infinite hidden angular measure) iff

$$1 - G \in RV_{-\alpha}.$$

In this case, for $i = 1, 2$, we have

$$\mathbb{P}[V^{(i)} > x] \sim \frac{1}{2(1-\alpha)}\bar{G}(x), \quad x \to \infty.$$

(ii) $\boldsymbol{V}$ is regularly varying on $\mathbb{E}$ with index $\alpha = 1$ and possesses hidden regular variation with infinite hidden angular measure iff

$$L(x) := \int_0^x \bar{G}(s)ds \in RV_0 \quad \text{and} \quad L(x) \uparrow \infty.$$

A sufficient condition is $\bar{G} \in RV_{-1}$ with $\int_0^\infty \bar{G}(s)ds = \infty$, in which case

$$\lim_{x\to\infty} \frac{\mathbb{P}[V^{(i)} > x]}{\mathbb{P}[R > x]} = \lim_{x\to\infty} \frac{\mathbb{P}[V^{(i)} > x]}{\mathbb{P}[\Theta^{(i)} > x]} = \infty$$

for $i = 1, 2$.

**9.11 ([162]).** Suppose $(U, V)$ has a distribution that is standard regularly varying on $[0,\infty)^2$, which means that there is a limit measure $\nu$ on $\mathbb{E}$ such that

$$t\mathbb{P}\left[\left(\frac{U}{t}, \frac{V}{t}\right) \in \cdot\right] \stackrel{v}{\to} \nu$$

in $M_+(\mathbb{E})$. Suppose $(U_i, V_i)$, $i = 1, 2$, are iid copies of $(U, V)$, and we define

$$(X, Y) = B(U_1, h(V_1)) + (1 - B)(h(U_2), V_2), \tag{9.106}$$

where $h \in RV_p$, $h(t)/t \to 0$, $B$ is a Bernoulli random variable independent of $\{(U_i, V_i); i = 1, 2\}$, and

$$\mathbb{P}[B = 0] = \mathbb{P}[B = 1] = \frac{1}{2}.$$

Then $(X, Y)$ is standard regularly varying and asymptotically independent and

$$t\mathbb{P}\left[\left(\frac{X}{h(t)}, \frac{Y}{t}\right) \in \cdot\right] \stackrel{v}{\to} \mu(\cdot), \tag{9.107}$$

where $\mu$ is given by

$$\mu([0, x] \times (y, \infty]) =: \frac{1}{2}\nu([0, x^{1/p}] \times (y, \infty]).$$

The condition $h(t)/t \to 0$ is necessary and sufficient for $(X, Y)$ to be asymptotically independent.

**9.12 (Different cone; products [215]).** Define the cone $\mathbb{D} := (0, \infty] \times [0, \infty]$. Assume that $X$ and $Y$ are strictly positive, finite random variables. Suppose that

$$t\mathbb{P}[X > b_X(t)] \to 1 \quad (t \to \infty).$$

Verify that the following are equivalent:

(i) $t\mathbb{P}[(\frac{X}{b_X(t)}, Y) \in \cdot] \stackrel{v}{\to} (\nu_\alpha \times G)(\cdot)$ on $\mathbb{D}$ for some $\alpha > 0$ and $G$ a probability measure satisfying $G(0, \infty) = 1$.

(ii) $t\mathbb{P}[\frac{(X, XY)}{b_X(t)} \in \cdot] \stackrel{v}{\to} \nu(\cdot)$ on $\mathbb{D}$, where $\nu(\{(x, y) : x > u\}) > 0$ for all $u > 0$.

In fact, $\nu$ is homogeneous of order $-\alpha$, i.e., $\nu(u\cdot) = u^{-\alpha}\nu(\cdot)$ on $\mathbb{D}$, and is given by

$$\nu = \begin{cases} (\nu_\alpha \times G) \circ \theta^{-1} & \text{on } (0, \infty) \times [0, \infty), \\ 0 & \text{on } \mathbb{D} \setminus ((0, \infty) \times [0, \infty)), \end{cases} \tag{9.108}$$

where $\theta(x, y) = (x, xy)$, if $(x, y) \in D \setminus \{(\infty, 0)\}$ and $\theta(\infty, 0)$ is defined arbitrarily.

**9.13 (Form of $b_n, \tilde{b}_n$).** Suppose that

$$\mathbb{P}[|Z_1| > x] \sim x^{-\alpha} \quad (x \to \infty).$$

Then with reference to Section 9.5.2 (p. 342), verify the following:

$$\mathbb{P}[|Z_1 Z_2| > x] \sim x^{-\alpha}(1 + \alpha \log x), \quad x \to \infty,$$

$$b_n \sim n^{1/\alpha}, \qquad \tilde{b}_n \sim (n \log n)^{1/\alpha},$$

$$\frac{b_n^2}{\tilde{b}_n} \sim \left(\frac{n}{\log n}\right)^{1/\alpha}.$$

**9.14.** Suppose that $(Z_1, Z_2)$ is standard bivariate normal with correlation $\rho$, and suppose for $\alpha > 0$ that $W$ is a nonnegative random variable with $\mathbb{P}[W > x] \in \mathrm{RV}_{-\alpha}$. Consider the distribution of $\sqrt{W}\mathbf{Z}$ and show that it possesses hidden regular variation in the first quadrant. If $\rho = -1$, the coefficient of tail dependence is 0, and if $\rho = 1$, it is 1; otherwise, it is strictly between 0 and 1.

Apply this to the multivariate $t$-density [218, p. 211].

# Part V

# Appendices

# 10

# Notation and Conventions

## 10.1 Vector notation

Vectors are denoted by bold letters, uppercase for random vectors and lowercase for nonrandom vectors. For example, $\boldsymbol{x} = (x^{(1)}, \ldots, x^{(d)}) \in \mathbb{R}^d$. Operations between vectors should be interpreted componentwise, so that for two vectors $\boldsymbol{x}$ and $\boldsymbol{z}$,

$$\boldsymbol{x} < \boldsymbol{z} \text{ means } x^{(i)} < z^{(i)}, \quad i = 1, \ldots, d, \qquad \boldsymbol{x} \leq \boldsymbol{z} \text{ means } x^{(i)} \leq z^{(i)}, \quad i = 1, \ldots, d,$$

$$\boldsymbol{x} = \boldsymbol{z} \text{ means } x^{(i)} = z^{(i)}, \quad i = 1, \ldots, d, \qquad \boldsymbol{z}\boldsymbol{x} = (z^{(1)}x^{(1)}, \ldots, z^{(d)}x^{(d)}),$$

$$\boldsymbol{x} \vee \boldsymbol{z} = (x^{(1)} \vee z^{(1)}, \ldots, x^{(d)} \vee z^{(d)}), \qquad \frac{\boldsymbol{x}}{\boldsymbol{z}} = \left(\frac{x^{(1)}}{z^{(1)}}, \ldots, \frac{x^{(d)}}{z^{(d)}}\right),$$

and so on. Also, define

$$\boldsymbol{0} = (0, \ldots, 0), \qquad \boldsymbol{1} = (1, \ldots, 1),$$

$$\boldsymbol{e}_i = (0, \ldots, 1, \ldots, 0), \qquad \boldsymbol{e}_i^{-1} = (\infty, \ldots, 1, \ldots, \infty),$$

where in $\boldsymbol{e}_i$ and $\boldsymbol{e}_i^{-1}$, the "1" occurs in the $i$th spot. For a real number $c$, write $c\boldsymbol{x} = (cx^{(1)}, \ldots, cx^{(d)})$, as usual. We denote the rectangles (or the higher-dimensional intervals) by

$$[\boldsymbol{a}, \boldsymbol{b}] = \{\boldsymbol{x} \in \mathbb{R}^d : \boldsymbol{a} \leq \boldsymbol{x} \leq \boldsymbol{b}\}.$$

Higher-dimensional rectangles with one or both endpoints open are defined analogously, for example,

$$(\boldsymbol{a}, \boldsymbol{b}] = \{\boldsymbol{x} \in \mathbb{R}^d : \boldsymbol{a} < \boldsymbol{x} \leq \boldsymbol{b}\}.$$

Suppose $\mathbb{E} = [\boldsymbol{0}, \boldsymbol{\infty}] \setminus \{\boldsymbol{0}\}$. Complements are taken with respect to $\mathbb{E}$, so that for $\boldsymbol{x} > \boldsymbol{0}$,

$$[\mathbf{0}, \boldsymbol{x}]^c = \mathbb{E} \setminus [\mathbf{0}, \boldsymbol{x}] = \left\{ \boldsymbol{y} \in \mathbb{E} : \bigvee_{i=1}^{d} \frac{y^{(i)}}{x^{(i)}} > 1 \right\}.$$

The axes originating at $\mathbf{0}$ are $\mathbb{L}_i := \{t\boldsymbol{e}_i, t > 0\}$, $i = 1, \ldots, d$. Then define the cone

$$\mathbb{E}^0 = \mathbb{E} \setminus \bigcup_{i=1}^{d} \mathbb{L}_i = \{\boldsymbol{s} \in \mathbb{E} : \text{for some } 1 \le i < j \le d, s^{(i)} \wedge s^{(j)} > 0\}.$$

If $d = 2$, we have $\mathbb{E}^0 = (0, \infty]^2$. The cone $\mathbb{E}^0$ consists of points of $\mathbb{E}$ such that at most $d - 2$ coordinates are 0.

## 10.2 Symbol shock

Here is a glossary of miscellaneous symbols, in case you need a reference guide.

| | |
|---|---|
| $C_K^+(S)$ | The space of continuous functions on the metric space $S$ with compact support equipped with the uniform topology. |
| $\mathcal{B}_+$ | Positive, bounded, measurable functions. |
| $D[0, \infty)$ | The Skorokhod space of real-valued càdlàg functions on $[0, \infty)$ equipped with the $J_1$-topology. |
| $D([0, \infty), \mathbb{R}^2)$ | The Skorokhod space of $\mathbb{R}^2$-valued càdlàg functions on $[0, \infty)$ equipped with the $J_1$-topology. |
| $D([0, \infty) \times S)$ | Same as above, but the range is the metric space $S$. |
| $D^{\uparrow}[0, \infty)$ | The subspace of $D[0, \infty)$ containing the nondecreasing functions $f$ such that $f(0) = 0$ and $f(\infty) = \lim_{x\to\infty} f(x) = \infty$. |
| $C[0, \infty)$ | The continuous functions on $[0, \infty)$. |
| $\epsilon_x$ | The probability measure consisting of all mass at $x$. |
| $f^{\leftarrow}$ | The left-continuous inverse of a monotone function $f$ defined by $f^{\leftarrow}(x) = \inf\{y : f(y) \ge x\}$. |
| $\mathbb{LEB}$ | Lebesgue measure. |
| $\mathbb{E}$ | Usually $[0, \infty]^d \setminus \{\mathbf{0}\}$, except when it isn't. |
| $M_+(\mathbb{E})$ | The space of nonnegative Radon measures on $\mathbb{E}$. |
| $M_p(\mathbb{E})$ | The space of Radon point measures on $\mathbb{E}$. |
| $\nu_\alpha$ | A measure on $(0, \infty]$ given by $\nu_\alpha(x, \infty] = x^{-\alpha}$, $\alpha > 0, x > 0$. |
| $\text{PRM}(\mu)$ | Poisson random measure on $\mathbb{E}$ with mean measure $\mu$. |
| $\Rightarrow$ | Convergence in distribution. |
| Nice | Locally compact with countable base. "Nice" sounds less ferocious. |
| $\mathcal{K}(S)$ | The compact subsets of the metric space $S$. |
| $\mathcal{G}(S)$ | The open subsets of the metric space $S$. |

| | |
|---|---|
| $\mathcal{F}(S)$ | The closed subsets of the metric space $S$. |
| $b(t)$ | The quantile function of a distribution function $F(x)$, defined by $b(t) = F^{\leftarrow}(1 - \frac{1}{t})$. |
| $\|\boldsymbol{x}\|_p$ | $(\sum_{i=1}^{d} \|x^{(i)}\|^p)^{1/p}$, $1 \le p$. |
| $\|\boldsymbol{x}\|_\infty$ | $\bigvee_{i=1}^{d} \|x^{(i)}\|$. |
| $\Gamma_k$ | $k$th Poisson point; sum of $k$ iid unit exponential random variables. |
| $\aleph$ | The unit sphere with respect to a chosen norm. |
| $\mathcal{C}(f)$ | The points at which the function $f$ is continuous. |
| $\mathcal{D}(f)$ | The points at which the function $f$ is discontinuous. |
| $\mathrm{PR}(\mathbb{S})$ | The metric space of probability measures on the Borel $\sigma$-algebra $\mathcal{S}$ of subsets of the metric space $S$ metrized by weak convergence. |

# 11

# Software

Here we collect some general comments about software and visualization tools useful for exploring data from heavy-tailed models. The software world changes rapidly and there is no guarantee that these comments are accurate by the time you read this.

Splus or R are convenient environments for graphical and exploratory data analysis. We will give some routines which reflect the way we think about heavy-tailed analysis. These have been written for R but should run in Splus. The routines are only intended to be an outline of the logic behind the semiparametric analysis of heavy-tailed data.

Other software that is helpful comes from the extreme-value world. Alexander McNeil (`http://www.ma.hw.ac.uk/~mcneil/`) has a very professional compilation of Splus routines for performing extreme-value analysis that can be adapted to heavy-tailed analysis. A version is now marketed as part of the *Finmetrics* module by Insightful. A port for R called EVIR, by Alec Stephenson, is available at `http://cran.r-project.org/src/contrib/Descriptions/evir.html`. As of this writing, McNeil advertises on his web site a package called QRMlib for risk analysis, which incorporates the extreme values routines; the package is intended for use with [218].

Stuart Coles has notes explaining his Splus routines currently linked to `http://homes.stat.unipd.it/coles/public_html/ismev/summary.html`. His routines are used in his interesting extreme-values book [50]. There is a menu-driven program called XTREMES that R. Reiss and M. Thomas packaged with the book [238], and more information is available at `www.xtremes.math.uni-siegen.de`.

*Warning*: When I was an undergraduate, Queens College did not have a single computer and I did not absorb computing with mother's milk. A charitable characterization of my programming skills would be "primitive." If one of the routines looks somewhat polished, it was probably written by a valued and skilled coworker such as Paul Feigin, Catalin Stărică, Krishanu Maulik, or Jan Heffernan.

## 11.1 One dimension

### 11.1.1 Hill estimation

We give three plots corresponding to the Hill, altHill, and smoothed Hill plot. Because of the difficulty of choosing $k$, the number of upper-order statistics used in the estimation, we have not included confidence intervals in the plots but this could be done based on the asymptotic normality of the estimators. (See Proposition 9.3 (p. 302) and also [58, 59, 62, 72, 90, 113, 154, 155].) The packages listed above typically do include normal confidence intervals and XTREMES even includes parametric bootstrap confidence intervals.

#### Hillalpha

The routine Hillalpha($x$), where $x$ is the name of the data set, produces the Hill plot for estimating $\alpha$.

```
Hillalpha<-function(x)
{
ordered <- rev(sort(x))
ordered <- ordered[ordered[] > 0]
n <- length(ordered)
loggs <- log(ordered)
hill <- cumsum(loggs[1:(n - 1)])/(1:(n - 1)) - loggs[2:n]
hill <- 1/(hill)
plot(1:length(hill), hill, type = "l",
        xlab = "number of order statistics",
        ylab = "Hill estimate of alpha", main="Hill plot")
}
```

#### altHillalpha

The routine altHillalpha($x$, theta1, theta2), where $x$ is the name of the data set, produces the Hill plot for estimating $\alpha$, but on a logarithmic scale. The plot is for the points

$$\left\{\left(\frac{\log i}{\log n}, H_{i,n}\right); n^{\text{theta1}} \leq i \leq n^{\text{theta2}}\right\}.$$

Parameters theta1 and theta2 can be invoked to eliminate estimates based on very small or very large numbers of order statistics which would distort the vertical scaling of the plot and produce uninformative pictures.

```
altHillalpha<-function(x,theta1, theta2)
{
        ordered <- rev(sort(x))
        ordered <- ordered[ordered[] > 0.]
        n <- length(ordered)
        loggs <- log(ordered)
        hill <- cumsum(loggs[1.:(n - 1.)])/(1.:(n - 1.))
               - loggs[2.:n]
        hill <- 1./hill
        s<-log((1:(n-1)))/log(n-1)
        plot(s[n^theta1:n^theta2], hill[n^theta1:n^theta2],
             type ="l", xlab = "theta",
             ylab = "Hill estimate of alpha",
             main="altHill")
}
```

Of course, the two Hill plots can be displayed side by side for comparison which is what twoHillalpha accomplishes.

```
twoHillalpha<-function(x,theta1,theta2)
{
        par(mfrow=c(1,2))
        Hillalpha(x)
        altHillalpha(x,theta1,theta2)
        par(mfrow=c(1,1))
}
```

**smooHillalpha**

This function computes the estimates $\text{smooH}_{k,n}$ of (4.25) and then plots the vector corresponding to $1/\text{smooH}_{k,n}$. For this function $x$ is the data set, $r$ is the amount of smoothing to be applied, $(l, \text{up})$ restrict the plot to serial indices between these bounds.

```
smooHillalpha<-function(x,r,l,up)
{
ordered <- rev(sort(x))
ordered <- ordered[ordered[] > 0]
n <- length(ordered)
loggs <- log(ordered)
hill <- cumsum(loggs[1:(n - 1)])/(1:(n - 1)) - loggs[2:n]
smoo<-hill[1]
for(k in (1:floor(n/r))){smoo=c(smoo,mean(hill[k+1:r*k]))}
plot((1:length(smoo))[l:up],1/smoo[l:up],type="l",xlab="",
              ylab="smoo est of alpha", main="smooHill")
}
```

The three plots can be combined into the Hill, altHill, and smooHill plots.

```
threeHillalpha<-function(x,theta1,theta2,r,l,up)
{
par(mfrow=c(1,3))
Hillalpha(x)
altHillalpha(x,theta1,theta2)
smooHillalpha(x,r,l,up)
par(mfrow=c(1,1))
}
```

### 11.1.2 QQ plotting

When the data from some point on are approximately a realization from a Pareto distribution, one takes logarithms to convert from Pareto to exponential distributions and then computes the slope of the least-squares line through a user-specified number of order statistics to estimate and report on the estimate of $\alpha$. One has to confirm that the data are approximately Pareto beyond a threshold and then decide on the threshold.

#### ppPareto

To decide that approximate Pareto is the right choice past a threshold, one can match all the sample quantiles of the log-transformed data with theoretical quantiles of the exponential density. The plot should look linear from some point onwards. This gives an idea of how many upper-order statistics to use when fitting the LS line (or, equivalently, what threshold to pick if you are a POT head). This was how Figure 4.11 was produced and it used the routine pppareto. In this function $x =$ the data set.

```
pppareto<-function(x)
{
l <- length(x)
s <- seq(1./(l + 1.), l/(l + 1.), length = l)
y <- - log(1. - s)
plot(y, log(sort(x)), pch = ".",
     xlab = "quantiles of exponential",
     ylab = "log-sorted data")
}
```

#### parfit

The R function parfit applied to approximate Pareto data takes logarithms to convert from Pareto to exponential distributions and then computes the slope of the least-squares line through a user-specified number of order statistics to estimate and report on the estimate of $\alpha$. Using ppPareto helps decide on the user-specified number of order statistics. This function was used to produce Figure 4.12. The function takes as inputs $x$, the name of the data, and $k$, the number of order statistics or that part of the plot to which lsfit fits a line and computes the slope. Outputs include the estimate of $\alpha$ which is the reciprocal slope and the intercept.

Here is the text of parfit:

```
parfit<-function(x, k)
{
l <- length(x)
s <- seq(1/(l + 1), l/(l + 1), length = l)
y <-  - log(1 - s)
plot(y[(l - k + 1):l], log(sort(x)[(l - k + 1):l]), pch ="*",
     xlab = "quantiles of exponential",
     ylab = "log-sorted data")
coeffs <- lsfit(y[(l - k + 1):l],
            log(sort(x)[(l - k + 1):l]))$coef
abline(coeffs[1], coeffs[2])
names(coeffs) <- NULL
list(logxl = coeffs[1], alpha = 1/coeffs[2],
bn = exp(coeffs[1] + log(l) * coeffs[2]))
}
```

## QQ estimator plot

The QQ estimator of Section 4.6.6 computes an estimate of $\alpha$ based on $k$ order statistics. This represents the reciprocal slope in parfit. To assess the sensitivity to the choice of $k$, it is wise to plot the estimates as a function of $k$ to check the stability of the plot. This is performed by qqest. Generally this plot looks smoother than the Hill plot.

```
qqest<-function(y, list = F)
{
        n <- length(y)
        x <-  - log((1.:n))
        z <- rev(sort(log(y)))
        sumx <- cumsum(x)
        sumz <- cumsum(z)
        sumxz <- cumsum(x * z)
        sumxx <- cumsum(x^2.)
        alphainv <- ((1.:n) * sumxz - sumx * sumz)/
         ((1.:n) * sumxx - (sumx)^2.)
        alpha <- 1./alphainv
```

```
        plot((5:n),alpha[5:n],type="l",
                 xlab="number of order statistics",
                 ylab="qq est of alpha",
                 main="QQ estimator")
        if(list) {
                alpha[50.:n]
        }
}
```

Comparison with the Hill plot is easy with qqHill.

```
qqHill<-function(x)
{
        par(mfrow = c(1., 2.))
        Hillalpha(x)
        qqest(x)
        par(mfrow = c(1., 1.))
}
```

### 11.1.3 Estimators from extreme-value theory

As discussed in Section 4.5, estimators from extreme-value theory are useful for rejecting the suitability of a heavy-tail model. In our experience, two in particular have been helpful: the Pickands estimator [85, 112, 235, 236] and the moment estimator [85–87, 90, 251] of Dekkers, Einmahl, and de Haan.

#### The Pickands estimator

The Pickands plots in Section 4.5 were produced with the R function Pickands, which estimates $\gamma = 1/\alpha$. The function requires $x$, the data set, as input.

```
pickands<-function(x)
{
ordered <- rev(sort(x))
n <- length(ordered)
ordered2k <- ordered[seq(2, (n/4), by = 2)]
ordered4k <- ordered[seq(4, (n/4), by = 4)]
l <- length(ordered4k)
gammak <- (1/log(2)) * log((ordered[1:l] - ordered2k[1:l])/(
ordered2k[1:l] - ordered4k[1:l]))
plot(1:length(gammak), gammak, type = "l", xlab =
"number of order statistics", ylab =
"Pickands estimate of gamma")
}
```

**The moment estimator**

Ignoring location and scale, the extreme-value distributions [41, 102, 197, 260] can be parameterized as a one-parameter family

$$G_\gamma(x) = \exp\{-(1+\gamma x)^{-\gamma^{-1}}\}, \quad \gamma \in \mathbb{R}, \quad 1+\gamma x > 0.$$

The moment estimator $\hat{\gamma}^{(\text{moment})}$ [85–87] is designed to estimate $\gamma$ for a random sample in the domain of attraction of $G_\gamma$. When $\gamma > 0$, this is the same thing as estimating $\gamma = 1/\alpha$. Recall that the exponential, normal, log-normal, gamma densities, and many other distributions with exponentially bounded tails are in the $D(G_0)$, the domain of attraction of the Gumbel distribution $G_0(x) = \exp\{-e^{-x}\}$, $x \in \mathbb{R}$. In cases where one is deciding whether heavy-tailed analysis is appropriate, this estimator provides another method for deciding when a distribution is heavy tailed or not. If $\hat{\gamma}$ is negative or very close to zero, there is considerable doubt that heavy-tailed analysis should be applied.

The moment estimator is defined as follows: Let $X_1, \ldots, X_n$ be the sample and let $X_{(1)} \geq X_{(2)} \geq \cdots \geq X_{(n)}$ be the order statistics. Define for $r = 1, 2$,

$$H_{k,n}^{(r)} = \frac{1}{k}\sum_{i=1}^{k}\left(\log \frac{X_{(i)}}{X_{(k+1)}}\right)^r,$$

so that $H_{k,n}^{(1)}$ is the Hill estimator. Define

$$\hat{\gamma}_n^{(\text{moment})} = H_{k,n}^{(1)} + 1 - \frac{1/2}{1 - \frac{(H_{k,n}^{(1)})^2}{H_{k,n}^{(2)}}}. \tag{11.1}$$

Then assuming $F \in D(G_\gamma)$, we have consistency

$$\hat{\gamma}_n \xrightarrow{P} \gamma,$$

as $n \to \infty$ and $k/n \to 0$. Furthermore, under an additional condition and a further restriction on $k$,

$$\sqrt{k}(\hat{\gamma} - \gamma) \Rightarrow N,$$

where $N$ is a normal random variable with 0 mean and variance

$$\sigma(\gamma) = \begin{cases} 1+\gamma^2 & \text{if } \gamma \geq 0, \\ (1-\gamma)^2(1-2\gamma)\left(4 - 8\frac{1-2\gamma}{1-3\gamma} + \frac{(5-11\gamma)(1-2\gamma)}{(1-3\gamma)(1-4\gamma)}\right) & \text{if } \gamma < 0. \end{cases}$$

The asymptotic variance of the moment estimator exceeds that of the Hill estimator when $\gamma > 0$, so from the point of view of asymptotic variance, there is no reason to prefer it. However, the moment estimator discerns a light tail more effectively than the Hill estimator, and thus it is often useful to apply the moment estimator to see if $\gamma \leq 0$, which would rule out heavy-tail analysis.

The moment estimator may be plotted to make a plot as a function of the number of upper-order statistics used. This is comparable to the Hill plot and the Pickands plot. The following R routine does this, taking as input the name of the data set. The output is the plot of the estimate of $\gamma$ as a function of $k$, the number of upper-order statistics used in the estimation. Recall that if $\gamma > 0$, then $\gamma = 1/\alpha$.

```
moment<-function(x)
{
ordered <- rev(sort(x))
ordered <- ordered[ordered[] > 0]
n <- length(ordered)
loggs <- log(ordered)
sqloggs <- ((loggs)^2)
hill <- cumsum(loggs[1:(n - 1)])/(1:(n - 1)) - loggs[2:n]
one <- cumsum(sqloggs[1:(n - 1)])/(1:(n - 1))
two <- (2 * loggs[2:n] * cumsum(loggs[1:(n - 1)]))/(1:(n -
1))
three <- sqloggs[2:n]
square <- one - two + three
gammahat <- hill[2:(n - 1)] + 1 - (0.5)/(1 - ((hill[2:(n -
1)])^2)/square[2:(n - 1)])
plot(5:length(gammahat), gammahat[5:length(gammahat)],
                type = "l",
                xlab = "number of order statistics",
                ylab = "Moment estimate of gamma",
                main= "Moment")
}
```

The plot can also be smoothed as with the smooHillalpha plot. See [251].

## 11.2 Multivariate heavy tails

Analysis of multivariate data involves the one-dimensional techniques to estimate marginal parameters. Dependence structure can be obtained by trying to estimate the angular measure after transformation to the standard case.

### 11.2.1 Estimation of the angular distribution

After transforming to the standard case to get $\nu_*$, we estimate the angular distribution $S$ or assume $S$ has a density and do a density estimation.

### Rank transform

This is simple enough that there may be a temptation to avoid a separate subroutine. This routine carries out the transformation of Section 9.2.3 (p. 310). Note the $k$ in (9.45) is omitted.

```
ranktransform<-function (x)
{
tr<-(length(x)-rank(x)+1)^(-1)
#invisible(list(tr=tr))
}
```

### Estimate the angular density using ranks

There are two routines for estimating the angular density. The first uses the rank transform and the norm is the $L_2$-norm.

```
angulardensityrank<-function(x, y, k)
{
#x <- x-vector
#y <- y-vextor
#k <- no of upper order statistics to be used
n <- min(length(x), length(y))
rx <- n - rank(x[1:n])+1 #compute anti ranks
ry <- n - rank(y[1:n])+1
theta <- atan(rx, ry)
rad <- k * sqrt((rx)^(-2) + (ry)^(-2))
plot(density(theta[rad > 1]), type = "b",
             xlab = "theta", ylab = "angular measure density")
abline(v = pi/4.)
}
```

### Estimate the angular density using power transforms

This method uses the technique of powering up components to achieve the standard case as in Section 9.2.3 (p. 310). Inputs include two data vectors, choice of $k$ and estimates of $\alpha$s.

```
angulardensityEstAlpha<-function(x, y, k, alpha1, alpha2)
{
n <- min(length(x), length(y))
x <- (x/(rev(sort(x)))[k])^(alpha1)
y <- (y/(rev(sort(y)))[k])^(alpha2)
```

```
r <- sqrt(x^2. + y^2.)
theta <- atan(x, y)
plot(density(theta[r > 1.]), type = "b",
xlab = "theta", ylab = "angular measure density")
abline(v = pi/4.)
}
```

### Estimate the angular distribution using the rank transform

If one is skeptical that a density exists, or skeptical of density estimation, one can estimate $S[0, \theta]$ as a distribution function.

```
angularDF<-function(x, y, k)
{
#x <- x-vector
#y <- y-vector
#k <- no of upper order statistics to be used
if((n <- length(x)) != length(y)) {
stop("The lengths of the data vectors do not match")
}
rx <- n - rank(x)
ry <- n - rank(y)
theta <- atan(rx, ry)
ord <- order(theta)
l <- rx[ord] < k | ry[ord] < k
stheta <- theta[ord]
cl <- cumsum(l)
sl <- sum(l)
plot(stheta[l], cl[l]/sl, xlim = c(0., pi/2.),
        ylim = c(0., 1.), type = "l",
xlab = "theta\n\t\t", ylab = "S(theta)",
main = "Spectral Distribution Function",
sub = paste(
"number of upper order statistics used",
as.character(k)), font.sub = 3.)
abline(h = 0.5, lty = 3.)
points(stheta[l], rep(0.5, sl), pch = "+")
}
```

### 11.2.2 The Stărică plot

The following function takes bivariate vectors $(\boldsymbol{x}, \boldsymbol{y}) = \{(x_i, y_i), 1 \leq i \leq n\}$ from a distribution $F$ whose tail satisfies standard regular variation, performs the power transform to approximate the standard case (see Section 9.2.3 (p. 310)), computes the radius vectors, and then creates the Stărică plot with a user-specified value of $k$. The plot draws horizontal and vertical lines through (1, 1). Abscissa values are restricted to being less than 10. This scheme is outlined in Section 9.2.4. The user can choose between the $L_1$- and $L_2$-norms.

### Norms

```
L1norm<-function (x)
{
    sum(x)
}

L2norm<-function (x)
{
    sqrt(sum(x^2))
}
```

### Stărică plot using the power transform

```
Starica2dPlot<-function (x, y, k, PlotIt = TRUE, norm = L1norm)
{
    if (length(x) != length(y)) {
        stop("x and y different lengths\n")
    }
    n <- length(x)
    alpha1 <- 1/EstimateEtaHills(k = k, data = x)
    alpha2 <- 1/EstimateEtaHills(k = k, data = y)
    x <- (x/rev(sort(x))[k])^(alpha1)
    y <- (y/rev(sort(y))[k])^(alpha2)
    r <- apply(cbind(x, y), 1, norm)
    u <- rev(sort(r))
    ratio <- (u * (0:(n - 1)))/(length(r[r > 1]))
    if (PlotIt) {
        plot(u[u < 10], ratio[u < 10], xlim = c(0.1, 10),
    type = "l",
            xlab = "scaling constant", ylab = "scaling ratio",
            col = "blue")
        abline(h = 1, col = "red")
        abline(v = 1, lty = 2, lwd = 0.5, col = "red")
        title(paste("k =", k))
    }
    invisible(list(r = u, ratio = ratio, k = k, norm = norm))
}
```

### Stărică plot using the rank transform

The following function uses the rank method. It requires user inputs of two data sets $x$, $y$ and a value of $k$ and produces the Stărică plot.

```
Starica2dplotrank<-function (x, y, k, PlotIt=TRUE,
          norm=L1norm)
{
    if (length(x) != length(y)) {
        stop("x and y different lengths\n")
    }
    n <- length(x)
rx<-ranktransform(x)
ry<-ranktransform(y)
R <- apply(cbind(k*rx, k*ry), 1, norm)
u <- rev(sort(R))
    ratio <- (u * (1:n))/(length(R[R > 1]))
    if (PlotIt) {
        plot(u[u < 5], ratio[u < 5], xlim = c(0.1, 5),
    type = "l",
            xlab = "scaling constant", ylab = "scaling ratio",
            col = "blue")
        abline(h = 1, col = "red")
        abline(v = 1, lty = 2, lwd = 0.5, col = "red")
        title(paste("k =", k))
    }
    invisible(list(r = u, ratio = ratio, k = k, norm = norm))
}
```

### Allowing the Stărică plot to choose $k$

This routine uses the rank transform and then computes the distance of the points to the horizontal line. It then cycles through $k$ and chooses the $k$ which achieves minimum distance. The procedure allows the distance to be computed between two abscissa values (Lower, Upper). It is sometimes finicky and sometimes gives ridiculously low values of $k$.

```
ChooseKrank<-function (x, y, PlotIt = TRUE,  norm = L2norm,
Lower,Upper)
{
    if (length(x) != length(y)) {
        stop("x and y different lengths\n")
    }
    n <- length(x)
```

```
rx<-ranktransform(x)
ry<-ranktransform(y)
R<-apply(cbind(rx,ry),1,norm) #norms of pairs
                               #after rank transform
R <- rev(sort(R)) #ordered norms; biggest first
nk <- min(c(500, ceiling(.5*n)))
Kseq <- (1:nk) #round(exp(seq(log(10), log(n/2), len = nk)))
dist<-rep(0,nk) #vector of length nk of zeros
for (i in 1:nk) {
dist[i]<-L2norm(
(  i*(1:n)*R/length(i*R[i*R>=1])  )*(i*R>Lower & i*R <=Upper)-
(i*R>Lower & i*R <=Upper)
)
}
 k <- (Kseq[dist == min(dist)])
if (PlotIt) {
 u <- k*R
  ratio <- k*R*(1:n)/length(k*R[k*R>=1])
 plot(u[u <= Upper & u>Lower], ratio[u <= Upper & u>Lower],
type = "l", xlab = "scaling constant", ylab = "scaling ratio",
col="blue")
    abline(h = 1,col="red")
    abline(v = 1,lty=2,lwd=0.5,col="red")
    title(paste("k =",k))
  }
k
}
```

# References


[1] R. J. Adler, R. E. Feldman, and M. S. Taqqu, eds., *A Practical Guide to Heavy Tails: Statistical Techniques and Applications*, Birkhäuser Boston, Cambridge, MA, 1998.

[2] J. Anděl, Nonlinear nonnegative AR(1) processes, *Comm. Statist. Theory Methods*, **18**-11 (1989), 4029–4037.

[3] J. Anděl, Nonnegative autoregressive processes, *J. Time Series Anal.*, **10**-1 (1989), 1–11.

[4] D. Applebaum, *Lévy Processes and Stochastic Calculus*, Cambridge Studies in Advanced Mathematics, Vol. 93, Cambridge University Press, Cambridge, UK, 2004.

[5] D. Applebaum, Lévy-type stochastic integrals with regularly varying tails, *Stochastic Anal. Appl.*, **23**-3 (2005), 595–611.

[6] M. Arlitt and C. L. Williamson, Web server workload characterization: The search for invariants (extended version), in *Proceedings of the ACM Sigmetrics Conference, Philadelphia, PA*, ACM, New York, 1996; available from `{mfa126,carey}@cs.usask.ca`.

[7] M. Arlitt and C. Williamson, *Web Server Workload Characterization: The Search for Invariants*, Master's thesis, University of Saskatchewan, Saskatoon, SK, Canada, 1996.

[8] S. Asmussen, *Applied Probability and Queues*, 2nd ed., Applications of Mathematics: Stochastic Modelling and Applied Probability, Vol. 51, Springer-Verlag, New York, 2003.

[9] K. B. Athreya, Bootstrap of the mean in the infinite variance case, in *Proceedings of the 1st World Congress of the Bernoulli Society* (*Tashkent*, 1986), Vol. 2, VNU Sci. Press, Utrecht, The Netherlands, 1987, 95–98.

[10] K. B. Athreya, Bootstrap of the mean in the infinite variance case, *Ann. Statist.*, **15**-2 (1987), 724–731.

[11] A. A. Balkema and S. I. Resnick, Max-infinite divisibility, *J. Appl. Probab.*, **14**-2 (1977), 309–319.

[12] B. Basrak, R. A. Davis, and T. Mikosch, The sample ACF of a simple bilinear process, *Stochastic Process. Appl.*, **83**-1 (1999), 1–14.

[13] B. Basrak, R. A. Davis, and T. Mikosch, A characterization of multivariate regular variation, *Ann. Appl. Probab.*, **12**-3 (2002), 908–920.

[14] B. Basrak, R. Davis, and T. Mikosch, Regular variation of GARCH processes, *Stochastic Process. Appl.*, **99**-1 (2002), 95–115.

[15] B. Basrak, *The Sample Autocorrelation Function of Non-Linear Time Series*, Ph.D. thesis, Rijksuniversiteit Groningen, Groningen, The Netherlands, 2000.

[16] J. Beirlant, Y. Goegebeur, J. Teugels, and J. Segers, *Statistics of Extremes: Theory and Applications*, Wiley Series in Probability and Statistics, Wiley, Chichester, UK, 2004 (with contributions from D. De Waal and C. Ferro).

[17] J. Beirlant, P. Vynckier, and J. Teugels, Tail index estimation, Pareto quantile plots, and regression diagnostics, *J. Amer. Statist. Assoc.*, **91**-436 (1996), 1659–1667.

[18] J. Beran, Statistical methods for data with long-range dependence, *Statist. Sci.*, **7**-4 (1992), 404–416 (404–427 including discussions and rejoinder).

[19] J. Bertoin, *Lévy Processes*, Cambridge Tracts in Mathematics, Vol. 121, Cambridge University Press, Cambridge, UK, 1996.

[20] R. N. Bhattacharya, V. K. Gupta, and E. Waymire, The Hurst effect under trends, *J. Appl. Probab.*, **20** (1983), 649–662.

[21] P. J. Bickel and D. A. Freedman, Some asymptotic theory for the bootstrap, *Ann. Statist.*, **9**-6 (1981), 1196–1217.

[22] P. Billingsley, *Convergence of Probability Measures*, Wiley, New York, 1968.

[23] P. Billingsley, *Weak Convergence of Measures: Applications in Probability*, CBMS–SIAM Series in Applied Mathematics, Vol. 5, Society for Industrial and Applied Mathematics, Philadelphia, 1971.

[24] P. Billingsley, *Probability and Measure*, Wiley Series in Probability and Mathematical Statistics, 3rd ed., Wiley–Interscience, New York, 1995.

[25] P. Billingsley, *Convergence of Probability Measures*, 2nd ed., Wiley–Interscience, New York, 1999.

[26] N. H. Bingham, C. M. Goldie, and J. L. Teugels, *Regular Variation*, Cambridge University Press, Cambridge, UK, 1987.

[27] D. C. Boes and J. D. Salas-La Cruz, On the expected range and expected adjusted range of partial sums of exchangeable random variables, *J. Appl. Probab.*, **10** (1973), 671–677.

[28] D. C. Boes, Schemes exhibiting Hurst behavior, in J. Srivastava, ed., *Essays in Honor of Franklin Graybill*, North-Holland, Amsterdam, 1988, 21–42.

[29] O. J. Boxma and J. W. Cohen, Heavy-traffic analysis for the GI/G/1 queue with heavy-tailed distributions, *Queueing Systems Theory Appl.*, **33**-1–3 (1999), 177–204.

[30] L. Breiman, On some limit theorems similar to the arc-sin law, *Theory Probab. Appl.*, **10** (1965), 323–331.

[31] P. J. Brockwell and R. A. Davis, *Time Series: Theory and Methods*, 2nd ed., Springer-Verlag, New York, 1991.

[32] P. J. Brockwell, N. Pacheco-Santiago, and S. I. Resnick, Weak convergence and range analysis for dams with Markovian input rate, *J. Appl. Probab.*, **19** (1982), 272–289.

[33] M. Brown, An invariance property of Poisson processes, *J. Appl. Probab.*, **6** (1969), 453–458.

[34] M. Brown, Some results on a traffic model of Rényi, *J. Appl. Probab.*, **6** (1969), 293–300.

[35] M. Brown, A property of Poisson processes and its application to macroscopic equilibrium of particle systems, *Ann. Math. Statist.*, **41** (1970), 1935–1941.

[36] M. Brown, Invariant of traffic streams, *Cahiers Centre Études Rech. Opér.*, **14** (1972), 23–26.

[37] M. Brown, Low density traffic streams, *Adv. Appl. Probab.*, **4** (1972), 177–192.

[38] F. H. Campos, J. S. Marron, C. Park, S. I. Resnick, and K. Jaffay, Extremal dependence: Internet traffic applications, *Stoch. Models*, **21**-1 (2005), 1–35.

[39] J. Cao, W. Cleveland, D. Lin, and D. X. Sun, On the nonstationarity of internet traffic, in *Proceedings of the ACM SIGMETRICS* 2001, ACM, New York, 2001, 102–112; available online from `http://cm.bell-labs.com/cm/ms/departments/sia/InternetTraffic/webpapers.html`.

[40] J. Cao, W. Cleveland, D. Lin, and D. X. Sun, The effect of statistical multiplexing on internet packet traffic: Theory and empirical study, 2001, `http://cm.bell-labs.com/cm/ms/departments/sia/InternetTraffic/webpapers.html`.

[41] E. Castillo, *Extreme Value Theory in Engineering*, Academic Press, San Diego, 1988.

[42] T. L. Chow and J. L. Teugels, The sum and the maximum of i.i.d. random variables, in *Proceedings of the Second Prague Symposium on Asymptotic Statistics* (*Hradec Králové*, 1978), North-Holland, Amsterdam, 1979, 81–92.

[43] E. Çinlar, *Random Measures and Dynamic Point Processes* II: *Poisson Random Measures*, Discussion Paper 11, Center for Statistics and Probability, Northwestern University, Evanston, IL, 1976.

[44] D. B. H. Cline, *Estimation and Linear Prediction for Regression, Autoregression, and ARMA with Infinite Variance Data*, Ph.D. thesis, Colorado State University, Fort Collins, CO, 1983.

[45] J. W. Cohen, *Heavy-Traffic Theory for the Heavy-Tailed M/G/*1 *Queue and ν-Stable Lévy Noise Traffic*, Technical Report PNA-R9805, Centrum voor Wiskunde en Informatica, Amsterdam, 1998.

[46] J. W. Cohen, *The ν-Stable Lévy Motion in Heavy-Traffic Analysis of Queueing Models with Heavy-Tailed Distributions*, Technical Report PNA-R9808, Centrum voor Wiskunde en Informatica, Amsterdam, 1998.

[47] J. W. Cohen, A heavy-traffic theorem for the GI/G/1 queue with a Pareto-type service time distribution, *J. Appl. Math. Stochastic Anal.*, **11**-3 (1998), 247–254.

[48] J. Cohen, S. Resnick, and G. Samorodnitsky, Sample correlations of infinite variance time series models: An empirical and theoretical study, *J. Appl. Math. Stochastic Anal.*, **11**-3 (1998), 255–282.

[49] S. G. Coles, J. E. Heffernan, and J. A. Tawn, Dependence measures for extreme value analyses, *Extremes*, **2**-4 (1999), 339–365.

[50] S. G. Coles, *An Introduction to Statistical Modeling of Extreme Values*, Springer Series in Statistics, Springer-Verlag, London, 2001.

[51] M. Crovella, A. Bestavros, and M. S. Taqqu, Heavy-tailed probability distributions in the world wide web, in M. S. Taqqu, R. Adler, R. Feldman, eds., *A Practical Guide to Heavy Tails: Statistical Techniques for Analyzing Heavy Tailed Distributions*, Birkhäuser Boston, Cambridge, MA, 1999.

[52] M. Crovella and A. Bestavros, *Explaining World Wide Web Traffic Self-Similarity*, Technical Report TR-95-015, Computer Science Department, Boston University, Boston, 1995; available from `{crovella,best}@cs.bu.edu`.

[53] M. Crovella and A. Bestavros, Self-similarity in world wide web traffic: Evidence and possible causes, in *Proceedings of the ACM SIGMETRICS* 1996 *International Conference on Measurement and Modeling of Computer Systems*, Vol. 24, 1996, ACN, New York, 160–169.

[54] M. Crovella and A. Bestavros, Self-similarity in world wide web traffic: Evidence and possible causes, *Performance Evaluation Rev.*, **24** (1996), 160–169.

[55] M. Crovella and A. Bestavros, Self-similarity in world wide web traffic: evidence and possible causes, *IEEE/ACM Trans. Networking*, **5**-6 (1997), 835–846.

[56] M. Crovella, G. Kim, and K. Park, On the relationship between file sizes, transport protocols, and self-similar network traffic, in *Proceedings of the* 4th *IEEE Conference on Network Protocols* (*ICNP*'96), IEEE Press, Piscataway, NJ, 1996, 171–180.

[57] S. Csörgő, P. Deheuvels, and D. Mason, Kernel estimates for the tail index of a distribution, *Ann. Statist.*, **13** (1985), 1050–1077.

[58] S. Csörgő and D. Mason, Central limit theorems for sums of extreme values, *Math. Proc. Cambridge Philos. Soc.*, **98** (1985), 547–558.

[59] S. Csörgő, E. Haeusler, and D. M. Mason, The asymptotic distribution of extreme sums, *Ann. Probab.*, **19**-2 (1991), 783–811.

[60] S. Csörgő, E. Haeusler, and D. M. Mason, The quantile-transform approach to the asymptotic distribution of modulus trimmed sums, in *Sums, Trimmed Sums and Extremes*, Progress in Probability, Vol. 23, Birkhäuser Boston, Cambridge, MA, 1991, 337–353.

[61] S. Csörgő, E. Haeusler, and D. M. Mason, The quantile-transform–empirical-process approach to limit theorems for sums of order statistics, in *Sums, Trimmed Sums and Extremes*, Progress in Probability, Vol. 23, Birkhäuser Boston, Cambridge, MA, 1991, 215–267.

[62] S. Csörgő and L. Viharos, On the asymptotic normality of Hill's estimator, *Math. Proc. Cambridge Philos. Soc.*, **118**-2 (1995), 375–382.

[63] C. Cunha, A. Bestavros, and M. Crovella, *Characteristics of WWW Client–Based Traces*, Technical Report BU-CS-95-010, Computer Science Department, Boston University, Boston, 1995; available from `{crovella,best}@cs.bu.edu`.

[64] M. Dacorogna, H. A. Hauksson, T. Domenig, U. Müller, and G. Samorodnitsky, Multivariate extremes, aggregation and risk estimation, *Quantitative Finance*, **1**-1 (2001), 79–85.

[65] D. J. Daley and D. Vere-Jones, *An Introduction to the Theory of Point Processes, Vol.* I: *Elementary Theory and Methods*, 2nd ed., Probability and Its Applications, Springer-Verlag, New York, 2003.

[66] J. Danielsson, L. de Haan, L. Peng, and C. G. de Vries, Using a bootstrap method to choose the sample fraction in tail index estimation, *J. Multivariate Anal.*, **76**-2 (2001), 226–248.

[67] S. Datta and W. P. McCormick, Bootstrap inference for a first-order autoregression with positive innovations, *J. Amer. Statist. Assoc.*, **90**-432 (1995), 1289–1300.

[68] B. D'Auria and S. I. Resnick, Data network models of burstiness, *Adv. Appl. Probab.*, **38**-2 (2006), 373–404.

[69] R. A. Davis and W. P. McCormick, Estimation for first-order autoregressive processes with positive or bounded innovations, *Stochastic Process. Appl.*, **31**-2 (1989), 237–250.

[70] R. A. Davis and T. Mikosch, The sample autocorrelations of heavy-tailed processes with applications to ARCH, *Ann. Statist.*, **26**-5 (1998), 2049–2080.

[71] R. A. Davis and T. Mikosch, Point process convergence of stochastic volatility processes with application to sample autocorrelation, *J. Appl. Probab.*, **38A** (2001), 93–104.

[72] R. A. Davis and S. I. Resnick, Tail estimates motivated by extreme value theory, *Ann. Statist.*, **12** (1984), 1467–1487.

[73] R. A. Davis and S. I. Resnick, Limit theory for moving averages of random variables with regularly varying tail probabilities, *Ann. Probab.*, **13**-1 (1985), 179–195.

[74] R. A. Davis and S. I. Resnick, More limit theory for the sample correlation function of moving averages, *Stochastic Process. Appl.*, **20**-2 (1985), 257–279.

[75] R. A. Davis and S. I. Resnick, Limit theory for the sample covariance and correlation functions of moving averages, *Ann. Statist.*, **14**-2 (1986), 533–558.

[76] R. A. Davis and S. I. Resnick, Extremes of moving averages of random variables from the domain of attraction of the double exponential distribution, *Stochastic Process. Appl.*, **30**-1 (1988), 41–68.

[77] R. A. Davis and S. I. Resnick, Limit theory for bilinear processes with heavy tailed noise, *Ann. Appl. Probab.*, **6** (1996), 1191–1210.

[78] R. Davis, E. Mulrow, and S. Resnick, The convex hull of a random sample in $\mathbf{R}^2$, *Comm. Statist. Stochastic Models*, **3**-1 (1987), 1–27.

[79] P. Deheuvels, D. M. Mason, and G. R. Shorack, Some results on the influence of extremes on the bootstrap, *Ann. Inst. H. Poincaré Probab. Statist.*, **29**-1 (1993), 83–103.

[80] P. Deheuvels and D. M. Mason, A tail empirical process approach to some nonstandard laws of the iterated logarithm, *J. Theoret. Probab.*, **4**-1 (1991), 53–85.

[81] P. Deheuvels and D. M. Mason, Functional laws of the iterated logarithm for the increments of empirical and quantile processes, *Ann. Probab.*, **20**-3 (1992), 1248–1287.

[82] P. Deheuvels, A construction of extremal processes, in *Probability and Statistical Inference* (*Bad Tatzmannsdorf*, 1981), Reidel, Dordrecht, The Netherlands, 1982, 53–57.

[83] P. Deheuvels, Spacings, record times and extremal processes, in *Exchangeability in Probability and Statistics* (*Rome*, 1981), North-Holland, Amsterdam, 1982, 233–243.

[84] P. Deheuvels, The strong approximation of extremal processes II, *Z. Wahrscheinlichkeitstheor. Verw. Gebiete*, **62**-1 (1983), 7–15.

[85] A. L. M. Dekkers and L. de Haan, On the estimation of the extreme-value index and large quantile estimation, *Ann. Statist.*, **17** (1989), 1795–1832.

[86] A. L. M. Dekkers and L. de Haan, Optimal choice of sample fraction in extreme-value estimation, *J. Multivariate Anal.*, **47**-2 (1993), 173–195.

[87] A. L. M. Dekkers, J. H. J. Einmahl, and L. de Haan, A moment estimator for the index of an extreme-value distribution, *Ann. Statist.*, **17** (1989), 1833–1855.

[88] E. del Barrio and C. Matrán, The weighted bootstrap mean for heavy-tailed distributions, *J. Theoret. Probab.*, **13**-2 (2000), 547–569.

[89] L. de Haan and J. de Ronde, Sea and wind: Multivariate extremes at work, *Extremes*, **1**-1 (1998), 7–46.

[90] L. de Haan and A. Ferreira, *Extreme Value Theory: An Introduction*, Springer-Verlag, New York, 2006.

[91] L. de Haan, E. Omey, and S. I. Resnick, Domains of attraction and regular variation in $\mathbb{R}^d$, *J. Multivariate Anal.*, **14**-1 (1984), 17–33.

[92] L. de Haan and E. Omey, Integrals and derivatives of regularly varying functions in $\mathbb{R}^d$ and domains of attraction of stable distributions II, *Stochastic Process. Appl.*, **16**-2 (1984), 157–170.

[93] L. de Haan and L. Peng, Comparison of tail index estimators, *Statist. Neerlandica*, **52**-1 (1998), 60–70.

[94] L. de Haan, S. I. Resnick, H. Rootzén, and C. G. de Vries, Extremal behaviour of solutions to a stochastic difference equation with applications to ARCH processes, *Stochastic Process. Appl.*, **32**-2 (1989), 213–224.

[95] L. de Haan and S. I. Resnick, Limit theory for multivariate sample extremes, *Z. Wahrscheinlichkeitstheor. Verw. Gebiete*, **40** (1977), 317–337.

[96] L. de Haan and S. I. Resnick, Conjugate $\pi$-variation and process inversion, *Ann. Probab.*, **7**-6 (1979), 1028–1035.

[97] L. de Haan and S. I. Resnick, Derivatives of regularly varying functions in $\mathbb{R}^d$ and domains of attraction of stable distributions, *Stochastic Process. Appl.*, **8**-3 (1979), 349–355.

[98] L. de Haan and S. I. Resnick, On regular variation of probability densities, *Stochastic Process. Appl.*, **25** (1987), 83–95.

[99] L. de Haan and S. I. Resnick, Estimating the limit distribution of multivariate extremes, *Stochastic Models*, **9**-2 (1993), 275–309.

[100] L. de Haan and S. I. Resnick, On asymptotic normality of the Hill estimator, *Stochastic Models*, **14** (1998), 849–867.

[101] L. de Haan and U. Stadtmueller, Generalized regular variation of second order, *J. Australian Math. Soc. Ser.* A, **61**-3 (1996), 381–395.

[102] L. de Haan, *On Regular Variation and Its Application to the Weak Convergence of Sample Extremes*, Mathematisch Centrum, Amsterdam, 1970.

[103] L. de Haan, Weak limits of sample range, *J. Appl. Probab.*, **11** (1974), 836–841.

[104] L. de Haan, An Abel–Tauber theorem for Laplace transforms, *J. London Math. Soc.* (2), **13**-3 (1976), 537–542.

[105] L. de Haan, von Mises-type conditions in second order regular variation, *J. Math. Anal. Appl.*, **197**-2 (1996), 400–410.

[106] J. Dieudonné, *Foundations of Modern Analysis*, enlarged and corrected printing, Pure and Applied Mathematics, Vol. 10-I, Academic Press, New York, 1969.

[107] J. L. Doob, *Stochastic Processes*, Wiley, New York, 1953.

[108] G. Draisma, L. de Haan, L. Peng, and T. T. Pereira, A bootstrap-based method to achieve optimality in estimating the extreme-value index, *Extremes*, **2**-4 (1999), 367–404 (2000).

[109] H. Drees, Refined Pickands estimators of the extreme value index, *Ann. Statist.*, **23**-6 (1995), 2059–2080.

[110] H. Drees, L. de Haan, and S. I. Resnick, How to make a Hill plot, *Ann. Statist.*, **28**-1 (2000), 254–274.

[111] H. Drees, Refined Pickands estimators with bias correction, *Comm. Statist. Theory Methods*, **25**-4 (1996), 837–851.

[112] H. Drees, *Estimating the Extreme Value Index*, Habilitationsschrift (Ph.D. thesis), Mathematics Faculty, University of Cologne, Köln, Germany, 1998.

[113] H. Drees, A general class of estimators of the extreme value index, *J. Statist. Plann. Inference*, **66**-1 (1998), 95–112.

[114] H. Drees, On smooth statistical tail functionals, *Scandinavian J. Statist.*, **25**-1 (1998), 187–210.

[115] H. Drees, Optimal rates of convergence for estimates of the extreme value index, *Ann. Statist.*, **26**-1 (1998), 434–448.

[116] R. M. Dudley, *Real Analysis and Probability*, Wadsworth and Brooks/Cole, Belmont, CA, 1989.

[117] N. G. Duffield, J. T. Lewis, N. O'Connell, R. Russell, and F. Toomey, Statistical issues raised by the Bellcore data, in *Proceedings of the* 11*th IEE UK Teletraffic Symposium, Cambridge UK*, IEE, London, 1994, 23–25.

[118] D. E. Duffy, A. A. McIntosh, M. Rosenstein, and W. Willinger, Analyzing telecommunications traffic data from working common channel signaling subnetworks, *Comput. Sci. Statist.*, **25** (1993), 156–165 (special issue: M. E. Tarter and M. D. Lock, eds., *Statistical Applications of Expanding Computer Capabilities: Proceedings of the* 25*th Symposium on the Interface between Statistics and Computer Science*).

[119] R. Durrett and S. I. Resnick, Functional limit theorems for dependent variables, *Ann. Probab.*, **6**-5 (1978), 829–846.

[120] M. Dwass, Extremal processes, *Ann. Math. Statist.*, **35** (1964), 1718–1725.

[121] M. Dwass, Extremal processes II, *Illinois J. Math.*, **10** (1966), 381–391.

[122] M. Dwass, Extremal processes III, *Bull. Inst. Math. Acad. Sinica*, **2** (1974), 255–265 (collection of articles in celebration of the 60th birthday of Ky Fan).

[123] B. Efron and R. J. Tibshirani, *An Introduction to the Bootstrap*, Monographs on Statistics and Applied Probability, Vol. 57, Chapman and Hall, New York, 1993.

[124] B. Efron, The bootstrap and modern statistics, *J. Amer. Statist. Assoc.*, **95**-452 (2000), 1293–1296.

[125] B. Efron, Second thoughts on the bootstrap, *Statist. Sci.*, **18**-2 (2003), 135–140 (issue on the silver anniversary of the bootstrap).

[126] J. H. J. Einmahl and D. M. Mason, Generalized quantile processes, *Ann. Statist.*, **20**-2 (1992), 1062–1078.

[127] J. Einmahl, L. de Haan, and V. Piterbarg, Nonparametric estimation of the spectral measure of an extreme value distribution, *Ann. Statist.*, **29**-5 (2001), 1401–1423.

[128] P. Embrechts and C. M. Goldie, On closure and factorization properties of subexponential and related distributions, *J. Australian Math. Soc. Ser.* A, **29**-2 (1980), 243–256.

[129] P. Embrechts, C. Kluppelberg, and T. Mikosch, *Modelling Extreme Events for Insurance and Finance*, Springer-Verlag, Berlin, 1997.

[130] P. D. Feigin, M. F. Kratz, and S. I. Resnick, Parameter estimation for moving averages with positive innovations, *Ann. Appl. Probab.*, **6**-4 (1996), 1157–1190.

[131] P. Feigin and S. I. Resnick, Estimation for autoregressive processes with positive innovations, *Stochastic Models*, **8** (1992), 479–498.

[132] P. Feigin and S. I. Resnick, Limit distributions for linear programming time series estimators, *Stochastic Process. Appl.*, **51** (1994), 135–165.

[133] P. Feigin and S. I. Resnick, Linear programming estimators and bootstrapping for heavy tailed phenomena, *Adv. Appl. Probab.*, **29** (1997), 759–805.

[134] P. Feigin and S. I. Resnick, Pitfalls of fitting autoregressive models for heavy-tailed time series, *Extremes*, **1** (1999), 391–422.

[135] W. Feller, *An Introduction to Probability Theory and Its Applications*, Vol. 2, 2nd ed., Wiley, New York, 1971.

[136] M. I. Fraga Alves, M. I. Gomes, and L. de Haan, A new class of semi-parametric estimators of the second order parameter, *Port. Math. (N.S.)*, **60**-2 (2003), 193–213.

[137] H. Furrer, *Risk Theory and Heavy-Tailed Lévy Processes*, Diss. ETH 12408, Ph.D. thesis, Eidgenössische Technische Hochschule, Zurich, 1997.

[138] H. Furrer, Risk processes perturbed by $\alpha$-stable Lévy motion, *Scandinavian Actuarial J.*, **1998**-1 (1998), 59–74.

[139] R. Gaigalas and I. Kaj, Convergence of scaled renewal processes and a packet arrival model, *Bernoulli*, **9**-4 (2003), 671–703.

[140] J. Galambos, *The Asymptotic Theory of Extreme Order Statistics*, Wiley Series in Probability and Mathematical Statistics, Wiley, New York, Chichester, UK, Brisbane, Australia, 1978.

[141] M. W. Garrett and W. Willinger, Analysis, modeling and generation of self similar vbr video traffic, in *Proceedings of the ACM SigComm, London*, 1994, ACM, New York, 1994.

[142] J. Geffroy, Contribution à la théorie des valeurs extrêmes, *Publ. Inst. Statist. Univ. Paris*, **7**-3–4 (1958), 37–121.

[143] J. Geffroy, Contribution à la théorie des valeurs extrêmes II, *Publ. Inst. Statist. Univ. Paris*, **8** (1959), 3–65.

[144] J. L. Geluk and L. de Haan, *Regular Variation, Extensions and Tauberian Theorems*, CWI Tracts, Vol. 40, Stichting Mathematisch Centrum, Centrum voor Wiskunde en Informatica, Amsterdam, 1987.

[145] J. L. Geluk and L. Peng, An adaptive optimal estimate of the tail index for MA(1) time series, *Statist. Probab. Lett.*, **46**-3 (2000), 217–227.

[146] J. Geluk, L. de Haan, S. I. Resnick, and C. Stărică, Second-order regular variation, convolution and the central limit theorem, *Stochastic Process. Appl.*, **69**-2 (1997), 139–159.

[147] E. Giné and J. Zinn, Necessary conditions for the bootstrap of the mean, *Ann. Statist.*, **17**-2 (1989), 684–691.

[148] C. M. Goldie, Implicit renewal theory and tails of solutions of random equations, *Ann. Appl. Probab.*, **1**-1 (1991), 126–166.

[149] M. I. Gomes, F. Caeiro, and F. Figueiredo, Bias reduction of a tail index estimator through an external estimation of the second-order parameter, *Statistics*, **38**-6 (2004), 497–510.

[150] M. I. Gomes, L. de Haan, and L. Peng, Semi-parametric estimation of the second order parameter in statistics of extremes, *Extremes*, **5**-4 (2002), 387–414.

[151] M. I. Gomes and M.J. Martins, "Asymptotically unbiased" estimators of the tail index based on external estimation of the second order parameter, *Extremes*, **5**-1 (2002), 5–31.

[152] M. I. Gomes and M. J. Martins, Bias reduction and explicit semi-parametric estimation of the tail index, *J. Statist. Plann. Inference*, **124**-2 (2004), 361–378.

[153] C. A. Guerin, H. Nyberg, O. Perrin, S. I. Resnick, H. Rootzén, and C. Stărică, Empirical testing of the infinite source poisson data traffic model, *Stochastic Models*, **19**-2 (2003), 151–200.

[154] E. Haeusler and J. L. Teugels, On asymptotic normality of Hill's estimator for the exponent of regular variation, *Ann. Statist.*, **13**-2 (1985), 743–756.

[155] P. Hall, On some simple estimates of an exponent of regular variation, *J. Roy. Statist. Soc. Ser.* B, **44**-1 (1982), 37–42.

[156] P. Hall, Asymptotic properties of the bootstrap for heavy-tailed distributions, *Ann. Probab.*, **18**-3 (1990), 1342–1360.

[157] J. M. Harrison, *Brownian Motion and Stochastic Flow Systems*, Wiley, New York, 1985.

[158] D. Heath, S. I. Resnick, and G. Samorodnitsky, Patterns of buffer overflow in a class of queues with long memory in the input stream, *Ann. Appl. Probab.*, **7**-4 (1997), 1021–1057.

[159] D. Heath, S. I. Resnick, and G. Samorodnitsky, Heavy tails and long range dependence in on/off processes and associated fluid models, *Math. Oper. Res.*, **23**-1 (1998), 145–165.

[160] D. Heath, S. I. Resnick, and G. Samorodnitsky, How system performance is affected by the interplay of averages in a fluid queue with long range dependence induced by heavy tails, *Ann. Appl. Probab.*, **9** (1999), 352–375.

[161] J. E. Heffernan and S. I. Resnick, Hidden regular variation and the rank transform, *Adv. Appl. Probab.*, **37**-2 (2005), 393–414.

[162] J. E. Heffernan and S. I. Resnick, Limit laws for random vectors with an extreme component, *Ann. Appl. Probab.*, submitted, 2005; available online from `http://www.orie.cornell.edu/~sid`.

[226] T. Mikosch and C. Stărică, Change of structure in financial time series and the GARCH model, *REVSTAT Statist. J.*, **2** (2004), 16–41.

[227] T. Mikosch, Modeling dependence and tails of financial time series, in B. Finkenstadt and H. Rootzén, eds., *SemStat: Seminaire Europeen de Statistique: Extreme Values in Finance, Telecommunications, and the Environment*, Chapman and Hall, London, 2003, 185–286.

[228] I. Molchanov, *Theory of Random Sets*, Probability and Its Applications, Springer-Verlag, London, 2005.

[229] J. Neveu, *Discrete-Parameter Martingales*, North-Holland Mathematical Library, Vol. 10, North-Holland, Amsterdam, 1975 (translated from the French by T. P. Speed).

[230] J. Neveu, Processus ponctuels, in *École d'Été de Probabilités de Saint-Flour* VI—1976, Lecture Notes in Mathematics, Vol. 598, Springer-Verlag, Berlin, 1977, 249–445.

[231] J. Neyman and E. L. Scott, Statistical approach to problems of cosmology, *J. Roy. Statist. Soc. Ser.* B, **20** (1958), 1–43.

[232] K. Park and W. Willinger, Self-similar network traffic: An overview, in K. Park and W. Willinger, eds., *Self-Similar Network Traffic and Performance Evaluation*, Wiley–Interscience, New York, 2000, 1–38.

[233] M. Parulekar and A. M. Makowski, Tail probabilities for a multiplexer with a self-similar traffic, in *Proceedings of the* 15*th Annual IEEE INFOCOM*, IEEE Press, Piscataway, NJ, 1996, 1452–1459.

[234] M. Parulekar and A. M. Makowski, Tail probabilities for M/G/$\infty$ input processes I: Preliminary asymptotics, *Queueing Systems Theory Appl.*, **27**-3–4 (1997), 271–296.

[235] L. Peng, *Second Order Condition and Extreme Value Theory*, Ph.D. thesis, Tinbergen Institute, Erasmus University, Rotterdam, 1998.

[236] J. Pickands, Statistical inference using extreme order statistics, *Ann. Statist.*, **3** (1975), 119–131.

[237] D. Radulović, On the bootstrap and empirical processes for dependent sequences, in H. Dehling, T. Mikosch, and M. Sorensen, eds., *Empirical Process Techniques for Dependent Data*, Birkhäuser Boston, Cambridge, MA, 2002, 345–364.

[238] R.-D. Reiss and M. Thomas, *Statistical Analysis of Extreme Values*, 2nd ed., Birkhäuser Verlag, Basel, Switzerland, 2001.

[239] A. Rényi, *Foundations of Probability*, Holden–Day, San Francisco, 1970.

[240] A. Rényi, *Probability Theory*, North-Holland Series in Applied Mathematics and Mechanics, Vol. 10, North-Holland, Amsterdam, 1970 (translated by L. Vekerdi).

[241] S. I. Resnick and P. Greenwood, A bivariate stable characterization and domains of attraction, *J. Multivariate Anal.*, **9**-2 (1979), 206–221.

[242] S. I. Resnick and H. Rootzén, Self-similar communication models and very heavy tails, *Ann. Appl. Probab.*, **10** (2000), 753–778.

[243] S. I. Resnick and R. Roy, Leader and maximum independence for a class of discrete choice models, *Econom. Lett.*, **33**-3 (1990), 259–263.

[244] S. I. Resnick and R. Roy, Multivariate extremal processes, leader processes and dynamic choice models, *Adv. Appl. Probab.*, **22**-2 (1990), 309–331.

[245] S. I. Resnick and R. Roy, Superextremal processes, max-stability and dynamic continuous choice, *Ann. Appl. Probab.*, **4**-3 (1994), 791–811.

[246] S. I. Resnick and R. Roy, Super-extremal processes and the argmax process, *J. Appl. Probab.*, **31**-4 (1994), 958–978.

[247] S. I. Resnick and M. Rubinovitch, The structure of extremal processes, *Adv. Appl. Probab.*, **5** (1973), 287–307.

[248] S. I. Resnick, G. Samorodnitsky, and F. Xue, How misleading can sample acf's of stable MA's be? (Very!), *Ann. Appl. Probab.*, **9**-3 (1999), 797–817.

[249] S. I. Resnick, G. Samorodnitsky, and F. Xue, Growth rates of sample covariances of stationary symmetric $\alpha$-stable processes associated with null recurrent markov chains, *Stochastic Process. Appl.*, **85** (2000), 321–339.

[250] S. I. Resnick and G. Samorodnitsky, A heavy traffic approximation for workload processes with heavy tailed service requirements, *Management Sci.*, **46** (2000), 1236–1248.

[251] S. I. Resnick and C. Stărică, Smoothing the moment estimator of the extreme value parameter, *Extremes*, **1**-3 (1998).

[252] S. I. Resnick and C. Stărică, Smoothing the Hill estimator, *Adv. Appl. Probab.*, **29** (1997), 271–293.

[253] S. I. Resnick and E. van den Berg, Sample correlation behavior for the heavy tailed general bilinear process, *Stochastic Models*, **16**-2 (2000), 233–258.

[254] S. I. Resnick and E. van den Berg, Weak convergence of high-speed network traffic models, *J. Appl. Probab.*, **37**-2 (2000), 575–597.

[255] S. I. Resnick and E. Willekens, Moving averages with random coefficients and random coefficient autoregressive models, *Comm. Statist. Stochastic Models*, **7**-4 (1991), 511–525.

[256] S. I. Resnick, Tail equivalence and its applications, *J. Appl. Probab.*, **8** (1971), 136–156.

[257] S. I. Resnick, Inverses of extremal processes, *Adv. Appl. Probab.*, **6** (1974), 392–406.

[258] S. I. Resnick, Weak convergence to extremal processes, *Ann. Probab.*, **3**-6 (1975), 951–960.

[259] S. I. Resnick, Point processes, regular variation and weak convergence, *Adv. Appl. Probab.*, **18** (1986), 66–138.

[260] S. I. Resnick, *Extreme Values, Regular Variation and Point Processes*, Springer-Verlag, New York, 1987.

[261] S. I. Resnick, Point processes and Tauberian theory, *Math. Sci.*, **16**-2 (1991), 83–106.

[262] S. I. Resnick, *Adventures in Stochastic Processes*, Birkhäuser Boston, Cambridge, MA, 1992.

[263] S. I. Resnick, Discussion of the Danish data on large fire insurance losses, *Astin Bull.*, **27** (1997), 139–151.

[264] S. I. Resnick, *A Probability Path*, Birkhäuser Boston, Cambridge, MA, 1998.

[265] S. I. Resnick, Why non-linearities can ruin the heavy tailed modeler's day, in M. S. Taqqu, R. Adler, and R. Feldman, eds., *A Practical Guide to Heavy Tails: Statistical Techniques for Analyzing Heavy Tailed Distributions*, Birkhäuser Boston, Cambridge, MA, 1998, 219–240.

[266] S. I. Resnick, Hidden regular variation, second order regular variation and asymptotic independence, *Extremes*, **5**-4 (2002), 303–336.

[267] S. I. Resnick, Modeling data networks, in B. Finkenstadt and H. Rootzén, eds., *SemStat: Seminaire Europeen de Statistique: Extreme Values in Finance, Telecommunications, and the Environment*, Chapman and Hall, London, 2003, 287–372.

[268] S. I. Resnick, On the foundations of multivariate heavy tail analysis, *J. Appl. Probab.*, **41A** (2004), 191–212 (special volume: J. Gani and E. Seneta, eds., *Stochastic Methods and Their Applications: Papers in Honour of C. C. Heyde*).

[269] E. L. Rvačeva, On domains of attraction of multi-dimensional distributions, in *Selected Translations in Mathematics, Statistics, and Probability*, Vol. 2, American Mathematical Society, Providence, RI, 1962, 183–205.

[270] J. D. Salas, D. Boes, and V. Yevjevich, Hurst phenomenon as a pre-asymptotic behavior, *J. Hydrology*, **44**-1–2 (1979), 1–15.

[271] J. D. Salas and D. Boes, Expected range and adjusted range of hydrologic sequences, *Water Resoures Res.*, **10**-3 (1974), 457–463.

[272] J. D. Salas and D. Boes, Nonstationarity of mean and Hurst phenomenon, *Water Resources Res.*, **14**-1 (1978), 135–143.

[273] G. Samorodnitsky and M. Taqqu, *Stable Non-Gaussian Random Processes: Stochastic Models with Infinite Variance*, Stochastic Modeling, Chapman and Hall, New York, 1994.

[274] K. Sato, *Lévy Processes and Infinitely Divisible Distributions*, Cambridge Studies in Advanced Mathematics, Vol. 68, Cambridge University Press, Cambridge, UK, 1999 (translated from the 1990 Japanese original; revised by the author).

[275] E. Seneta, *Regularly Varying Functions*, Lecture Notes in Mathematics, Vol. 508, Springer-Verlag, New York, 1976.

[276] R. Serfozo, Functional limit theorems for extreme values of arrays of independent random variables, *Ann. Probab.*, **10**-1 (1982), 172–177.

[277] R. W. Shorrock, On discrete time extremal processes, *Adv. Appl. Probab.*, **6** (1974), 580–592.

[278] R. W. Shorrock, Extremal processes and random measures, *J. Appl. Probab.*, **12** (1975), 316–323.

[279] M. Sibuya, Bivariate extreme statistics, *Ann. Inst. Statist. Math.*, **11** (1960), 195–210.

[280] G. Simmons, *Topology and Modern Analysis*, McGraw–Hill, New York, 1963.

[281] U. Stadtmüller and R. Trautner, Tauberian theorems for Laplace transforms, *J. Reine Angew. Math.*, **311–312** (1979), 283–290.

[282] U. Stadtmüller and R. Trautner, Tauberian theorems for Laplace transforms in dimension $D > 1$, *J. Reine Angew. Math.*, **323** (1981), 127–138.

[283] U. Stadtmüller, A refined Tauberian theorem for Laplace transforms in dimension $d > 1$, *J. Reine Angew. Math.*, **328** (1981), 72–83.

[284] A. Stam, Regular variation in $\mathbb{R}_+^d$ and the Abel–Tauber theorem, Technical report, Mathematisch Instituut, Rijksuniversiteit Groningen, Groningen, The Netherlands, 1977.

[285] A. Stegeman, Modeling traffic in high-speed networks by on/off models, Master's thesis, Department of Mathematics, Rijksuniversiteit Groningen, Groningen, The Netherlands, 1998.

[286] C. Stărică, Multivariate extremes for models with constant conditional correlations, *J. Empirical Finance*, **6** (1999), 515–553.

[287] C. Stărică, Multivariate extremes for models with constant conditional correlations, in P. Embrechts, ed., *Extremes and Integrated Risk Management*, Risk Books, London, 2000, 515–553.

[288] M. S. Taqqu, W. Willinger, and R. Sherman, Proof of a fundamental result in self-similar traffic modeling, *Comput. Comm. Rev.*, **27** (1997), 5–23.

[289] W. Vervaat, Functional central limit theorems for processes with positive drift and their inverses, *Z. Wahrscheinlichkeitstheor. Verw. Gebiete*, **23** (1972), 245–253.

[290] W. Vervaat, *Success Epochs in Bernoulli Trials* (*with Applications in Number Theory*), Mathematical Centre Tracts, Vol. 42, Mathematisch Centrum, Amsterdam, 1972.

[291] W. Vervaat, Random upper semicontinuous functions and extremal processes, in *Probability and Lattices*, CWI Tracts, Vol. 110, Stichting Mathematisch Centrum, Centrum voor Wiskunde en Informatica, Amsterdam, 1997, 1–56.

[292] R. von Mises, La distribution de la plus grande de $n$ valeurs, *Rev. Math. Union Interbalcanique*, **1** (1936), 141–160.

[293] R. von Mises, *Selected papers of Richard von Mises, Vol.* 2: *Probability and Statistics, General*, American Mathematical Society, Providence, RI, 1964.

[294] I. Weissman, Extremal processes generated by independent nonidentically distributed random variables, *Ann. Probab.*, **3** (1975), 172–177.

[295] I. Weissman, Multivariate extremal processes generated by independent non-identically distributed random variables, *J. Appl. Probab.*, **12**-3 (1975), 477–487.

[296] I. Weissman, On weak convergence of extremal processes, *Ann. Probab.*, **4**-3 (1976), 470–473.

[297] L. Weiss, Asymptotic inference about a density function at an end of its range, *Naval Res. Logist. Quart.*, **18** (1971), 111–114.

[298] W. Whitt, The continuity of queues, *Adv. Appl. Probab.*, **6** (1974), 175–183.

[299] W. Whitt, Heavy traffic limit theorems for queues: A survey, in *Mathematical Methods in Queueing Theory* (*Proceedings of a Conference at Western Michigan University, May* 10–12, 1973), Lecture Notes in Economics and Mathematical Systems, Vol. 98, Springer-Verlag, Berlin, 1974, 307–350.

[300] W. Whitt, Some useful functions for functional limit theorems, *Math. Oper. Res.*, **5**-1 (1980), 67–85.

[301] W. Whitt, *Stochastic Processs Limits: An Introduction to Stochastic-Process Limits And their Application to Queues*, Springer-Verlag, New York, 2002.

[302] D. V. Widder, *The Laplace Transform*, Princeton Mathematical Series, Vol. 6, Princeton University Press, Princeton, NJ, 1941.

[303] W. Willinger, V. Paxson, and M. S. Taqqu, Self-similarity and heavy tails: Structural modeling of network traffic, in R. J. Adler, R. E. Feldman, and M. S. Taqqu, eds., *A Practical Guide to Heavy Tails: Statistical Techniques and Applications*, Birkhäuser Boston, Cambridge, MA, 1998, 27–53.

[304] W. Willinger and V. Paxson, Where mathematics meets the Internet, *Notices Amer. Math. Soc.*, **45**-8 (1998), 961–970.

[305] W. Willinger, M. S. Taqqu, M. Leland, and D. Wilson, Self-similarity in high-speed packet traffic: Analysis and modelling of ethernet traffic measurements, *Statist. Sci.*, **10** (1995), 67–85.

[306] W. Willinger, M. S. Taqqu, M. Leland, and D. Wilson, Self-similarity through high variability: Statistical analysis of ethernet lan traffic at the source level (extended version), *IEEE/ACM Trans. Networking*, **5**-1 (1997), 71–96.

[307] W. Willinger, *Data Network Traffic: Heavy Tails Are Here to Stay*, presentation at Extremes—Risk and Safety, Nordic School of Public Health, Gothenberg, Sweden, 1998.

[308] A. L. Yakimiv, *Probabilistic Applications of Tauberian Theorems*, Modern Probability and Statistics, VSP, Leiden, The Netherlands, 2005.

[309] A. L. Yakimiv, Multidimensional Tauberian theorems and their application to Bellman–Harris branching processes, *Mat. Sb.* (*N.S.*), **115 (157)**-3 (1981), 463–477, 496.

[310] A. L. Yakimiv, Asymptotics of the probability of nonextinction of critical Bellman–Harris branching processes, *Trudy Mat. Inst. Steklov.*, **177** (1986), 177–205, 209; Probabilistic problems of discrete mathematics, *Proc. Steklov Inst. Math.*, **1988**-4 (1988), 189–217,

[311] A. L. Yakimiv, The asymptotics of multidimensional infinitely divisible distributions, *J. Math. Sci.*, **84**-3 (1997), 1197–1207 (special issue: D. H. Mushtari, ed., *Proceedings of the* XVII *Seminar on Stability Problems for Stochastic Models, Kazan′, Russia, June* 19–26, 1995, *Part* III).

[312] A. L. Yakimiv, Tauberian theorems and the asymptotics of infinitely divisible distributions in a cone, *Teor. Veroyatnost. i Primenen.*, **48**-3 (2003), 487–502.

[313] M. Zarepour and K. Knight, Bootstrapping unstable first order autoregressive process with errors in the domain of attraction of stable law, *Comm. Statist. Stochastic Models*, **15**-1 (1999), 11–27.

[314] V. M. Zolotarev, *One Dimensional Stable Distributions*, Translations of Mathematical Monographs, Vol. 65, American Mathematical Society, Providence, RI, 1986 (translated from the original 1983 Russian edition).

[315] V. M. Zolotarev, The first passage time of a level and the behaviour at infinity of a class of processes with independent increments, *Theory Probab. Appl.*, **9** (1964), 653–661.

# Index

Printed in the United States of America.